PRECIOUS DUST

ALSO BY PAULA MITCHELL MARKS

Turn Your Eyes Toward Texas
And Die in the West:
The Story of the O.K. Corral Gunfight

PRECIOUS DUST

The American Gold Rush Era: 1848 - 1900

Paula Mitchell Marks

William Morrow and Company, Inc. / New York

It is the policy of William Morrow and Company, Inc., and its
imprints and affiliates, recognizing the importance of preserving
what has been written, to print the books we publish on acid-free
paper, and we exert our best efforts to that end.

Library of Congress Cataloging-in-Publication Data
Marks, Paula Mitchell, 1951–
Precious dust : the American gold rush era: 1848–1900 / Paula
Mitchell Marks.
p.   cm.
Includes bibliographical references and index.
ISBN 0-688-10566-1
1. North America—Gold discoveries. 2. Gold mines and
mining—North America—History—19th century.   I. Title.
E45.M27   1994
970.04—dc20   93-28196   CIP

Printed in the United States of America

First Edition

1 2 3 4 5 6 7 8 9 10

BOOK DESIGN BY LINEY LI

TO CARRIE,
golden girl,
and in memory of
Dick Hughes,
gallant traveling companion

Acknowledgments

In the process of researching and writing this book, I often felt like a solitary gold-seeker struggling across the western landscape and despairing at the magnitude of the journey. Yet a fine "traveling company" eased and sped my steps. Thanks first to my New York companions—to my agent, Robert Gottlieb, to William Morrow's Will Schwalbe and Zachary Schisgal, who helped guide the book to publication; and especially to my editor, Harvey Ginsberg. His encouragement always renewed my energy and resolve, and the book is stronger for his astute editing suggestions.

Unlike the argonauts, I had reliable information along the way, provided by the staffs of a number of libraries. Special thanks are due to the librarians at the Western History Collection of the Denver Public Library, and to those at the Bancroft Library, University of California at Berkeley. At Yale's Beinecke Library, George Miles, curator of the Western Americana Collection, graciously directed me to some delightful gold rush materials. I am also grateful to the Bancroft and Beinecke curators for permission to quote from their manuscript collections. Closer to home, Allison Carpenter at the St. Edward's University Library obtained valuable books and articles for me through interlibrary loan.

The St. Edward's University community accompanied me in this undertaking with encouragement and interest. Dr. Joseph

O'Neal, Dean of New College, and Dr. Donna Jurick, Academic Dean, were flexible and supportive, as were the New College faculty and staff. My fall 1992 women's history class provided feedback on some last-minute considerations, and my advanced writing students shared in the hazards and satisfactions of the writing process with me.

Among my colleagues, special thanks go to Dr. Kay Sutherland, with whom I had a number of discussions on this topic, and to Dr. John Houghton, who provided criticisms of selected chapters. Dr. Richard B. Hughes kindly joined me on this adventure from the beginning, becoming my first audience for the whole manuscript. Warmest thanks also go to Dr. Malcolm Rohrbough of the University of Iowa, who provided commentary during the revising and editing stage. All of these people offered perceptive suggestions, and any inaccuracies or inadequacies in the text are mine and mine alone.

Appreciation is also due three fellow writers and esteemed friends—Dr. Jeanell Buida Bolton, Dr. Bill Goetzmann, and Kathi Jackson—who helped me both gain and keep perspective during the researching and writing process. Joe Toyoshima supplied me with the music of the rushes, eloquently reinforcing the themes of stampeder letters and memoirs, while Elaine Bargsley cheerfully and professionally mapped their journeys.

My family, as always, brought their generous support to the quest. Glenna Marks cooked and baby-sat and waited on computer deliveries. Ivan Marks and Phillip Mitchell smoothed my way through more than one computer crisis with their expertise, and friend Ed Schultz served as a dependable backup. Paul Mitchell plied me with nourishment from his bountiful garden and continued to believe I could do anything. Mark Mitchell listened to my ideas. Anna Beth Mitchell did all of the above *and* helped research and prepare the manuscript.

Alan and Carrie loved me, put up with me, and celebrated with pride—and relief—when each chapter was finished. To these two I owe the greatest debt of all. Their sustaining presence truly made "the crooked [places] straight and the rough places plain."

Contents

Introduction

"Put forth thy hand, reach at the glorious gold," the Duchess of Gloucester urges her husband in Shakespeare's *Henry VI*. From the California gold rush that ignited in 1848 to the frenzied scramble on the gold-laden beaches of Nome, Alaska, at century's end, hundreds of thousands of American and foreign adventurers reached for the gold hidden—or believed hidden—in the far reaches of the American and Canadian West. Gold-seekers reached with every sinew in their bodies and every ounce of their wills, making their way across sun-scorched deserts, through dense forests and swollen streams, and over frigid precipices, then adapting to a rough and chaotic frontier environment where many worked harder than they ever had before in order to extract the "glorious gold" from soil, sand, gravel, and rock.

This is, first and foremost, their story—a chronicle of what propelled them westward, how they lived, how they met the challenges of the journey and search, what kept them going or separated them from their dreams, and what sense they made of the whole enterprise. Those who joined the rushes to "mine the miners," by providing goods and services, also enter the drama, but of primary interest here are the experiences of those who went west and north with the intention of finding gold and who persevered to mount at least one mining effort in the "diggins."

In tracing the impact of this experience on the seekers, the book also traces some of the broader implications of the gold rushes that galvanized western communities, whole regions, and the nation itself for over half a century. It demonstrates how the rushes provided *the* major impetus for the initial development of western regions in the mid- to late-nineteenth century and how they served as a "safety valve" for restless dreamers and a laboratory for the American democratic experiment.

In addition, this narrative shows how the rushes both contributed to a distinctive frontier culture and exposed some of the tensions and paradoxes in American culture. The old preindustrial ideal of moderate success built on frugality, industry, and perseverance clashed with a late-nineteenth-century model of grand fortune based at least in part on luck, cunning, and avarice. The American myth of abundance for all stood in contrast to the reality of wealth for the few. Frictions surfaced between an American belief in democratic fair play and the presence of virulent racism, between romantic appreciation of the land and greedy exploitation of it, and between the American celebration of individualism and freedom and two quite different challenges to it: the need for community and commitment and the demands of a changing economic and social order.

For American enterprise was altering dramatically during this period, as "big business" and wage labor replaced old agrarian and entrepreneurial patterns in which the worker had charted an independent course, however humble. In large part, then, the rushes provided an opportunity for people to reassert their individuality and resist the loss of autonomy imposed by the encroaching industrial age. For this reason, the presence or rumored presence of "placer" (or "poor man's") gold, requiring simple individual effort for retrieval, generated more fevered widespread interest than did the promise of rock-bound quartz gold or silver, whose mining required financing, long-term development, and industrial extraction methods. Thus, the reader will find the emphasis here on the placer gold rushes, with occasional reference to silver booms and with limited attention paid to the quartz gold excitements that tended to swell as placer opportunities diminished.

The seekers themselves comprised a regional, ethnic and socioeconomic cross section of the American male population, with the greater number being middle-class Euro-Americans. Among the foreign-born and the foreign adventurers who came to America or Canada specifically to hunt gold, there was also great variety. The scales were tipped toward the young, the adventurous, and the desperate, but otherwise, the stampeders defy easy classification. Most planned to go home when they had "made their pile," and most found themselves headed home—or into the oblivion of the whiskey bottle—without one. Some stayed where they were and channeled their energies into other business endeavors, while a significant minority embraced the life of the long-term prospector.

The rushes in which they participated varied, too, from brief flurries of limited local activity to massive long-term migrations impacting whole regions. The California rush set the tone and in fact continues to overshadow even the Yukon stampede, for California's "mother lode" lured the most people westward over the longest rush period, was based on a genuine treasure trove, and produced not only "precious dust" but a wealth of letters, journals, and memoirs. Further, it proved a huge stimulus for far western development and established patterns for subsequent rushes—for example, in the codification of "miners' law." For all these reasons, the reader will find this rush emphasized both as a phenomenon in itself and as a point of comparison.

Each rush, large or small, did have a dynamic life of its own, and the West itself clearly changed tremendously in this half century. I have pointed out a number of these changes, yet it is my contention that the gold rush experience remained essentially the same from year to year and from place to place, that despite developments in transportation and communication or modifications in mining techniques and life-style, the California argonaut of 1849 and the Klondike gold-seeker of 1898 shared basically the same tribulations and pleasures in traveling to and working and living in the gold regions.* Whatever social, economic, and political upheavals were occurring back in the East or in the ill-defined, barely governed chunk of the western frontier

---

* William Greever in *The Bonanza West* similarly concluded that mining frontier conditions remained remarkably consistent.

in which they pitched their tents, the gold-seekers in a sense stepped outside history even as they made it, choosing, at least temporarily, an alternative to the more circumscribed, structured, and connected lives of most Americans.

Of course, men could abandon the structures of civilization more readily than could women, who often proved a novel sight in the rush areas. Because the rush experience was overwhelmingly masculine, I occasionally use the term "men" to refer to the gold-seekers, and "he" or "him" to identify a typical seeker. At the same time, I have attempted to show both in the general text and in a special chapter that women were involved, in varied and intriguing ways, in gold rush society.

A note on the book's organization is in order. The first chapter provides a chronological overview of the rush decades. Then, because the journeys to the goldfields loomed large in many stampeders' experience, the next four chapters focus on the arduous "getting there"; for clarity's sake in describing the various trails taken, I have made those chapters chronological as well. In grouping all the pilgrimages together in order of their occurrence, I am aware that I have dragged the reader along far more trails than even the most obsessed prospector ever traveled, and all without pausing in any "diggins."

.This situation is finally rectified in Chapter 6, where the general approach shifts from the chronological to the thematic in order better to explore the gold rushes as a whole in relation to specific topics: the challenge of gleaning the gold and of life in the diggings; the growth of gold rush urban areas; the problems of building communities, especially in regard to caring for the ill and destitute and establishing and enforcing law and order; the distinct treatment of and experiences of minorities; the complex home ties most stampeders maintained, and the presence, effect, and experience of women in the rushes.

The last chapter briefly explores the significance of the stampedes in terms of who "won" and who "lost" in the great fortune hunt, in terms of how participants' experiences both reflected and rejected the larger American culture of the period, and in relation to both "triumphalist" and more critical interpretations of western history.

Throughout the book, I have made use of the gold-seekers'

own words as much as possible, retaining some of the charming misspellings and other grammatical idiosyncrasies but editing to standard English format in most cases for the sake of readability and consistency. The voices provide a vivid testimony to the nature of the adventures that tempered some, liberated others, and destroyed still others in the mania for gold-hunting that periodically swept the nation in general and the West in particular during this era of frontier opportunity and expansion.

... like everyone else in those days, we were bitten by the gold bug, and mine we must, and mine we did. . . .

—Granville Stuart

I had a day-dream, you know, that when I got my shaft down to bedrock it might be like the streets of the new Jerusalem.

—Robert Henderson

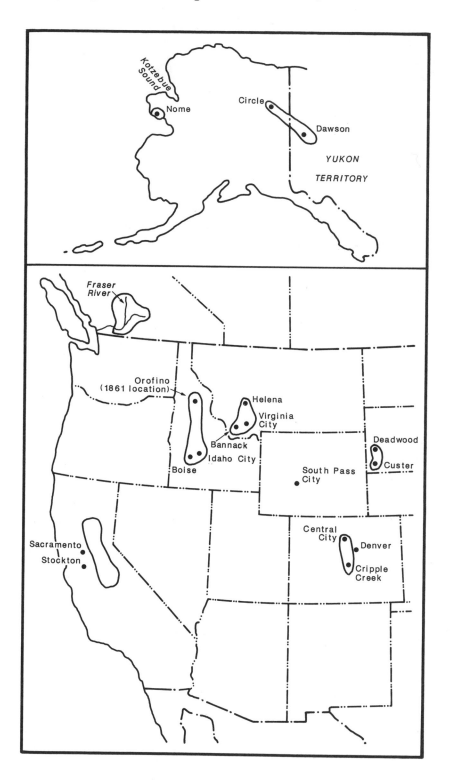

Major gold rush regions and sites, 1848–1900

Chapter One

"The Age of Gold"

It is coming—nay at hand—there is no doubt of it. We are on the very brink of the Age of Gold.

—*Newspaper editor Horace Greeley, on California discoveries*

What a mad rush this is to a land nobody knows anything about.

—*Alaska stampeder Joseph Grinnell*

In January 1849, E. L. Cleaveland of New Haven, Connecticut, was a worried man. As pastor of a church planted firmly on a tranquil green beside the Yale University campus, Cleaveland perceived a danger to his orderly New England universe and to the character formation of his spiritual charges, the young men attending Yale. That danger was gold, gold in the distant territory of California, and the students and citizens of New Haven, like their contemporaries across the United States, seemed to talk of, think of, dream of, nothing else.

California gold fever had coursed across the settled half of the nation since the preceding autumn, when fantastic reports from the western lands newly acquired from Mexico accumulated until they could no longer be ignored. Far out there across the Sierra Nevada, the foothills were said to be brimming with gold—shimmering flakes, dust, even large nuggets—that a man could easily pick out of a crevice with a knife, pluck from a streambed, or capture with the aid of water in a simple washing pan.

Affirmation of these reports had come in early December 1848 from President James Polk, who announced in his farewell address to Congress that the amazing stories had been "corroborated by authentic reports of officers in the public service." By January, hastily organized eastern companies—made up of farmers, merchants, professional men, youths fresh out of school or still in it—raced to book passage on a motley collection of brigs, ships, barks, and schooners pressed into service for a long voyage to El Dorado. Southerners and midwesterners began organizing and planning spring overland journeys. As groups formed to travel together, often making plans to dig together as well, E. L. Cleaveland mounted his pulpit to deliver a sermon titled "Hasting to Be Rich."

Before his listeners decided to join a gold-seeking company, Cleaveland urged them to consider whether their financial need was severe enough to warrant such a speculation, what the effects would be on family and friends left behind, what physical dangers—from illness, accident, or violence—adventurers would face, and, most important, what this greedy pursuit of material wealth would do to their moral and spiritual health and development. Solemnly, he quoted Proverbs 28:20: "He that maketh haste to be rich shall not be innocent." Gold-seekers would enter a land, he warned, "where you will find no Sabbath, unless you find it in your own heart."

Other religious and secular leaders voiced similar moral warnings, invoking the Puritan work ethic, reinforced with the admonition that a precipitate path to riches did not guarantee happiness, that calm, steady labor remained the most virtuous course. Practical critics joined in with sobering assessments of men's chances for success, and out-and-out cynics lampooned as gullible dreamers the "fools of '49."

Many of the gold-seekers did feel a dissonance between the wonderful stories and their own common sense and between the values their culture had instilled in them and their headlong dash for gold. But gold *did* exist in great quantities in California, and as for values, ministers were not above dropping everything to join the rush. Even the financially strapped University of

Notre Dame was sending four Holy Cross brothers and three laymen to dig for gold. If representatives of religious groups could succumb to temptation or justify their actions as means to a greater good, how easy it was for any individual with a thimbleful of imagination to jettison all those teachings about prudence and frugality and character-building in the splendiferous clamor over California gold!

Young Yale graduate John Angell was preparing for his trip to the goldfields about the time Cleaveland delivered his sermon. To his mother, Angell admitted he wanted to get rich, but insisted that that was not his only motivation. No, he was also going to improve his health (this from a man who would become so seasick he would vomit for twenty-seven straight days on the outward voyage), to "see more of the world," and—an ingenious touch designed to melt a God-fearing parent's heart—to distribute Bibles and religious literature obtained from a New York tract house. He, too, clinched his arguments with a Bible verse: "All things shall work together for good to those that put their trust in Him."

Some mothers could be convinced on more practical grounds. George Hearst, destined to become one of the mining kings of the West, found his mother resisting his desire to depart Missouri for California until he emphasized that miners were making forty and fifty dollars a day there and "it seemed to me it was by far the best thing to do, as it was pretty hard pulling where we were."

The departure scenes, whether at inland communities or eastern ports, reverberated with excitement and gaiety. In New York City, docks were "crowded with fathers and mothers, brothers and sisters and sweethearts," the well-wishers engaged in much "embracing and waving of handkerchiefs" amid cries of "I say Bill! If you send me a barrel of Gold Dust don't forget to pay the freight on it." Such send-offs obscured any scoffing and warnings, confirming for the travelers what they were embarking on a grand adventure in which each was "more or less a hero."

An estimated ninety thousand stampeders left the settled states for the California goldfields in 1849 alone. They were followed the next year by almost as many—an estimated eighty-five thousand. Over the next few years, the heavy stream of westering gold-hunting humanity would continue. But even before the forty-niners loaded their wagons or boarded their ships, the California stampede had become the first great American gold rush.

California at the time was inhabited by Indians, by landowning rancheros and a small urban population (both overwhelmingly of Spanish-Mexican origins), and by various isolated American immigrants—the latter primarily single male adventurers, some of whom had intermarried with the Mexican and Indian natives. When the first reports of gold began rippling outward from the Coloma Valley, the news nearly drained San Francisco, Sonoma, and Monterey of men. Upon seeing a bag of gold from the Yuba River diggings, Monterey carpenters cast aside their saws and planes and "shouldered their picks," while an entire platoon of soldiers from the nearby fort deserted for the goldfields. The news had also already drawn most of the male settler population of Oregon Territory southward and attracted adventurers from the Pacific ports of Hawaii, Chile, and Peru.

This was to be the granddaddy of all the rushes from the Fraser River to the Yukon, from Pikes Peak to Nome. A chance discovery of gold in California's picturesque Coloma Valley in early 1848 had inaugurated more than fifty years of American gold excitement, creating a glistening lure of independent wealth for those willing to brave the rigors of a frontier journey and frontier living. As many as a million people hoping to find the gold or to glean it from those who did would join the successive gold rushes of the late nineteenth century. In the process they would not only stimulate the exploration and initial development of large portions of the American and Canadian West but would share in a series of experiences that indelibly marked their lives and helped shape western frontier culture.

The stampedes, and the fever that engendered them, had precedents in American history. After all, Columbus had "discovered" the Americas in part out of a burning desire for gold. Whoever had it, he reasoned, "can do whatever he likes in the world ... even bring souls into paradise." From the Spanish conquistadors onward, immigrants and wanderers from European soil had dreamed of such power, of wonderful, lustrous, malleable gold in the far reaches of the New World. For the Spaniards, the dream had been partially realized in Mexico, with its gold and silver mines. And always, there were tales told by natives—or told by people who claimed to have learned them from natives.

Besides, even without stories of distant gleaming cities and their gold-bedecked inhabitants, men would dig for the precious metal. "No talk, no hope, no work but to dig gold, wash gold, refine gold, load gold," lamented a resident of Jamestown, Virginia, when colonists mistakenly thought they were unearthing riches in the English settlement in 1608, the year after its founding. The Anglo-American settlers who quickly established ethnic dominance along the eastern seaboard and began moving inland were just as susceptible to the presence or rumored presence of gold as were Spanish adventurers clattering across the Southwest looking for Cibola.

The first Anglo-American finds of any significance occurred in the Piedmont region of central North Carolina in the first decade of the nineteenth century, as farmers turned prospectors with the news that one family's rock doorstop had yielded a gold ingot. Those drawn to the strikes followed the Piedmont Plateau southwestward, extending the goldfields into South Carolina and then to northern Georgia, site of the first rush of any consequence.

The Georgia rush exploded in 1829, initially attracting an estimated five thousand seekers. It seemed "as if the world must have heard of it," men arriving "from every state . . . afoot, on horseback, and in wagons, acting more like crazy men than anything else." They panned the streams and branches and pocked the earth with holes. By the next year, as many as twenty thousand fellow prospectors had joined them.

The Georgia rush foreshadowed stampedes to come. Prospectors totally disregarded Indian claims to land—in this case, the rights of the Cherokee for whom the area had been reserved. There was frenetic initial activity, the influx of too many seekers for the claims available, an exploitative attitude that made ecological havoc and mobility the norm, a lack of social and cultural restraint that led observers to judge mining communities "immoral" in the extreme, and a dizzying regional and ethnic diversity that jammed not only Anglo-Americans but Germans, Mexicans, Cornishmen, Scotsmen, Spaniards, and African-American freedmen together in the "diggins." Here, as well, some of the basic gold-mining techniques that would be used in the western diggings were tested. And here, as in subsequent rushes, the emphasis was on placer mining.

Placering, the collecting of gold already released from rock through erosion, at its most basic required only a shovel, a washing pan, water, muscle, and a willingness to work. As the last two ingredients indicate, it was not necessarily easy labor for the gold still mixed with the gravel and soil near bedrock. Many a man fantasizing of easy riches received a rude awakening in the diggings as he toiled down through feet of dirt and sand and rock and gravel to locate the "pay streak," repeatedly hauling up and panning the earth to see what he had. Soon a persistent miner was graduating to homemade rockers and sluices that sped the washing operation but did not lessen the amount of effort.

The Georgia excitement fed the dream of golden riches for Americans from the East Coast to the midwestern frontier. Virginians dug up their farmland looking for gold; Missourians listened to mountain men's tales of glinting treasure troves in the Rockies. But few American leaders seemed to realize the potential for either placer or quartz mining. In 1790, Benjamin Franklin had declared flatly, "Gold and silver are not the produce of North America, which has no mines." Nor did Thomas Jefferson's purchase of the vast trans-Mississippi region in 1803 stimulate prominent American thinkers and dreamers to give much consideration to the presence of gold, silver, or other mineral resources in the new lands.

Instead, the prairies, plains, deserts, mountains, and valleys of the West improbably gave rise in the early nineteenth century to visions of a mercantile "passage to India" or an agricultural empire. Disgusted with the impracticality of the latter, in 1843 United States Senator George McDuffie of South Carolina could only see the West, a land he "wish[ed] to God we did not own," as a giant buffer "to secure [Americans] against the intrusion of others."

At the time McDuffie voiced these sentiments, the only American economic activity of any significance in the West—furtrapping—was dying out with the decline in value of beaver pelts, Mexico still claimed the Southwest and California, and Mexican prospectors had already made a few small placer gold strikes in present-day New Mexico and in California's San Fernando Valley. American farm families had commenced immigration to the fertile Willamette Valley of Oregon, and a few farm families and other American adventurers were trickling into California.

The gold lay waiting. It dotted the river bars and dry riverbeds along the western fringe of the Sierra Nevada, hid in Rocky Mountain canyons and gulches, mixed with the rock and soil of the Sioux holy ground in the Black Hills, eroded into swift Alaskan waterways, reposed in the cold creeks flowing downward from King Solomon's Dome in the Canadian Yukon. Wherever the land was most challenging and remote, the gold waited.

The boom era began with former New Jersey carpenter James Marshall's Coloma Valley discovery. One of the restless Americans who had wandered all the way into California, Marshall was building a mill on the American River in partnership with John Sutter, a Swiss emigrant who had used Mexican land grants to establish an agricultural empire in the Sacramento Valley. In January 1848, the carpenter spotted a golden particle "about half the size and the shape of a pea" in the millrace he and his crew had just constructed. "It made my heart thump, for I was certain it was gold," he would recall. To his workmen, he announced, "Boys by G[o]d, I believe I have found a gold mine."

Within two weeks, before the news had spread beyond Sutter and his workers, Mexico ceded California—along with New Mexico Territory and the Rio Grande area of Texas—to the United States in the post-Mexican-War Treaty of Guadalupe Hidalgo. Soon it would appear to Mexicans that the Americans had been damned lucky in their timing. It was almost as if God or fate were on the Yankees' side, especially when forty-niners crossing the harsh southern California desert encountered a river that reportedly had not flowed for ninety years. Some Mexicans saw its presence as a "miracle . . . specially designed for American emigrants on whose side Providence had arrayed itself."

Americans were not averse to reading the signs that way, either. Most shared an ebullient national optimism, one part youthful vigor, two parts confidence in American resourcefulness, and three parts a belief in unlimited opportunity, or abundance for all. Nothing seemed too grand for the American imagination as the population and settled territory burgeoned and as the nation muscled its way to world prominence—especially with the glowing reports from California in 1848 and 1849. "Plenty for all, for years to come," wrote one early arrival at the diggings. "Oh, the gold is there, most anywhere," sang the forty-niners.

Tied in with this unbridled optimism about the multiplying California gold discoveries was a general faith in "manifest destiny," in Euro-American superiority to other ethnic and racial groups, and in the grandiose idea that, with the course of empire surging westward, world civilization would culminate in the New World, specifically in the American West. Prophetic tales of a "city to come, all framed in gold" were linked in gold-seekers' minds with "the day now dawning in California."

Indeed, Marshall had stumbled upon a portion of "perhaps the greatest trove of undiscovered placer gold in history," a series of deposits that by fall 1848 were being worked "from the Trinity River in the north to the Tuolumne in the south," along a four-hundred-mile ribbon. At a time when many eastern workers were receiving a dollar for a day's labor, prospectors were panning an ounce or two of gold—at fifteen or sixteen dollars an ounce—in one day's effort, and some were working claims yielding hundreds of dollars' worth of dust and nuggets daily. Furthermore, the holders of rich claims were paying twelve dollars a day for laborers.

The excitement such deposits produced would draw argonauts from around the globe, while in the American East the hullabaloo reached such a crescendo that men "decided to go to California as they might get religion at a revival, or volunteer at the outbreak of a war."

Most of the argonaut companies consisted of middle-class tradesmen and farmers and young men of their families—in short, those who could afford the journey or borrow the needed funds. Even upper-class professionals joined in. "Every day men of property and means are advertising their possessions for sale in order to furnish them with means to reach that golden land," reported the *New York Herald*.

Others had the added impetus of declining family fortunes. California-bound Alonzo Hill wrote to his financially distressed father, "Remember Father, this step is taken more for you than for me, yet I am pleased to think that I am he who can cheerfully & fearlessly take it." Ohio native Charles Churchill wrote that his widowed mother's shaky finances made it necessary "that I should be in a situation to contribute something to the support of the family." Besides, he noted, "at my time of life, it is time I was accumulating something."

Yet few were in the economic straits that would help propel their fellows into the subsequent large rushes, when national economic panics fed the numbers headed for Pikes Peak, the Black Hills, and the Yukon. Philosopher Josiah Royce, son of a couple who joined the 1849 rush, would later characterize the participants as Jonahs, and many were temporarily or permanently escaping home responsibilities or expectations—or consequences of previous actions. As a popular ditty asked, "Oh, what was your name in the States?/Was it Thompson, or Johnson or Bates?/Did you flee for your life or murder your wife?/Say, what was your name in the States?"

But most resembled Huck Finn more nearly than Jonah or a criminal fugitive, for the majority were young, excited by the possibilities of the American West, and looking for adventure and its freedoms as much as for the gold. "So good bye to all our friends and relations," sang the forty-niners. "And if we fail to get the gold, we'll see most all creation." They were, like Huck, "light[ing] out for the Territory," and the quest had all the elements of a rite of passage, a test of their abilities to survive and prosper on their own in a new and challenging environment.

Whatever the travelers' motivations, merchants and hucksters were not slow in exploiting this migration. The would-be miners bought "Captain Sutter's long mining waistcoats"; pills guaranteed to guard against the dreaded cholera; and large, elaborate, totally useless contraptions for washing gold, blissfully unaware that they would be forced to abandon both the ludicrous and the practical on the plains and deserts.

They also purchased a guidebook—one of the more than thirty rushed into print by the spring of 1849, many of them grossly unrealistic in their descriptions of routes, distances, and trail conditions. Lansford W. Hastings's *The Emigrants' Guide to Oregon and California,* by minimizing the difficulties of crossing the Great Basin and the Sierra Nevada, had already contributed to the Donner Party debacle in which snowed-in Sierra travelers were reduced to eating the flesh of their dead companions. Now Hastings quickly concocted a travel book that indicated argonauts could journey from New Orleans across Mexico and northward to the California goldfields in no more than thirty-six days, when the actual length of the trip for those who accomplished it was six times that.

Communities along the various routes added to the spate of misinformation in their attempts to promote their services as outfitting centers. A Corpus Christi newspaper touting the Texas coastal town as a starting point recommended a "trail" through arid, Apache-scourged, nearly uninhabited Mexican terrain and equally isolated southwestern landscapes with the blithe assurance that the harrowing route passed through "no wilderness," all of it leading instead through "thickly settled country."

The ignorance of the information givers and the self-interest of outfitting the transportation companies would plague all of the subsequent rushes as well as contribute to high turnaround rates. Of the ninety-thousand gold-seekers heading west in 1849, only an estimated forty thousand completed their journey to the Sierra Nevada mining districts. There most found the best claims already taken by the early-comers of '48.

This cruel discovery at the end of an arduous journey would become a gold rush norm. If, as the rushers themselves reported, ninety-nine out of a hundred men in the California placer diggings were lucky to make expenses, what of the rushes based on small finds magically inflated by optimistic promoters, or, even worse, of complete chimeras fueled by garbled rumors, dreamed up by an individual seeking the limelight, or calculated by transporters or suppliers?

For fifty years, stampeders would gamble their fortunes and their very lives on the strength of a whispered rumor, of an overenthusiastic newspaper account, of a glittering nugget pulled from the pocket of a shabby, grizzled prospector. Some would learn to live for the excitement of the gamble, so that even a prosperous working claim could not hold them for long. In fact, observers found, the greater the volume of gold, "the more unsettled the population."

That made the California diggings the most unsettled of all. Despite the poor fortunes of most gold-seekers, the rush itself is usually considered a phenomenon that lasted at least through 1852, and perhaps as late as 1857. California catapulted directly into statehood in 1850 on the basis of its gold resources, bypassing organized territorial status. As the mining population exploded, men rushed from the northern mines to the southern at the same time their fellows were abandoning the southern mines for the northern.

They also followed rumors to new California locations. In 1852, the conjecture that in the Sierra Nevada range near Downieville existed a lake, "evidently the crater of a large volcano," its shores strewn with gold, led as many as five thousand miners, some with lucrative working claims, to stream from the established sites. They paid exorbitant prices for pack mules and horses, "so great was the desire to be the first in the new diggings." Some lost their dear-bought mounts to horse thieves, who hurried the animals back to market in the new mining supply centers of Sacramento and Stockton. Many of the stampeders spent the whole summer searching for Gold Lake, locating only "mountains of snow" and returning "ragged and footsore" in the fall to find their abandoned claims taken up by others.

In addition, "old-timers" and newcomers combined to pursue stories of gold strikes beyond the state's borders. An 1851 strike in southern Oregon Territory drew prospectors and initiated a small but genuine series of Oregon discoveries. In 1852, an estimated 600 to 700 prospectors from both Oregon and California converged on Queen Charlotte Island off the coast of British Columbia. One later flatly summarized the experience: "Somebody had started the excitement, and we went. When we did not find anything of course we all had to go away." Other seekers were drawn to an Australian discovery in 1851–52 and to fanciful reports of a strike in Peru in 1854. The grandest rush— and the grandest humbug—involving Golden State miners ig-

Canada's Fraser River, a Rocky Mountain steam bisecting the rugged western terrain of what was about to become British Columbia.

Californians regarded the reports from Fraser River warily at first. By this time, too many tales had proven false, too many men had invested all their funds and risked their health in ill-fated gold expeditions. One 1854 "Miners' Commandment" warned, "Thou shalt not tell any false tales about good diggings in the mountains to thy neighbor, that thou mayest benefit thy friend who hath mules and provisions and tools and blankets he cannot sell," for the neighbor might return with his rifle and present the rumormonger "with the contents thereof."

Further, in the case of Fraser River, gold reports were muted. Prospectors returning from the shallow, limited placers on and around the Fraser displayed their gold dust in San Francisco, but "without at all exaggerating the resources of the land from which they had come."

Then the *Sacramento Union* published the account of one of these returnees. He talked of forbidding mountains, treacherous rivers, hostile Indians, bitter cold, and lack of timber, but also of gold "lying thick as pebbles" on the bottom of one stream. Perhaps his acknowledgement of the difficulties made the report more believable. More likely, the image of the gold winking from the waters crumbled the fragile defenses of men wanting desperately to believe. By June 1858, claim-seekers crowded the trails from the California diggings to Fraser River, and each steamer out of San Francisco carried a thousand or more.

Many soon found themselves wandering penniless along the shores of Bellingham Bay just south of the Canadian border in Washington Territory. In July, despite the news of men digging clams for food in order to survive, of flooding in the placer mines, and of restrictions in the form of licensing by the British government (anathema to Americans used to exploiting public land freely and proceeding strictly on their own initiative), stampeders continued to crowd the steamers bound for the new El Dorado. The British government hurried to create the crown colony of New Caledonia, which would quickly become British Columbia.

By fall, however, the rush had fallen flat as a flapjack. Some miners remained to establish diggings to the north, on the Horse-fly River, on the forest-bordered Quesnel River, and eventually on Antler Creek and William's Creek in the "Cariboo" gold district, focus of a rush in the early 1860s. But most returned to California in 1858 or '59 and considered who really got rich out of the experience. The answer was clear: the transportation companies and suppliers. "We are again quiet at least until the Steamboat Companies can find another new Eldorado," wrote one San Franciscan. California prospectors who had joined a rush to New Mexico Territory's Gila River as their fellows tackled Fraser River reached a similar conclusion: Those stimulating and profiting from the rush were merchants, the stage company, and territorial promoters.

Meanwhile, another apparent El Dorado was in the making in the American Rockies. In 1850, en route to the California goldfields, a prospecting party had located a small amount of gold on Ralston Creek, near the future site of Denver, but had resolutely pressed onward. The knowledge of this discovery, and the severe national financial Panic of 1857, spurred a handful of new prospecting companies to venture into Colorado's craggy Front Range. As the Fraser River excitement swelled and died to the northwest, one persistent thirteen-man group under the leadership of Georgia and California veteran Green Russell made a small strike on the South Platte River, then on its Cherry Creek tributary.

Their summer's work panning and prospecting up and down the South Platte yielded $800 worth of gold dust—about $62 apiece—not the kind of riches that should lead dreamers to drop everything and head west. But word of gold in the Rockies pulsed eastward, suffusing the whole Front Range in a golden glow. A Kansas City paper published a hyperbolic report stating that men had gleaned "over $600 per week, with nothing but their knives, tomahawks, and frying pans," the tomahawks being a fanciful reference to the Indians who had supposedly been smitten with gold fever as well.

The reaction of frontier folk and of the economically distressed farmers of the Midwest was swift and uncontrolled as a prairie fire. Despite the arrival of fall with snows descending on the Rockies, in September a number of parties from the frontier towns set off for the eastern slopes. Most received the news too late to mount a journey to and prepare for a winter in the mountain. They dreamed of spring and talked gold, "Pikes Peak gold," despite the fact that the diggings lay eighty miles to the north of the peak. In Kansas City that autumn, "every man [had] gold on his tongue, if none in his pocket." One could not enter a Leavenworth barbershop without hearing, "Have it done in Pikes Peak style, sah?" nor a Leavenworth restaurant without being asked to select beef cooked "a la mode Pikes Peak" or pudding "with Pikes Peak sauce."

Again, those who stampeded were primarily motivated by money and adventure, with Pikes Peak promising "gold, and plenty of it . . . influence, power and position in our middle age, and ease and comfort in our decline." And again, argonauts felt the need for justification of such a risky venture. "You might consider this a foolish move for me," Minnesotan James Fergus wrote to his brother, "but I am doing little or nothing here for my self. There I will see the country, probably lay the foundation for a future business, and if nothing else, I may be able with my own hands to dig the gold to pay you for the money you sent me." Like previous seekers, Fergus also perceived and presented his departure as a moral duty that fell to him as provider for his wife and four children, reading them a poem in which he promised to "wend my way through wet and cold to dig for you, the hidden gold."

Although nobody had proven there were abundant deposits, throngs of eager Americans were preparing to move 600 to 700 miles from the Missouri River frontier, grabbing up the new spate of gold rush handbooks, in which "distances were shrunk by two and three hundred miles; non-existent bridges over rivers were solemnly listed; deserts disappeared; grass grew where it had never grown before" and expenses "plummeted to ridiculous proportions."

A certain nostalgia already pervaded the rush experience; one man wrote of seeing a "rocker" built for gold washing sitting in front of a Kansas City store. "This brings to memory some scenes buried with my California life," he wrote, "and almost makes me 'sigh for the mines once more.' "

Because settlement had extended along the Missouri and had spilled onto the prairies in the decade since the California gold migration began, the towns vying for outfitting status now included raw Kansas Territory settlements eager to provide supplies. An outfit for a party of four for four to six months would run about five hundred dollars. However, the buyer had to beware; his meager travel fund could quickly disappear as he piled up "Pikes Peak hats, Pikes Peak boots, Pikes Peak guns, Pikes Peak picks and shovels, Pikes Peak cure-all pills," and other sundry items.

Negative reports of the diggings soon elbowed their way into the newspapers, but once set in motion a rush took on a heady, inexorable life of its own. Early spring 1859 found thousands on the trails to "Pikes Peak," and thousands becoming disillusioned with the journey itself and the dwindling hopes in the diggings.

Then, in early 1859, a few genuine strikes seemed to confirm the faith of those who believed in the Rocky Mountains as a source of golden wealth. At Idaho Springs, George Jackson thawed a frozen patch of ground by building a fire (a technique that would become popular in the Yukon forty years later) and promptly panned ten dollars in gold from it. "I've got the diggings at last," Jackson confided to his journal and to friend Tom Bolden, whose mouth was "as tight as a No. 4 beaver trap." Jackson returned to the site with a group of grubstakers from Chicago in April, the party netting almost two thousand dollars' worth of gold in their first week of digging. Meanwhile, other prospectors located the placering and quartz camp of Gold Hill and John Gregory, a former Georgia gold-miner working a short distance north of Jackson, made a significant find in what became known as Gregory Gulch.

These three strikes "saved the rush from collapse." In massive numbers—with estimates of up to one hundred thousand in 1859—the seekers poured into the Cherry Creek area, now the site of the rival towns of Denver City and Auraria. Dick Whitsitt, Denver City town recorder and real estate speculator, proclaimed, "The Rocky Mountains are full of gold and there are fortunes for all in this country. Come and if you can't come on two legs come on one."

Those who responded were, like their forty-niner predecessors, predominantly farmers, craftsmen, laborers, and clerks under thirty, residents of all sections of the nation and world, but with the largest native numbers from New England, Virginia, and Georgia and the largest foreign-born from Canada, Ireland, and Germany. Most had only the slimmest notion of how to find and mine gold, although many companies contained a forty-niner or two. These men had returned home after the big rush, but more in body than in spirit; rushes were addictive affairs from which men often extracted themselves with difficulty and to which some often returned. The "fierce, riotous, wearing, fearfully excited life" of the stampeder led naturally to "restlessness, craving for stimulant, unscrupulousness, hardihood, impulsive generosity, and lavish ways," all, except perhaps the hardihood, decidedly out of place on a Missouri farm or in a Iowa dry-goods store.

Fifty-niners hurtling into the unknown confidently, defiantly emblazoned their wagon covers with the slogan "Pikes Peak or Bust." A Lawrence, Kansas, newspaper noted 450 wagons crossing the Des Moines River at Des Moines in one day, 12,000 crossing the Missouri River at St. Joseph on one of the three ferries within a space of ten days. At the Arkansas River, a traveler wrote, "Jerusalem, what a sight! Wagons—wagons—Pikes Peak wagons." Ominously, some of the wagons were those of disillusioned returnees.

By 1860, many busted stampeders were looking to hang a guidebook author or two. There was even talk of burning down one of the outfitting towns. "Busted, by God!" men wrote ruefully across the wagon covers. But still the hopeful adventurers flowed into Denver City. And gold fever spiraled once more when Abe Lee in April 1860 scooped up a panful of dirt on the Arkansas River near the Continental Divide, yelling to his companions, "By God, I've got all of California in this here pan!" Four months later, ten thousand or more people were jammed into the narrow ravine where Lee had hit pay dirt, now christened California Gulch.

The series of strikes and the possibility of more caused even determinedly realistic mining camp commentators to vacillate between somber assessments and seductive lures in their advice to would-be prospectors. *New York Tribune* editor Horace Greeley, touring the Rocky Mountain goldfields in the boom summer of 1859, joined with two traveling companions to warn would-be prospectors away from the overcrowded diggings, predicting much suffering and many dashed hopes. But the three undermined their own advice even as they offered it by admitting that there were likely many rich gold ravines awaiting discovery in the area. No wonder Greeley's compelling command "Go west, young man," overshadowed his cautions.

Of the estimated 100,000 people who scurried to Colorado in 1859 in response to the news of gold, fully half turned around before getting there or almost immediately after arriving, and only about half of the remainder—25,000—actually did any mining. The percentage of miners among the 60,000 to 70,000 who arrived in summer 1860 was probably the same or less. Even the most optimistic eventually had to admit that the Colorado placers did not have the long-term richness of the California ones. But the rush had carried with it settlers as well as prospectors, had turned some prospectors into settlers, and had led others to join the diehard gold-seekers fanning out across the mountain West. Further, it had given birth to Denver, "a great city built up . . . on a barren plain, nearly a thousand miles away from water or the trail of the iron horse," and it had propelled Colorado into being and statehood, granted in 1861.

Meanwhile, Californians had been busy with the West's first large-scale silver boom, occurring in Nevada, then part of Utah Territory. Silver would never exert the appeal that gold did, for as the Mexicans succinctly explained, "it required a gold mine to open a silver mine." In fact, the prospectors who in 1859 began taking silver from their strike on "Gold Hill" didn't even recognize it; they responded with nothing but obscenities to the blue substance mixed in with their precious gold finds. Yet they were perched on a portion of the incredible Comstock silver lode, "the richest silver strike in American history," and when they finally figured things out, other Californians came rushing in to help them locate more of what they had so derisively termed "damned blue stuff."

The mining camps of the northern Sierra Nevadas "pour[ed] their population back over the Sierra emigrant trails" to the "Washoe Diggings" in the summer and fall of 1859, with at least ten thousand following in 1860, and Gold Hill's successor Virginia City quickly became the hub of an industrial mining empire.

But industrial mining empires were not the stuff of which most stampeders' dreams were made. In the early 1860s, prospectors spilled out in every direction across the hills, mountains, and deserts of the West, including among their number "the backwash of the Civil War—young men who had been thrown out of employment by the war, discharged troopers from both sides, and a great crowd of deserters" and "low characters." In the Southwest, prospectors fanned outward from an 1859 strike at Pinos Altos, New Mexico Territory. In Oregon, granted statehood in 1859, gold strikes on the John Day River and Powder River in 1861 drew more than a thousand seekers to jockey for claims. Washington Territory, too, lured gold-seekers into its remote reaches, one adventurer in 1861 finding the Cascade Mountains and their streams "rarely, if ever, without a number of *gold hunters.*"

But to enhance one's chances of "making a pile" with a gold claim, the best places to search in the early 1860s were the future states of Idaho and Montana. Idaho, then a section of Washington Territory, possessed a tortured, wild, remote grandeur, as reflected in the names of its topographical features—the Bitterroot Range, the Sawtooth Mountains, the Snake River Plain, the Salmon River country, the last an "uninhabitable wasteland, with precipitous cliffs and yawning ravines, that looked like the ruins of a world." Yet gold would make the cliffs less precipitous, the ravines less daunting to the diehard seekers.

The first Idaho strike occurred north of Salmon River country, on Orofino Creek, a tributary of the Clearwater River, in 1860. This was Nez Percé land, but Indian trader Elias Pierce and his party dodged the Nez Percé to locate their claim and carried enough dust back to Walla Walla, Washington, to stimulate interest there. In a situation foreshadowing events in the Black Hills fifteen years later, the second group departing Walla Walla had to dodge not only the Indians but United States troops sent to keep them off Indian property. Soon the mining camp of Orofino had been established, and prospectors were ranging north and south, discovering placers in both directions. Most excitement centered on the Salmon River placers—limited in size, but very rich—and soon prospectors from California, Oregon, and Colorado were flocking into the Salmon River camp of Florence. "It is now estimated that there will be twenty thousand people in the Mountains so high is the gold fever," wrote one 1862 traveler.

The discoveries spiraled southward to the Boise Basin in 1863; once these goldfields had been defined, a series of small Idaho placer strikes continued throughout the 1860s, each precipitating a minor rush and whetting American interest. An 1864 argonaut found "nearly every third person" in Chicago apparently "bound for Idaho."

Meanwhile, on the northeastern side of the Bitterroot Range in the Rockies, pioneer Montana prospectors had located placer and quartz gold in the 1850s, the boom beginning in 1862 with the discovery by an Idaho-bound party of placer gold—including, gloriously, gold that could be shaken from the roots of sagebrush—on Grasshopper Creek in the Deer Lodge Valley. Others who had been headed for Idaho, along with scattered area prospectors and anyone else close enough to take advantage of the find, congregated to create the town of Bannack.

When the placers played out—as they did very quickly—a group of six prospectors stumbled upon extremely rich Alder Gulch, only seventy-five miles away, and Virginia City, Montana, was born.* Virginia City became a magnet for western prospectors and entrepreneurs in general, as well as greenhorns from the East, with an estimated 10,000 to 15,000 people located in and around the town within a few months of its founding.

Then, as too many people crowded into Virginia City, discoveries at Emigrant Gulch, at Confederate Gulch, and at Last Chance Gulch sparked continued movement. The discovery at

---

* The presence of two prominent nineteenth-century mining towns named Virginia City has proven confusing even to researchers. Because Virginia City, Montana, was a gold town established on placer gold, it enters into this narrative much more frequently than the primarily silver town of Virginia City, Nevada. Thus, in subsequent references, I simply use "Virginia City" for the Montana location but identify Virginia City, Nevada, by state.

Last Chance Gulch also gave rise to the town of Helena and to subsequent discoveries in the immediate area—New York Gulch, First Chance Gulch, Montana Bar.

Scenes in the Montana camps amply demonstrated the volatility of a gold rush population. When the Alder Gulch discoverers traveled into Bannack for supplies, determined to keep their find a secret, something tipped the many idle men in town. Perhaps it was the visitors' demeanor; as one of the six wrote upon reaching the town, he was "too tired and too glad!" The news leapfrogged with dizzying speed through the existing diggings, the discoverers unable to depart without being trailed by a mob. After Alder Gulch was established and other hopefuls poured in, one man "spent the day looking for a lost pig, followed by a crowd who were convinced he was leading them to Golconda." And an "old mountaineer" in newly established Virginia City buying supplies was heard to tell an acquaintance, "I have got as good a thing as I want." He was referring to his Blackfoot Indian wife, but an estimated twelve hundred goldseekers trailed him out of town in snow and bitter cold.

Early 1864 brought "a regular stampede craze" to Virginia City: "Somebody would say that somebody said, that somebody had found a good thing and without further inquiry a hundred or more men would start out for the reported diggings." Four hundred men stampeded through the snow to a vaguely reported find on the Gallatin River. A subsequent rush to Wisconsin Creek nearly depopulated Virginia City. Despite the fact that neither rush had proven of the slightest substance, a third rush, to Boulder Creek, drew in even those who had resisted the first two.

Then, in an idea common to mining camps, someone deduced that the town itself was sitting on golden treasure. Streets and backyards were attacked with gusto, with businesses in danger of being "dug up and washed out without ceremony." When this fruitless effort subsided, six hundred people took off for the Prickly Pear Valley, where there actually *were* paying gold claims, but, of course, far too few to go around.

A similar restlessness manifested itself all over the West during the 1860s and 1870s, keeping the mining frontier extremely transient and unstable while ironically paving the way for settlement and development. An 1867–68 rush to the area of the South Pass of the Rocky Mountains, a prominent landmark on the Overland Trail, gave birth to South Pass City, a "City of Rumors," where each day brought a new promise of elusive placers. Here residents slept "with their blankets tied in a roll on their backs, and their coffee-pot in one hand and a tin cup in the other," ready for "the first whisper of a new thing being struck." The town and all the excitement fizzled quickly, but they had served as an impetus in Wyoming's organization as a territory in 1868.

During the early 1870s, it might have seemed that the era of large gold excitements had expired naturally. While some had hit pay dirt in Colorado, Montana, and Idaho, nowhere had the gold finds proven as extensive or as accessible as they had in early California. Too, westerners had become wiser about the difficulties of living in new and remote locations and digging gold. Further, mining had come to be synonymous with industrial gold and silver lode development, and was associated in the public mind with eastern engineers and financiers, with stamp mills and multilevel mines worked by wage-earning laborers. Then, too, silver was stealing the luster from gold, as evidenced by the 1873 discovery of the Comstock Lode's incredible "Big Bonanza" over a thousand feet below ground and by other silver bonanzas that would continue to be uncovered from Arizona to northern Idaho through the 1870s and 1880s.

In addition, it was becoming harder to find a relatively unexplored western area in which to prospect. Gold or no gold, Americans had long followed a cultural imperative to press westward. More than any other economic lure—farming, ranching, trade, urban development—the rushes of the 1840s, 1850s, and 1860s had hastened this migration, and completion of the transcontinental railroad in 1869 had further reduced the West's remoteness.

No wonder, then, that stampeders rushed northward into a relatively unexplored section of British Columbia in 1874—and that the next big excitement invaded an off-limits region: the pine-covered Black Hills of Dakota Territory, land considered sacred by the Sioux and officially deeded to them in 1868.

The Sioux had been guarding this region against gold-hungry interlopers for a long time, decreeing death for any Indian who revealed the presence of gold in the hills to one of the encroachers, with the same fate awaiting the prospector who was caught. The U.S. Army had been generally successful in the 1860s and early 1870s in barring prospecting parties from the region. But as other western gold finds played out or settled into their industrial development phase, more and more westerners coveted a chance to mine the Black Hills. With the severe national economic Panic of 1873, easterners, too, were looking for another instant El Dorado as a solution to their financial woes. Unchecked railroad speculation had hurt many businesses, large and small, and midwestern farmers had had about all they could stand from four years of a grasshopper plague; they had "raised three crops for the grasshoppers" and were ready to conclude that mining couldn't be "as risky as farming."

Enter Lieutenant Colonel George Armstrong Custer, commissioned by the United States War Department in 1874 to explore the Black Hills with an eye to locating a military post as a way station on a route to Montana and as a monitoring post for Sioux hunting and war trails. Custer's party panned a moderate amount of gold in "Custer's Gulch" without succumbing to gold fever. His report reflected other cautioning-yet-encouraging accounts in the Horace Greeley vein. He hadn't been able to "determine in any satisfactory degree the richness or extent of the gold deposits," and a miner would probably not "in one panful of earth find nuggets of large size or deposits of astonishing richness." But in some places the prosecutor could "reasonably expect . . . to realize from every panful of earth a handsome return for his labor."

What more did people in financial straits need to hear? And again, restless Americans had an excuse for an adventure; a man could not "set comfortable by the fire when there's gold in the hills only five hundred mile[s] from his door." Newspapers blared the cry of "Gold!" and prospecting companies were formed in cities from Boston to New Orleans, from Memphis to San Francisco. Army commanders sighed and waited for the onslaught.

"The Age of Gold"

In the fall, gold-seekers were poised and chafing for action at the towns bordering the Black Hills; some slipped through the army lines, with more following in the spring. By July 1875, according to one "conservative estimate," 600 to 800 prospectors were clambering down limestone and sandstone cliffs into Black Hills canyons. The number would have been much higher, of course, if not for the threat of removal by the army—and of death at the hands of the enraged Sioux, with whom the United States government was now agitatedly negotiating for the hills.

The Indians' real nemesis at this point was probably U.S. Attorney General Edward Pierrepont, who ruled that the law excluding trespassers from the Sioux lands did not apply to United States citizens but only "to those who owed their allegiance to other nations." The ruling removed the army's ability to oust the prospectors. And the Sioux, forced into a dependence on the federal government for their very sustenance, were losing any bargaining power they once had.

The results were predictable: a spring 1876 rush of approximately ten thousand people to Deadwood Gulch in the northern hills, site of the most promising 1875 find, and the Indians' official ceding of the hills in exchange for winter provisions.

For a while, Americans could pretend the old California days had returned. Prospectors spread over a fifty-mile vertical expanse containing thirty-eight gulches and creek valleys, "all of which contained some gold," then over the hundred-mile length and sixty-mile width of the hills, establishing placer mining districts. Meanwhile, town-lot and mining speculators, saloonkeepers, merchants, and other entrepreneurs created and maintained raw, fragile communities bursting with vitality and pie-eyed optimism. Deadwood, crammed into a gulch south of the Belle Fourche River, became the hub of Black Hills life, the quintessential wild and woolly frontier town, and the jumping-off point for a number of chaotic minirushes.

The boom continued until 1879, but, as with other mining frontiers, individual seekers faced a high failure rate. The goldfields were not as extensive or rich as they had appeared, and the luckiest prospectors enjoyed no great fortunes. The real mining bonanzas were accumulating in the hands of a small mining elite.

Former California argonaut and millionaire businessman George Hearst, who had amassed one "pile" in the Nevada silver mines, would buy into the single most promising property in the Black Hills—the Homestake gold lode mine at Lead near Deadwood—and make another. The ordinary American citizen during the gold rush era was repeatedly given evidence that the fortune went to the few, and that these few were usually tough, grasping empire builders, corporate capitalists whose wealth and business savvy were making them powers in American culture.

Nevertheless, the continuing dream of "poor man's gold" led an estimated five thousand seekers—farmers from eastern Washington and Idaho, rapidly aging California and Colorado veterans, and gamblers and merchants—into the Coeur d'Alene Mountains of northern Idaho in early 1884, men floundering through snow-packed gulches and wending through forests so dense that pack mules often stalled before a barrier of trees. A string of small strikes—placer and quartz—spawned a string of sad-looking mining outposts, but even the placers, embedded under timber, thick soil, gravel, and boulders, proved almost impossible to work with the rudimentary equipment at hand, and the seekers concluded dourly that "it was not a poor man's country."

The dream also led some into Baja California in 1889, amid reports of rich Santa Clara Valley placers. San Diego experienced a brief boom as a way station for eager seekers—at one point, more than nine hundred train passengers, most bound for the diggings, disembarked in the city in one half-hour period. The placer camp population was soon estimated at two thousand or more, but the rush deflated quickly, the money that seekers had expended to get to camp and outfit themselves far exceeding the amount of gold they extracted.

The dream received a boost and an initially satisfying twist in Colorado in the early 1890s. After the early excitement generated by the Pikes Peak rush, silver had become the real bonanza for the state in the 1870s and 1880s, with the biggest silver stampede bringing an average of a hundred people a day to Leadville. But unbeknownst to everyone except a hard-drinking, much-scorned ranch hand named Bob Womack, Colorado still harbored a golden bonanza, a "Bowl of Gold," in the form of a high-altitude cow pasture overlooked by Mount Pisgah a short distance west of Pikes Peak.

Womack found gold "float"—particles washed down from another location—in this basin in 1878, but nobody paid much attention. Colorado had already seen too many flimsy excitements, and besides, as the old prospectors were quick to point out, the terrain was all wrong for gold. Then a fellow named "Chicken Bill" made things worse for Womack in 1884 by perpetrating the "Mount Pisgah Hoax." Chicken Bill added gold to, or "salted," a worthless claim near Mount McIntyre, a short distance to the west, but somehow McIntyre and Pisgah got mixed up in everyone's minds, and a number of people were tricked into rushing to the fake gold mine. After that, Womack didn't have a prayer of convincing anyone with his stories of gold at the base of Mount Pisgah.

Then, in 1890, Womack located a lode from which the float might have come and staked his El Paso claim. Still, the gold was resolutely ignored. A few tenderfeet wandered up from Colorado Springs, as did a somber Springs part-time prospector and carpenter named Winfield Scott Stratton. They found little placer gold, but some promising lodes. Finally, in late 1891, a Colorado Springs businessman sparked a general excitement by agreeing to pay eighty thousand dollars for one of the claims, and by spring 1892 the rush was on. The mining community of Cripple Creek, "the world's greatest gold camp," sprawled suddenly across the center of the basin.

The twist here was that seemingly ordinary men could make extraordinary fortunes with quartz lode mines. In the past, lode discoverers had usually sold out early for paltry sums or had gone broke trying to develop their finds. This time around, some discoverers obtained top money for the lode locations—or held on and became successful industrial titans. Colorado druggists A. D. Jones and J. K. Miller, novices at gold-seeking, became rich by tossing a hat in the air, digging on the spot where it fell, and hitting a golden lode. Winfield Scott Stratton became the biggest success story in early Cripple Creek by establishing the Independence and Washington mines, then remaining in control during their lucrative development. The high-country Rocky Mountain cow pasture "created at least twenty-eight millionaires" by 1899, although most were speculators rather than prospectors.

Bob Womack was not among the number, having followed the earlier pattern of selling out cheap and fast. And as always, there were many more losers than winners. The economic Panic of 1893 led desperate seekers from both East and West to converge on a town that very quickly became just another industrial mining location, with serious labor unrest exploding only a few years after its founding. Stratton and other frontier seekers who had struck independent wealth and parlayed it into positions as ranking capitalists now were caught between their western independent, democratic ideals and the realities of corporate hierarchical enterprise in late-nineteenth-century America.

Thus the dream of independent gold seemed to die a harsh death. But nature had two more surprises for the gold dreamers of America—and the first of them, amazingly, would equal the initial California rush in size and in sheer human drama.

In 1896, the United States had possessed Alaska for almost thirty years, but few had ventured into its grand, inhospitable terrain, and fewer still had persevered through an arctic winter or two. Still, some American prospectors exhibited tremendous range, forging into Alaska and from there into the Canadian Yukon, then part of the Northwest Territories.

The 1880s and early 1890s saw a series of small strikes. Juneau owed its existence to an 1880 lode gold discovery; this Inland Passage town became "a springboard to Alaska and the Canadian Yukon" for a grab bag of determined long-term prospectors who minded not a whit getting away from the crowds and comforts of civilization. They were periodically joined by an occasional westerner plagued by the failure of all his enterprises and looking for "a life independent of business vagaries."

On the Fortymile River in the heart of Yukon River country, two prospectors in 1886 hit placer pay dirt, causing a small rush and giving birth to the town of Fortymile in Canadian territory. In 1892, ever farther to the north, two prospectors found gold on Yukon tributary Birch Creek on the Alaskan side of the as-yet-nebulous border, and Circle City sprang up.

News of the Yukon and Alaskan finds created a steady current of excitement in the States; some white-collar New Yorkers were even mushing bobsleds pulled by St. Bernards through city streets in preparation for an Alaskan gold adventure.

Meanwhile, in August 1896, native Californian, George Washington Carmack, son of a forty-niner, and his two Siwash Indian brothers-in-law Skookum Jim Mason and Tagish Charley turned from salmon fishing to prospecting between the Klondike and Indian rivers after a tip from long-term prospector Robert Henderson. On Rabbit Creek—soon to be rechristened Bonanza Creek—they hit gold at four dollars to the pan (or quarter ounce), far richer than previous Alaskan and Yukon yields. The largesse of Bonanza Creek—and nearby Eldorado Creek, "the richest placer creek in the world"—would soon make a four-dollar pan seem puny; one famous pan from a Yukon creek would yield eight hundred dollars' worth of gold.

Like Bob Womack before him, Carmack—known as "the all-firedest liar on the Yukon"—did not seem the most likely person to have located genuine gold. But when he arrived in Fortymile to record his claim carrying shimmering proof of the find, the town quickly emptied. News of the Rabbit Creek strike spread as if by some kind of telepathy among the scattered prospectors in the region. Within two weeks, all the best claims among Bonanza were staked, and gold-seekers were spreading out along the other creeks that, with Bonanza, "all led away in sinuous outlines like immense cobras" from an imposing protuberance called King Solomon's Dome.

In the first fever of discovery, men abandoned or traded for a pittance plots of ground that would bring in a half-million dollars or even more for their lucky second owners. A group of Scotch miners, anxious to move to another creek, relinquished a claim on Eldorado to former Seattle YMCA employee Tom Lippy and his wife Salome, who had migrated north on a hunch and hung on to the claim to realize more than a million and a half dollars in gold.

Confirmation of the dramas being enacted in the Far North did not reach the West Coast until mid-July 1897, but when it came, it touched off a euphoric excitement unequaled since the California rush. For the confirmation arrived in the form of two

Alaskan steamers bearing prospectors from Bonanza and Eldorado creeks and their gold. One ship, the *Excelsior,* docked at San Francisco without much fanfare, but the small crowd of onlookers was treated to the sight of Tom and Salome Lippy struggling to drag a suitcase loaded with gold down the gangplank. Even the least lucky of the men on the ship had fifteen thousand dollars' worth of gold.

By the time the *Portland* steamed into Seattle two days later, five thousand people had assembled on the docks. And they were not disappointed. Former Seattle bookseller William Stanley, who had gone Yukon-prospecting in an attempt to support his wife and seven children, walked down the gangplank with more than one hundred thousand dollars in gold. Another lucky miner had to hire two onlookers to help him transport the nuggets and dust he had bundled in a blanket.

Newspapers across the United States blazoned the story on their front pages, reporting gold "piled about the [*Portland*] staterooms like so much valueless hand baggage" and its source "the most marvelous placer digging the world has ever seen." Again, some offered cool counsel in light of such euphoric statements. A "Miner's Catechism" asked such pointed questions as "Can I leave home perfectly free, leaving no one dependent on me in any manner for support?" and such practical ones as "Am I willing to put up with rough fare, sleep anywhere and anyhow, do my own cooking and washing, mend my own clothes?"

But a nation still reeling from the financial Panic of 1893 embraced the cry of "Gold in the Klondike" with all the fervor of a desperate patient snatching a patent medicine. In particular, the news electrified the economically depressed Pacific Northwest. A large chunk of Seattle's male population—and some women and children—crammed onto northward-bound steamers, with similar scenes being enacted at other West Coast ports. At the same time, railway companies were deluged with business, flooding seekers into these ports from the East and Midwest; a Klondike-bound reporter in late July found Chicago "stirred from center to horizon" by the gold news and St. Paul "Klondikized to a remarkable extent."

Most travelers sailed Alaska's Inland Passage to Skagway or

Dyea, the two chaotic, muddy little towns that materialized as gateways to the Yukon. More than five hundred miles away, over the White and Chilkoot passes and down the treacherous Yukon River, near the diggings, the community of Dawson emerged, spilling across "frozen swampland" at the junction of the Klondike and Yukon.

By September 1897, an estimated 6,000 to 8,000 were in the Dawson area, with another 2,000 navigating and portaging down the Yukon and "other thousands" streaming upriver. In November, the Klondike and its tributaries were "staked from source to mouth."

Yet the stampede to Dawson had really just begun. The mass of greenhorns—now called "cheechakos"—filled the streets of Seattle and came crowding up the trails in 1898, the year Yukon Territory was split off from the Northwest Territories on the strength of its gold deposits.

Some, believing all of Alaska and the Canadian Yukon pregnant with gold, were simply "flying northward like geese in the springtime." They spread across Alaska's tundra following rumors of strikes. At least 1,200 men voyaged on a fruitless gold quest all the way to the Kobuk, or Kowak, River which emptied into Kotzebue Sound in the Arctic Circle.

Meanwhile, 40,000 of the estimated 100,000 bound for Dawson pushed all the way to their goal. Former Boston shoe clerks valiantly tackled the frigid, rugged passes and whipsawed the lumber to build boats to carry them down the Yukon, all in a very poor gamble for wealth. Those who succeeded in reaching the seat of gold excitement, as with previous rushes, succumbed immediately to new lures, everyone stampeding "to all points of the compass" at the slightest rumor. One rush started on the strength of a Swede's "bringing in a handkerchief of dirt which panned 85 [cents]."

As those who had arrived fanned out, other stampeders followed alternate routes to Dawson across glaciers, through near-impenetrable forests, across treacherous, icy lakes, in the process crashing "the barrier of the frontier and open[ing] up the northwest." Stampeders were still struggling into Dawson—some from an alternate murderous Canadian route—in 1899. By that time, most of the life had swept out of the town like an ebb tide; almost everyone had returned home, joined the prospecting across the Yukon and Alaska, or headed for Nome.

For Nome provided the last large-scale placer excitement, the last opportunity to claim poor man's gold. In 1898, the "three lucky Swedes"—Jafet Lindeberg, Eric Lindblom (actually a Norwegian), and John Brynteson—had staked wonderfully rich placers on Anvil Creek and on nearby streams in the Nome area. By the time the Yukon stampeders—and those from the rush to Kotzebue Sound—converged on frigidly desolate Nome in early 1899, all the best creek claims were taken. Men out of money, out of energy, and out of hope camped despondently among the driftwood on the Nome beach.

Then in July 1899, someone discovered that the sand of the cold, barren-looking beach was littered profusely with gold. The gold proved so easy to collect that digging it was said to be "easier than stealing it." Rejuvenated "beachcombers" lined the waterfront, covering an area about 150 feet wide and six miles long, many of the seekers averaging $20 to $100 a day in gold.

More people rushed in from Dawson and from other northern points, while a steamer into Seattle sparked frenzy in the States. Lindeberg and Brynteson disembarked amid rumors that they had already dug $400,000 worth of gold, but again, every man aboard carried a rich "poke" of precious dust. Word soon spread that the new diggings would accommodate "half the population of the United States."

However, the steamer had arrived after ice and cold had virtually sealed off the Nome diggings to long-distance travelers. Adventurers all over the United States chafed and awaited spring. Then the steamers chugged northward, swelling the population of the remote outpost of Nome to eighteen thousand.

The excitement was short-lived, the placer gold quickly gleaned. A few subsequent rushes did flame in the first few years of the new century, with Fairbanks, Alaska, and Tonopah and Goldfield, Nevada, originating in gold strikes and gold fever. But Nome had been the last big "poor man's rush." Americans entering the twentieth century faced a world vastly different from the one they had shared a half century before. The frontier no longer spread huge and inviting; patterns of movement and of economic advancement had changed.

Yet men and women carried with them to their graves memories of that golden era. Many had grown up, had grown bitter, had grown foolish, had grown wise, in this grand late-nineteenth-century treasure hunt. And their education had begun on the hard, hard trails. In particular, burned deep into many a stampeder's memory were the daunting overland trails to California in 1849 and the early 1850s.

Overland trails to California
(reflecting 1850 territorial boundaries)

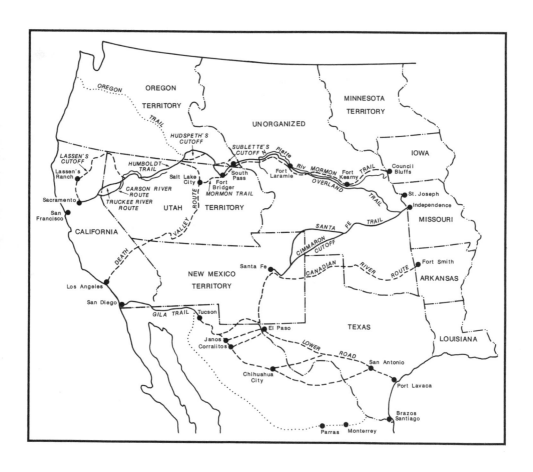

Chapter Two

Overland to California

This California Crusade will more than equal the great
military expeditions of the Middle Ages in magnitude,
peril, and adventure.

—*Journalist Bayard Taylor*

In the mid-nineteenth century, a vast, formidable frontier
stretched between the Missouri River and the California goldfields. Shallow western rivers and a few rough overland trails
provided the only transportation arteries. Way stations consisted
of a handful of Mexican and Indian villages and a few frontier
trading posts, the latter manned by only the hardiest entrepreneurial adventurers for trading with the Indians.

Furthermore, this western half of the nation little resembled
the eastern half. The West loomed as a "land of extremes"—here
were the country's "highest and lowest spots, its hottest and coldest, and its wettest and driest." Travelers could expect to encounter vast treeless tableaus, seemingly bottomless canyons and
gorges, vaulting peaks, strange prickly plants and equally strange
animals, including "the country's longest and deadliest snakes
and its fastest and largest mammals."

How was a gold-seeker to cover this expanse safely and as
quickly as possible? A steady trickle of agricultural immigrants
had managed the feat each year since 1841, most of them bound
for Oregon. Now suddenly, with the discovery of California gold,
the trickle promised to become a flood. In response, a few "pas-

senger line" enterprises were touted, including a dirigiblelike "aerial locomotive" that would whisk argonauts to the diggings in three days. To their credit, few forty-niners fell for this wistful scheme.

One advertised coach service, the St. Louis-based Pioneer Line, actually rolled 120 passengers in twenty vehicles out of Independence, Missouri, in May 1849 on a promised sixty-day jaunt to the diggings. Overloaded wagons, short food supplies, and other problems would cause the trip to stretch to more than twice the promised time, with many of the riders abandoning the enterprise and straggling westward on foot.

For most, going to California meant striking out on their own—but always by attaching to a company for protection from Indians and assistance with the difficulties of overland travel. A fellow would join a group of like-minded men from his town or area. If their number was small—say four or five—they might proceed to a gathering point on the Texas, Arkansas, or Missouri frontier and latch onto a larger company. However, if they could muster twenty or thirty men before departing home, they might form their own quasi-military company complete with a constitution, officers, and separate small "messes." Many, in fact, formed joint-stock companies, each member putting in a few hundred dollars or procuring the money from a local businessman who hoped to share in the venture's profits.

These fiercely democratic companies often cut across or even upended class distinctions; one New York doctor reportedly joined a party under the leadership of his former coachman. The name of a company usually reflected its members' common regional or urban origin; the participants were usually linked by family and community ties. But these organizations also pulled together dissimilar acquaintances or even strangers whose only connection might be the quest for gold.

Whatever the arrangement and company composition, it tended to fall apart early on the overland routes as men chose to shift to other companies—and sometimes other routes—or turned back in discouragement. Typical was one New Englander company that had barely reached Fort Smith, their starting point on one of the southern trails, when a "mess" of seven men from Philadelphia asked to withdraw. At first, the company voted against allowing this move; then it agreed to approve it if the men would settle all bills. The next day, the whole company voted to disband, then decided to continue under new leadership.

Sometimes, one man created problems early in the journey. A member of an Illinois company traveling by riverboat from St. Louis to the Overland Trail outfitting town of St. Joseph became fearful of the cholera then raging and cried all the way. At St. Joseph, he "pour[ed] down cholera drops, camphor, Laudanum, Brandy, or something all the time" and languished in his tent while the others labored to prepare for the trail beyond the settlement. "What we shall do with him I don't know," lamented a companion who soon joined with others in paying the fellow for his share of the supplies and sending him home.

Once the overlanders had launched onto the prairies, leaving their less ready companions behind, their experiences between the Arkansas, Missouri, and Texas frontier towns and the California diggings were so intense and rigorous that California itself has been called anticlimactic. "I would not take $10,000 for what I have learned. . . . It is the journey to learn human nature," wrote young Lucius Fairchild upon completing the trek in fall 1849. Another man wrote in a similar vein, but one weighted with weariness: "You may rest assured that I have an older head on my shoulders by about 1,000 years than when I left the states."

For many, the trip did overshadow the actual search for gold. Even when it did not, it assumed a tremendous significance. Whether overland gold-seekers were stumbling on foot across the Nevada desert or hoisting their wagons piece by piece over a Mexican precipice, they entered a series of intriguingly new and challenging physical environments, saw human nature at its best and worst, and discovered their own levels of endurance. In the parlance of the era, they "saw the elephant"—confronted the difficulties of the journey and emerged subtly but irrevocably changed.

Ironically, these difficulties were overshadowed or exacerbated by the scourge gold-seekers brought with them. The laudanum-dosing Illinois man had reason to cry. Overlanders in the early years could not escape the Asiatic cholera devastating the settled United States and stealing into Mexico—and in their crowding along the trails and unsanitary conditions, they provided an even more virulent breeding ground for the disease.

Cholera did not allow its victims a neat or noble death. Men suddenly bent double with stomach cramps, vomited and lost control of their bowels, alternately shook with cold and burned with fever. One formerly hearty member of a New York company following a Texas-Mexican route vomited so continuously that the veins in his face burst, his "broad forehead . . . marked with the blue and purple streaks of blood that stood under the skin and down both sides of the nose, stagnating in the delicate veins round the mouth and the large arteries of the neck." He soon occupied a Rio Grande grave.

The fatal sickness also lurked on the main route from St. Louis to one of the three Missouri River outfitting towns and westward across the prairies and plains along the artery used by Oregon farm emigrants, the Overland Trail (also known as the Oregon Trail or the Oregon-California Trail). In 1849, the steamboats between St. Louis and the outfitting towns stopped periodically to deposit blanket-wrapped corpses in hurriedly dug holes on river bars or in "lonely woodyards on the river bank." At the outfitting towns, even company doctors fell ill and died. Once beyond the Missouri frontier, it was common for companies to halt briefly on the prairies and plains to bury their cholera dead until at last the outbreak receded in the higher, drier climate west of Fort Laramie.

By one estimate, at least fifteen hundred cholera-stricken travelers died along the Overland Trail in 1849. A young Cherokee Indian in one gold party would recall a scene from Hades on the Platte River, its banks lined on both sides with "a solid mass of wagons" and "men digging graves on each side of the river; men dying in their wagons, hallooing and crying and cramping with the cholera, women screaming and hallooing and praying." Of himself and a youthful friend, he noted, "Oh! My God, if there were ever two boys that wanted to get back to their mothers, we did."

Illness created genuine conflicts for the sufferer's companions. If they were members of a close-knit company from the same town or rural region, perhaps consisting of brothers, fathers, cousins, or lifelong friends, they would be more likely to ease the fellow's pain, slow their pace, and give him, if necessary, as decent a burial as possible under the circumstances.

But often men struggled between humane impulses and the instinct of self-preservation. Many, not for the last time, had to ask themselves whether rendering a service to a fellow traveler would hamper or destroy their own chances of getting to the gold when they had families waiting and depending upon their safe and successful return. Others simply exhibited a selfish impatience stimulated by the quest itself, an impatience that led them to abandon cholera-stricken companions by the side of the road. All learned uncomfortable lessons about their own limitations and those of others. The trails were long and hard enough without the physical and moral burdens imposed by dying men.

Those trails snaked across the continent east-to-west along roughly parallel routes from the northern United States to northern Mexico. Most of the stampeders from the North and Midwest, along with some from the South, chose the best-known, the Overland Trail itself, while others elected to move southwest from Independence along the Santa Fe Trail, a trading avenue since the early 1820s. Still others—southerners, Texans, and northeastern companies that had sailed to Texas or Mexico—followed a series of alternate routes across Texas, Mexico, and the American Southwest, most sharing for part of the journey the Gila Trail through present-day Arizona.

Given the distances and slow modes of travel, the amount of activity on these trails was phenomenal. In 1849, an estimated ten thousand goldstruck travelers took the southern trails and as many as thirty-five thousand followed the Overland. The next year, the overall trail total for 1849 was met or surpassed. Although the numbers dipped in 1851, tens of thousands of gold-seekers continued to flow westward through the summer of 1852.

Each spring, the Overland Trail travelers arrived from towns, cities, and rural regions all over the nation to congregate outside the Missouri outfitting towns, their tents dotting the nearby meadows and riverbanks. Their presence conjured up images of a great army poised to move. "Neither the Crusades nor Alexander's expedition to India (all things considered) can equal this emigration to California," wrote one awed forty-niner.

While most arrived on steamers from St. Louis, some had already had a potentially daunting dose of trail life. In Iowa, men wrestled with wagons too heavily loaded, built rafts and repaired tenuous bridges in order to cross swollen streams, and navigated seemingly bottomless prairie sloughs by cutting and piling great slabs of sod. Sarah Royce, traveling with her husband and two-year-old daughter to the goldfields in a small company, reported that the wagons would sink nearly to the wheel hubs in Iowa mud, and while the party struggled to extricate them, their livestock would wander away, causing further delays.

Once at the Missouri River outfitting towns, the sojourners awaited the growth of prairie grasses, the essential fuel for the animals who would carry or pull their provisions across the West. Pack mules would spare travelers the expense and trouble of a wagon, but repeated packing and unpacking of a cantankerous mule was no easy task, and most travelers wanted a wagon to hold more supplies and to provide refuge when they desired privacy and rest, felt ill, or needed shelter from harsh weather. To pull the wagons, the companies could choose the relatively fast mules, or they could take oxen, cheaper and steadier, but—at about two miles an hour under optimum conditions—aggravatingly slow, given the throbbing impatience of the quest. Most nonetheless opted for oxen, the general wisdom being that "they can be managed with much more facility and with one-half the trouble that is required by mules and with far less damage."

The Overland Trail initially left Independence and cut northwest across the Kansas River, picking up the migration from St. Joseph about a hundred miles out, then running along the north bank of the Little Blue River almost all the way to Fort Kearny on the Platte. This stretch of the journey led over undulating prairies, bright during the peak travel season with wildflowers and pungent in spots with wild garlic. Pilgrims admired the vastness of this grass ocean, as well as the fertile farming soil and tree-lined streams that seemed to beckon settlement and agricultural development. They also paused to unload articles that already seemed an unnecessary drag on their progress and energies.

At Fort Kearny, established by the government in 1848 to provide support for western immigrants, travelers in early 1850 found "3 good-looking houses, a dozen mud or turf houses . . . and several tents." They then moved onto the plains of present-day Nebraska. Now the broad, brown Platte, so muddy "you had to chew it," became their great thoroughfare westward. "This strangest of all rivers," one gold-seeker called it, "with its yellow-ish, whitish water," its wide, shallow course, and its treacherous quicksands. Those who had passed through Independence and St. Joseph followed the south bank through the broad river valley; those who had journeyed along the third artery from Council Bluffs traveled the north bank along the "Mormon Trail." The paths spread as wide as the river, leading almost imperceptibly upward in a tilt toward the Rockies. Immigrants marveled at the already storied landmarks: the domed mass of Courthouse Rock, the fragile spire of Chimney Rock, the imposing wall of Scotts Bluff. The air grew drier, the soil sandier. The high plains rolled to the north and south. Artemisia and silver-green wild sage bordered the trail.

Near the intersection of the North Platte and the clear, sparkling Laramie River stood Fort Laramie, a trading post converted to a military site in June 1849. Westward beyond Laramie thrust the Black Hills of Wyoming, Laramie Peak "towering far above all others . . . like a giant among men to whom they dare not approach." Beyond Laramie, companies faced "a broken, rocky, mountainous country," with "broken ledges of bare rock and sparse, scrubby pine rising hill upon hill." Most continued to follow the North Platte, although now canyons and gorges made it difficult to do so. Their immediate goal was the Platte ferry crossing run by a few of the Mormons whose settlement of the Great Basin had barely predated the rush. From this dangerous crossing many of the travelers proceeded along an alkaline desert bordering the North Platte, joining their fellows on the Sweetwater River and moving steadily toward Independence Rock. Reaching the granite mass, they added their names to a surface already embellished with the scribbled and carefully lettered names of other western adventurers.

Now the route led gradually upward to South Pass, an unprepossessing indentation among arid hills bordered on the west by a broad plateau. The pass marked the halfway point of the journey from the Missouri, and across the plateau lay the Rocky Mountains. Streams provided no east-west thoroughfares through this rugged terrain, and a maze of trails and cutoffs competed for the traffic. A sizable minority of the gold-seekers on the Overland Trail (one third in 1849) dipped southward on the Mormon Trail, having decided to take advantage of the comforts and services of the settlement of Salt Lake over two hundred miles to the southwest. The stretch between South Pass and Salt Lake contained mountain man Jim Bridger's rude trading fort. Beyond it lay the torturous trail through the Wasatch Range. Travelers ascended and descended deep-gouged canyons on "the most miserable road ever traveled by civilized man" before emerging through an aperture resembling "an immense doorway unarched at the top" to a panoramic view of cultivated fields and adobe Mormon homes.

Those who took this route averaged about a week's stay in Salt Lake, devouring fresh vegetables, divesting themselves of wagons, livestock, and personal articles, reorganizing companies, and assuaging their curiosity about the sect with visits to church services and public entertainments.

Most of the gold-seekers, however, took Sublette's Cutoff directly westward, first hitting a fifty-mile waterless stretch on the plateau and eventually wending into the lush, picturesque Bear Valley. When the cutoff arched too far northward, following the route to Oregon, forty-niners forged a path that became known as Hudspeth's Cutoff and joined with the Salt Lake City sojourners at Steeple Rocks near the present Utah-Idaho border. All were bound for the Humboldt River, derisively termed the "Humbug" by those disgusted with its meager, muddy waters and the choking dust that rose along its banks. "The traveling on this river tries men's souls," wrote one man.

Beyond the three-week trek along the Humboldt, though, was greater misery: an expanse of desert so forbidding that many jettisoned the last of their cargo, often abandoning whole wagons, and desperately sought alternate routes. The first established trail, to the Carson River, contained a long stretch without water; the second, to the Truckee River, boasted one boiling spring. Stories of "wagon wheels hub-deep in the hot sands, hundreds of oxen and mules too weak to move, left to rot with other carcasses in the blazing sun, men with lips cracked and tongues swollen begging for water" on both these trails led some forty-niners to take a newly blazed third route, Lassen's Cutoff, a near-disastrous northerly detour marked by the stark, imposing Black Rock Desert.

All these routes led at last to the Sierra Nevada, but this range, too, had to be crossed to reach gold country—or any semisettled country, for that matter. Travelers wearily navigated along and over jagged mountain ridges; Sarah Royce would remember inclines so steep that the head of the animal she was riding "was even with my eyes" as she leaned forward clutching her toddler in one hand and holding on to the saddle pommel with the other. There were also deep, boulder-strewn ravines and swift mountain streams to cross before arriving at one of the California ranchos in the Sacramento Valley, in the rainy season a lush contrast to the arid and harsh environments the seekers had just braved.

Meanwhile, to the south, others on the gold quest were scrambling across landscapes just as intimidating and inhospitable. L. Dow Stephens's Jay Hawker Company had initially taken the Overland Trail but followed bad counsel west of Salt Lake and found themselves on a trail of sorts to Los Angeles, traversing salt-encrusted Death Valley and a great chunk of the southern California desert. This desert, in fact, would be the final trail challenge for all those traveling the southern routes.

Those who took the Santa Fe route traveled over the Kansas prairies, meeting and following the Arkansas River and passing the blackened sandstone of Pawnee Rock. At the big bend in the Arkansas, the trail split. One branch arched across the plains through present-day southeastern Colorado. Travelers enjoyed a respite at the thick-walled trading post of Bent's Fort, then crossed steep Raton Pass southward into New Mexico, bound for the old Mexican trading center of Santa Fe.

The second branch, called the Cimarron Cutoff, had the disadvantage of a sixty-mile desert stretch but was easier for wagon travel. It led along the Cimarron River through a corner of present-day Oklahoma, left the river and dipped gradually across northeastern New Mexico, crossed the Canadian and Mora Rivers, where the first route joined it, and led across Glorietta Pass to Santa Fe. Here the seekers picked up one of various alternate routes across present-day western New Mexico and Arizona, many following the Gila River through the deserts.

Joining them on the Gila would be the parties who departed Fort Smith, Arkansas, and took the Canadian River route across present-day Oklahoma and north Texas. Also traveling the Gila were those who had launched onto a spiderweb of faint trails through southern Texas and northern Mexico. Companies departing from San Antonio or from frontier Dallas journeyed across the arid unsettled half of the new state to El Paso along the Lower Road, "not much more than a hopeful line on a not too accurate map," then cut slightly southward to the Mexican outposts of Corralitos or Janos. Others paralleled the Rio Grande through southern Texas, then dipped across the border to Mexico's Chihuahua City, on the way to Janos. There they joined companies that had crossed through the state's southern extension and those that had sailed to the desolate tip-of-Texas military outpost of Brazos Santiago and then traveled overland through Monterrey and Parras. All these groups then forged through a stretch of northern Mexico as well, its settlements Apache-scourged, its mountains and deserts harsh and forbidding. Beyond Janos, the land itself, at least temporarily, ceased to be an enemy. The travelers found good grass and water and collected their strength for the Sierra Madre Occidental mountain range at Guadalupe Pass.

Yet even finding the pass among the jumbled, rugged ridges and cliffs of the Sierra Madre was not easy. And the road through, such as it was, contained yawning chasms and steep cliffs. Tired men were forced to break apart their wagons and haul them down the gorges and over the heights. They then made their way into the lush Santa Cruz Valley, recuperating briefly and nervously at Santa Cruz, for this beautifully situated settlement, too, had been devastated by Apaches. There followed a hundred miles of mesquite-studded Sonoran Desert to the Mexican frontier town of Tucson.

Beyond Tucson, the gold-seekers thrust their way toward the Gila River, then, swallowing ash-fine dust, followed this "meandering line of brown, liquid mud" westward to its meeting with the Colorado at the present juncture of Arizona, southern California, and Mexico. Now the Colorado Desert spread before them, achingly dry and hostile, its infrequent water holes often "covered with green scum and mosquitos" or "polluted by bloated and half-eaten carcasses and dead mules and oxen." Across it lay San Diego, where the travelers caught ship's passage or purchased new animals for the trip northward to the mines.

These trail descriptions only partially capture the demands of the overland routes, north and south. Gold-seekers daily struggled with climatic changes and other challenges. In a dry summer, grass for one's animals might disappear. In a wet one, streams would swell into monstrous obstacles. Then there were the normal vagaries of weather, heightened by the argonauts' increased exposure. Even on the relatively peaceful prairies just beyond the Missouri, prodigious storms bludgeoned caravans, drenching and pummeling everyone and everything, destroying whole wagons, sending terrific lightning bolts hurtling around companies, and frightening the mules and oxen into wild stampedes while lacerating their hides with hailstones. One Irish forty-niner buffeted by such a storm exclaimed, "Faith and by Jesus, all is lost; we are overboard and none of us will be saved!" There were also windstorms, dust storms, and extremes of heat and cold, with those on the Overland Trail painfully aware that they must get through the Sierra Nevada before the heavy snows set in or risk another Donner Party-type tragedy.

However trivial in comparison, insects proved a maddening accompaniment to the journey. Overlanders found mosquitoes ever "ready to take possession of us and all we possessed." On the Canadian River on the Santa Fe route, sand fleas "fill[ed] the atmosphere like dust," stinging "ears, nose, eyes and beard . . . face, neck, breasts . . . and wrists." Not only were the bites painful, but men's faces swelled so that they could barely see.

At water crossings, frustrations grew from delay, cost, and the dangers of navigation. West of crowded St. Joe, teams "formed in a line by the hundreds . . . awaiting their turns to cross the river," with quarrels over order of crossing "quite common." The cost of the hastily improvised ferries provided by frontier entrepreneurs—including the Mormons at the North Platte crossing on the Overland Trail and the Yuma Indians at the Colorado on the Gila—proved high. Thirty dollars for ferry services between the Missouri frontier and California was a substantial unanticipated cash outlay in an era when an eastern laborer was lucky to make fifteen or twenty dollars a month.

Yet work and skill were involved in moving people, livestock, wagons, and provisions across a swift stream, often with a shifting sand bottom. When ferries were unavailable, many travelers had to lash together their own makeshift rafts, try to swim their cattle across, and hope for the best.

Even with teamsters and ferrymen guiding recalcitrant livestock and travelers over, drowning deaths occurred frequently. Drownings "commenced at the crossing of the Missouri River even before the trip had fairly begun, and continued at virtually every stream and river crossing" on the Overland Trail. Daily fatalities were often reported on the North Platte, even superior swimmers succumbing in the icy, roiling currents. If a stream seemed fordable without swimming, men might still step into a deep current and disappear.

Some streams were less formidable but required constant crossings and recrossings. A New Yorker on the Gila Trail noted that his company had forded the same stream in the Sierra Madre fifty-two times in one day.

More threatening and tiring than the presence of water, though, was its lack. Every trail seemed to contain a stretch of hot, dry desert hell—or at least of bitter plains purgatory. An 1850 pilgrim wrote, "The desert! You must see it and feel it in an August day, when legions have crossed it before you, to realize it in all its horror. But heaven save you from the experience."

Of course, the primary sensation associated with the desert was thirst. L. Dow Stephens of the Jay Hawker Company would later assert that no punishment could compare with this: "Our tongues would be swollen, our lips crack, and a crust would form on our tongue that could not be removed. The body seemed to be dried through and through, and there wouldn't be a drop of moisture in the mouth." And Stephens had not yet even reached Death Valley!

In 1849, much uncertainty and misinformation existed about where the next water would be, especially on the trails only recently blazed or put into use by large numbers of thirsty humans and livestock. One company, stopping at "a fine water hole" in west Texas, met a man who told them that they would find "plenty of water" ahead. The group discarded its water, then found none to replace it. "I never saw the earth so dry as it is in this district, vegetation is completely burned up, and every water hole is dry as a bone yard," wrote a member. After reaching the Pecos River," he concluded, "I hope that I may never experience the want of water again, or witness the like suffering that our men and horses endured."

Another Texas company was too ignorant to fill their canteens completely before venturing onto a long stretch of high plains. Traveling "a region level as a table, destitute of vegetation except grass and weeds," the men came upon only a series of brackish salt lakes, the water undrinkable. Unlike many other overlanders, the thirsty and footsore men found the buzz of a mosquito one night "welcome singing," for it finally indicated the presence of good water holes.

Farther along the southern trails, another California-bound gold party, looking for the Gila River, changed from "hearty and well" men to exhausted, desperate souls within the course of a day, drinking cholera medicine simply for the liquid and chewing mescal plants to stay alive. "None of us cared but little for gold that night water, water was more precious than gold," one confided to his diary.

Of the three Nevada desert routes—Carson, Truckee, and Lassen—the Carson and Truckee were bad enough, especially if a party missed a water hole, as Sarah Royce's did. Surrounded by a sagebrush desert after unintentionally veering at night on the Carson route, the small party had to turn back: "*Turn back,* on a journey like that; in which every mile had been gained by most earnest labor . . . [and] it had seemed that the certainty of *advance* with every step, was all that made the next step possible."

Lassen's Cutoff was the worst for thirst, especially with its two-hundred-mile detour. Argonauts pushed eagerly toward Rabbit Hole Wells, expecting a green watered haven, getting instead "an abomination of desolation . . . ash heaps of hills into which slowly percolated filthy-looking, brackish water," many of the wells "filled with the carcasses of cattle which had perished in trying to get water." Then they encountered the Black Rock Desert, enduring a desperate trek across its expanse to the boiling Black Rock Spring.

So much food had been discarded by the time the overlanders got well into the deserts that hunger added to their fatigue. Some companies subsisted on the parched corn carried as animal feed, and many were lucky to have held on to moldy flour, coffee grounds, a bit of bread, and "rancid bacon with the grease fried out by the hot sun." In the Colorado Desert, two brothers from Ohio agreed to save their last slice of bacon and use it "only as a greaser for cooking cakes to be made in the frying pan of the meager remnant of flour remaining to them." Overcome with desire, one rose early, and fried and ate the "precious slice," causing "a small pandemonium" when his brother found out.

The desert crossings brought other sufferings, ranging from snakebite to severe sunburn. One traveler reached Janos on the Texas-Mexican route with his knees so swollen that he was unable to climb down from his mount and with his eyes "terribly inflamed from the constant glare of the sun."

At least on the deserts the paths were relatively level and broad; the mountains were a different matter. Gold-seekers hauled wagons up steep inclines only to stand on a summit and see spread before them a daunting series of corrugated ridges. They descended canyon after canyon, locking the wagon wheels and trying to "slide" the vehicles down. Where trees were available, men would cut one down and tie it to the rear axle in order to provide a drag. These descents were as difficult as ascents— sometimes more so. Lucius Fairchild judged the trail from the Sierra Nevada summit to a California rancho to be the "most *damniabl* road on the face of the earth," cluttered as it was with rocks "from the size of a teakettle up to that of a hogshead, over which we were obliged to drive, or rather lift the wagons." The mountains also contained the "stony and rough" streambeds that had to be crossed and recrossed.

The mundane yet crucial task of adapting to the daily demands of camp living provided another test of endurance. Multiple demands presented themselves—locating an acceptable campsite, setting up camp, standing night guard over the animals, repairing wagons, rounding up wayward livestock and rescuing them from poisonous alkali sloughs, finding campfire fuel, concocting meals from provisions or from sources at hand, washing and mending clothes, breaking camp—all to be accomplished again and again for the duration of a journey of three to seven months.

Even if a traveler came from a practical rural background, the work proved unusually demanding—especially since the lack of women forced men into the domestic sphere. By one estimate, women made up only about 2 percent of the travelers on the Overland Trail in the first two years of the gold migration. (Other estimates are higher, probably reflecting agricultural emigration.)

One result of this dearth of females was "a heightened sensitivity to the arduousness of women's work." Washing clothes proved such an unpleasant task that men frequently mentioned it as the worst chore of the trip. One fellow was almost relieved when he returned to the stream where he had left his wash and found that it had floated away. In fact, some cheerfully admitted to discarding dirty apparel as they went along—and three young men who reached the mining camp of Weaverville claimed they had made the whole journey from the Missouri frontier without ever removing an article of clothing—"not even their boots."

The other primary female activity, cooking, was not a task that could be jettisoned. Most of the small, military-style messes comprising the companies designated one man, either temporarily or for the length of the journey, as cook. This fellow's chores were "the hardest of all." While the others were releasing their teams to graze and pausing to rest, he was searching out water for coffee and fuel for the cooking fire. Getting the buffalo-dung or wood fire going "rain or shine," breathing smoke and swatting at insects, and working with limited supplies and utensils, the chef often produced disappointing and unappetizing results.

Some nonetheless took pride in their ability to assuage ravenous appetites. One member of a combined company from Arkansas and Indian Territory pronounced himself a "great cook, beat the company making biscuits," while a member of a New York company noted with satisfaction, "Made batter cakes and fried them in bear grease, they were first rate."

Sometimes the fare appealed only to the most adventurous taste. Brother Gatien Monsimer, one of the Holy Cross party charged with finding gold for the University of Notre Dame, reported that a student in the expedition had "mixed beans, pepper sauce, pork, molasses, tea & bread." Monsimer found the concoction "excellent," but the company captain "puked" on first taste.

When women *were* present, they usually proved up to the journey's challenges, "doing their full share of the work" and demonstrating a gritty fortitude at odds with the nineteenth-century image of the delicate, hearth-bound "lady." One company was none too happy when a man and woman and their three children asked to join them in the desert; the men "didn't want to be encumbered with any women," yet "hadn't the heart to refuse." Despite their qualms, they found the wife and mother "as plucky and brave as any woman that ever crossed the plains."

Further, women brought a certain comfort and order to the camps. At Council Bluffs, the wagons containing women and children were "distinguished by the greater number of conveniences, and household articles they carried" and had a more "home-like" look. A Texan journeying in company with a married couple credited the woman for his group's comfortable standard of camp life, concluding that "ladies' " presence was missed "as much . . . in the camp as in the palace."

Without ladies as carriers of the values of civilization, however, there was a certain rude freedom, including the freedom to curse the livestock to kingdom come. Although companies had already expended much effort in the outfitting towns breaking the mules and oxen, trail accounts are rife with references to hours lost in search of capricious mules or more time spent dealing with their tricks. A recalcitrant mule in one company, an animal "that should have been named Brimstone," ran a short distance from the caravan, throwing off his pack as the saddle slid under his belly. Then he lay down and "hammered the saddle with his forelegs" and "crunch[ed] the hollow-sounding wooden saddle" until "his mouth was bloody from laceration of his gums by saddle tacks, etc."

New Yorker Phineas Blunt reported a "ludicrous scene" with a horse on the Gila, the animal refusing to get up or not finding itself able to do so. "[Two] men took hold of its head one took hold its tail and one or two got behind him and then we lifted the animal up and got him once more on his feet," Blunt explained. He noted with pity the pack mules' backs, flesh rubbed and split to the bone, and judged that "many of our animals will give out before we get to our destination."

Blunt was right. While parties went to great lengths to keep their animals going, even ripping into the wagon mattresses to obtain hay for feed, many of these animals suffered greatly only to perish dehydrated and starving in the western deserts. In the crucible between Salt Lake and Death Valley, Stephens's Jay Hawkers stopped to butcher an ox as it expired, converting it to meat and using the hide to make moccasins for the remaining oxen, now without iron shoes. Stephens's laconic report of an incident in Death Valley says much about the difficulties of the trip and the matter-of-fact way in which companies met them: Climbing down a mountainside, "one of the best oxen went over the cliff and broke his back. We had to stop and make him into jerky."

Many animals were simply left to rot, their stench greeting the stream of later arrivals. On the Carson branch of the Overland Trail, one count yielded almost ten thousand livestock carcasses. For those who traveled at night to escape the sun's blistering rays, the bones and bodies loomed pale and ominous in the sagebrush. On the Gila route, engaging in a bit of macabre humor, desert travelers would prop dead, dried animals up on their withering legs along the trail.

The upright corpses symbolized all the gold rush waste—not only of life, animal or human, but of resources in general. Between Fort Laramie and the Sierra Nevada, one traveler found the road lined not only with dead cattle but with "wagons, and fragments of wagons," as well as "log chains in any number, trunks and clothing . . . and bacon and flour . . . in piles and piles." Sarah Royce felt as if she were "at the rear of a routed army."

In addition to provisions, the travelers hated to relinquish their practical and expensive wagons, with their seasoned hardwood axles, waterproofed canvas covers, and iron tires or rims. First, the contents were dumped; then the wagons themselves were pared, with the next wave of argonauts appropriating the abandoned parts for firewood or for modifications on their own stripped-down conveyances.

Some companies also had to relinquish the equipment that played a major role in their grand plans for California. A New York-based group, the Kit Carson Association, carried "a monumental five-story gold washer consisting of five sets of sieves" on a voyage to Galveston, Texas, and then overland westward as far as Laredo, where in desperation they bestowed it upon some no-doubt-bemused Mexican women on the street.

When some companies left supplies behind, they were known to burn them rather than allow them to be used by others—a mean-spirited act and not confined to the travelers' feeling toward any particular group who might make use of them. One 1849 immigrant between Fort Bridger and Salt Lake found many wagons burned, apparently to keep them "from being serviceable to anybody else." Another traveler witnessed a German breaking his tools with the angry comment, "Dey cosht me plendy of money in St. Louis, and nopoty shell have de goot of dem, py Got!"

Immigrants reportedly pursued and shot a man who fired dry grass in order to hinder the progress of his potential competitors in the goldfields, but the burning of bacon and other supplies was considered appropriate behavior when performed to keep Indians from obtaining the goods, for Indians were perceived as a tremendous threat. With varying degrees of trepidation, wariness, hostility, and curiosity, overlanders faced the possibility of interaction with three groups foreign to them: the Mexican populace—and Mexican bandits—on the southern routes, the various Indian tribes on all overland routes, and the Mormons on the Overland Trail route and Salt Lake Cutoff.

The argonauts who followed the southern routes viewed Mexicans as both exotic and inferior. To an extent, they were enchanted by Latino culture, by the elaborately dressed rancheros, the dark-skinned Indian laborers, the glossy-haired girls in embroidered blouses, the crisply uniformed military men. But to the Euro-Americans, all were ultimately "greasers," and they viewed the military in particular with a scorn stimulated by the United States' recent victory in the Mexican War, by the fact that Apaches continued to dominate northern Mexico, and by a complex of cultural assumptions. To mid-nineteenth-century Americans, Mexicans were dirty, lazy, and backward, and gold-seekers often blasted through their communities with all the arrogance of conquerors. Three Mexican War veterans in one Texas company proposed that the group capture the town of El Paso after an altercation over stolen horses there. Rejecting this plan, the company journeyed to Janos where the Mexican residents called upon them to enter into a plot to ambush a number of troublesome Apaches. Here, despite their obvious prejudices, the Texans expressed some commonality with the Mexicans, allowing that the latter were "civilized men" with whom they could sympathize. But the company's acquiescence seemed to have more to do with sheer eagerness for a fight than with anything else.*

All in all, the gold-seekers' attitude toward their Mexican hosts reflected strong racism but also some genuine sensitivity. American gold-seekers expressed shock and disgust when four of their number entered a Mexican church containing worshipers, strode to the front, and lit their cigars from the tapers on the altar. While the overwhelmingly Protestant travelers were vocal critics of Catholic "popery," most responded in awe to the grand Mexican churches and maintained some religious tolerance.

Away from the churches, away from the adobe villages, the travelers faced the threat of the ladrones, many of them former Mexican soldiers harboring special resentment against their Anglo opponents in the recent war.

If no one cared to meet a ladrone, many wanted at least to see

---

* The plan fizzled, and in a later meeting with the Apaches, the Indians tried to enlist the party against the Janos townspeople, claiming that the Mexicans were "damned Christians" but that Anglos and Indians were "*gringos* (heathen)."

a Plains or southwestern Indian. Some were eager to live out the western myth and do battle with them, but most combined their curiosity with intense fear. Members of gold companies were known to sit down and scribble a will when they saw Indians approaching. Steeped in a cultural tradition that could view Indians only as savages and enemies, many assumed that all Indian encounters would be hostile and fraught with danger.

Largely for this reason, men had loaded themselves with personal arsenals—rifles, pistols, knives—and had received considerable encouragement from the United States Congress, which subsidized a cut-rate price for weapons in 1849 and 1850. The travelers often panicked and fired away at imaginary Indians—a distant tree or a clump of suspicious sagebrush. One wagon party mistook prairie hens for Pawnee. Ironically, it was not uncommon for the inexperienced, easily spooked, and bullet-happy marksmen to shoot themselves or their companions, the woundings and killings by accidental discharge easily outnumbering those perpetrated by supposedly bloodthirsty Indians.

On the southern trails, the Apache and Comanche did pose a looming danger, as scourged Mexican villages testified. But travelers encountered members of both tribes without any hostilities—and sometimes men owed their lives to the natives. Apaches directed gold-seekers to water. On the Gila Trail, the Pima Indians, who had transformed the desolate earth bordering the Gila River into a bright patchwork of fruit and vegetable crops, hurried ten or fifteen miles into the desert to meet weary travelers and supply them with "gourds of water, roasted pumpkin, and green corn."

The Yuma maintained an essential Colorado River ferry service. Men complained that the Yuma stole stock and intentionally drowned travelers' animals in order to make dinner of them, but an army lieutenant sent to establish a post nearby found the Yuma—lithe, imposing figures with polished seashells ornamenting their features—more sinned against than sinning. Argonauts would refuse to pay for the dangerous river work and would even steal stock and blame the theft on the Indians.

Serious Indian trouble was seldom encountered on the prairies and plains. The Shawnee and Potawatomi proved "peaceful, poor, altogether a contrast to expectations," and the Pawnee, once fearsome, had been devastated by the Sioux and the cholera. Even in Sioux country in the valley of the Platte, the hordes of wild Indians that peopled many a forty-niner's imagination failed to materialize.

One company, the "Prairie Rovers," did encounter a strong Indian force barring a Platte River crossing and announcing a list of grievances, including the interlopers' destruction of buffalo and pasturage. "It is true that we felt somewhat guilty of those charges," wrote one Rover, "but [we] felt no disposition to turn back." Farther along the trail, the stealthy, secretive Digger Indians, a "most miserable set of beings," proved the biggest threat to immigrants, quietly stealing stock and occasionally killing individuals caught away from their companies.

Stampeders responded to the Indians they encountered along the trail with a mixture of feelings—wonder, interest, disgust, appreciation, ridicule. There are hints in their accounts of brief sexual liaisons, or at least flirtations, with Indian women. Contradictions and strains of romanticism waft through most accounts of Indian contact; one Texan waxed rhapsodic about a California Tejon Indian with whom he visited, invoking the image of the noble savage in a pure natural environment. Yet he also called "sound" the advice given him by a settler back in Texas: "Shoot at every Indian you see and save them a life of misery in subsisting on snakes, lizards, skunks, and other disgusting objects."

Unabashedly ethnocentric, the argonauts could never quite see the Indians as *people*. And in some ways, the Mormons seemed just as strange as the Indians. Their practice of polygamy, their leaders' public proclamations of enmity to the United States government, and their aggressive exercise of power in Salt Lake City did not sit well with most stampeders, but common language and cultural and ethnic roots and the fact that each group needed the other made relations generally agreeable. The travelers needed the Mormons to ferry them across the Platte and Green rivers, to provide information about what routes to pursue and what to expect, and to offer a way station in Salt Lake. For the Mormons, who had suffered under terribly deprived frontier conditions in establishing their colony, the migration hastened a far better standard of living as the gold-seekers "came powering in from Different parts of the Earth on their way for the Mines," leaving "Horses Oxen Waggons Cloathing Ploughs Spades Shovels Hoes Saws Augers Chisels Plains."

While relations with the Mormons, as with the Indians and Mexicans, turned out to be generally amicable, men traveling together often faced dissension or hostility within their own camp. Companies continued to break up, shift membership, and change captains all along the trails. Men found coordinating the movements of large groups and locating enough grass for the livestock to be exhausting. They also differed on which route to take or what pace to set, discovered personal incompatibilities, or simply lost their civilized veneer or good humor under the trying conditions: "If there is any inclination to shirk or do any little mean trick or the slightest tendency to hoggishness, it will soon develop." On the Gila Trail, one member of an Alabama company, sharing joint interest in a wagon with three or four companions, became displeased and determined to chop off a rear wagon wheel as his share. He was prevented only by the threat of a bullet through the heart.

Near the end of a hard journey, traveling northward from Los Angeles, Texan Benjamin Butler Harris tricked fellow company member William Kelly into thinking he had eaten poisoned bread. Kelly retaliated by secretly substituting horsemeat, which Harris had vowed never to eat, for the elk meat he was expecting at a meal. The jokes quickly turned sour, revealing the strain of the long-shared ordeal. In a dispute over which way to position meat to dry, the two almost came to a pistol duel. And the bitterness lingered; many years later, Harris accused Kelly of subsequently telling friends in California and back home in Texas "that, in California, my nature had become demonish and that I had, without cause, desired to murder him."

Perhaps the bloodiest and most bizarre example of violence within companies occurred after Parker French in spring 1850 convinced 187 men to buy tickets for French's Express Passenger Train across Texas and along the Gila to California. He lured his customers with promises of "large tents, portable stoves, an escort of Texas Rangers, and even two howitzers to intimidate any foolishly aggressive Comanches or Apaches." The promises turned out to be lies, and French proved to be an autocratic charlatan blustering his way west on fraudulent letters of credit. The "express train," already in severe disarray, fell apart with a thud in El Paso and when one group of "passengers" pushed onward, they were actually attacked by French and some of his cronies, with fatalities on both sides.

Even without such extreme animosities, gold-hungry men continued to foresake sick companions as cholera was replaced west of Fort Laramie with mountain fever (either Colorado tick fever or Rocky Mountain spotted fever), scurvy, and "bowel complaints." Sufferers were reported straggling into Salt Lake City, having been "left . . . to die by the way side." Overtaking an Indiana company with a sick member on the Humboldt River, a traveler "looked in one of the wagons and saw him still alive and smothered with dust and blankets." He noted, "They buried him in a most inhuman manner, perhaps alive."

Yet there were many positive aspects of the journey that added a sheen to the golden days of 1849. The camaraderie and shared purpose, the pride in proving one's mettle, the pleasures of a free outdoor life, the enjoyment of natural wonders along the way, the exuberant hope of gold—all combined to brighten the pilgrimage.

A strong spirit of community periodically manifested itself, including a willingness to practice the golden rule and a sense of shared adventure. One forty-niner insisted he had seen no man in trouble abandoned "without all the assistance [other immigrants] could render."

Strangers coming into a camp could usually expect a welcome and the best food available, and one company came across half of a "fat, fresh-killed elk" left by a preceding party, with a note "requesting our acceptance of same." At a water hole on the Overland Trail in 1852, another group found a notice put up by previous travelers warning that the water was poisonous. On the Truckee River route, when one party reached the boiling spring, they found water barrels filled by previous wayfarers so that the next arrivals could enjoy cool water. Even better, a forty-niner discovered that previous immigrants had taken the time and effort to bring kegs of water from the Truckee a distance of eight to ten miles along the desert road leading to the river.

Sometimes the service rendered demonstrated self-sacrifice and/or meant the difference between life and death for the recipient. One man twice made his way back across the Nevada desert to take water to the dog who had collapsed while accompanying him on the journey. And L. Dow Stephens, trudging endlessly through Death Valley with the Jay Hawkers, combined the minuscule provisions he and a messmate had been saving—"two or three of our little biscuits and a couple of spoonfuls of rice" each—into a "kind of stew" for an older, failing company member. The latter exclaimed, "Boys, you saved my life." Stephens recalled, "I knew we had. It did us more good, yes ten times over, than if we had eaten it ourselves."

Even when people could not share their scant supplies, they could share information, and a further spirit of communal caring was evidenced by the many messages left along the trails. Sometimes, of course, these messages served as not-so-subtle boasts: "We are ahead of you." But usually they established a vital cord of communication, drawing the travelers together and letting those behind know how their fellows were faring.

When the trails forked, the messages provided information on which one to take—information of limited value in 1849, since the writers themselves went only on guesswork, rumor, or the word of an occasional mountain man or disspirited backtracker. Still, men covered trading post walls with notices to friends and filled barrels at the trail forks with letters to those coming behind. They also wrote on every available surface along the way—"Not a buffalo skull or elk antler along the road but has a notice on it," one traveler observed. The effect of such missives was to make every emigrant feel himself "in the midst of a great company of friends and fellow-travelers."

Another pleasant feature of the journey was the satisfaction of adapting successfully to camp life, to its duties, rhythms, rough conditions, and freedoms. Especially during the easier early stages of the journey, men waxed enthusiastic. Charles Hinman acknowledged the "Burthens of the Journey" in a letter to his son, but added that "as it is I enjoy it. There is something exciting all the time." Lucius Fairchild wrote that it came "as natural as life to move on every day & and live out doors." And a Pennsylvanian wrote his brother that he found the trail experience "far more pleasing . . . than to sit daily locked up in a dirty office. Besides the pleasure of the thing, it gives us health and strength." He considered the tent or wagon a better place for repose than a bed, "no roaches to disturb us at midnight here and no bell to call us at breakfast."

Stampeders also enjoyed the sharing of entertainment after a hard day's travel. Almost inevitably, the entertainment consisted of music and dancing. Fiddles and violins made "many a gold hunter's heart . . . merry and soothed into forgetfulness of daily trials, perils, and pangs for those left behind." Men "danced cotillions upon the green prairie," taking turns playing the woman's part, and engaged in three-part sing-alongs on the open plains, setting the wolves to howling.

At whatever stage of their journey, the travelers also found much pleasure in observing and exploring the landscape. Many diarists revealed a practical botanical awareness born of agrarian living with the land; a typical entry noted "large quantities of wild rye of which our horses are very fond, plenty of pigweed, wild onions, yellow dock, wild peas and dandelions." And even though they were traveling westward to ravage the terrain, stampeders also showed a romantic appreciation for the environment. Men pressed flowers between book pages and attempted to collect seeds. On the Truckee River route, they exulted at the sight of a tree, and on the Gila Trail, they greeted their first saguaro cactus with excitement. The broad vistas of earth and sky aroused even more enthusiasm. Even a dissatisfied observer found on the Gila that "the whole scenery has been sublime, with a little aid of the imagination one could see any description of architecture from the tower of Babel to the most formidable mountains."

Having just survived the Black Rock Desert, William Swain took a mountain ramble: "had a hard tramp, found no game but was well-paid by the grand view of the mountains around." That Swain could take such pleasure in mountains he still had to cross says much about the stamina and basically optimistic outlook of many making the journey. And female immigrants shared the spirit of exploration; Sarah Royce climbed Independence Rock with her toddler in tow and pushed ahead of her party as they neared the summit of their Sierra Nevada crossing, being rewarded with a solitary view of the distant Sacramento Valley.

Wonders multiplied along the way, with the eerie mirages exciting much comment. On the Gila, one company saw fantastic apparitions, including one "resembling the city of Washington, showing the Capitol, White House and other features identified by those who had visited that city." In the Colorado Desert, there suddenly appeared before the same company "three or four hundred mounted and armed men, supposedly Indians." Most members panicked, but two rode forward to investigate, waving and receiving waves in return from the multitude. Then the phantoms "commenced vanishing at the rear, leaving only a dim, light mist where they had so lately stood."

Wild animals, too, excited comment. While wolves, snakes, and mountain lions naturally provoked wariness, men responded with pleased awe to the wild horses of the plains, their tails "sweeping the ground," their manes "heavy" and "flowing." Buffalo were an even more exhilarating sight to those who had only heard of them. At the first sign of the shaggy hosts, company members would race to a ridge to witness the roiling, earth-pounding passage. They also ran, guns in hand, with the hunter's instinct flaring within them. For American men at mid-century, particularly those on the farm frontier, hunting continued to be "a central aspect of male identity," and buffalo steak was repeatedly pronounced "the sweetest and tenderest meat I have ever eaten."

All in all, there was much to recommend this hearty outdoor life when the stampeder was healthy and resourceful, the environment generous. But even if some of the positive aspects faded as the journey became more demanding, occasional apparent confirmation of continuing California gold discoveries buoyed many a sinking spirit. On the southern trails, a steady stream of Sonorans carrying home gold dust and nuggets fueled hopes. On the Overland Trail, reports were limited, most returnees choosing to sail home via Panama instead. But as the Overland Trail travelers approached the diggings, they received encouraging reports from California ranchers and passers-by—

including one ludicrous tale that "no miners were taking less than $1,000 a day, the average mined being $5,000 to $20,000 per day."

Even when reports were disheartening, argonauts pressed onward enthusiastically. As one member of a gold-seeking company mused, "It is a remarkable and probably a commendable trait in the character of our go-ahead countrymen, to admit no statement contrary to their preconceived opinions, till by personal observation they have proved its truth or falsity."

Whatever hopes and pleasures the traveler experienced, the trek took its toll. Journalist Bayard Taylor, observing the first Gila Trail emigrants to reach San Diego, noted that even after a few days' recuperation they were "lank and brown 'as is the ribbed seasand'—men with long hair and beards and faces from which the rigid expression of suffering was scarcely relaxed." They were clad in tattered clothes and moccasins, and "except their rifles and some small packages rolled in deerskin, they had nothing left of the abundant stores with which they left home."

The next year, a woman on the Carson River found her fellow travelers "a woebegone, sorry-looking crowd—the men with long hair and matted beards, in soiled and ragged clothes covered with alkali dust" presenting "a half-savage appearance." And even in 1852, after the trails had been much-traveled, an observer described women at journey's end as "looking as haggard as so many Endorean witches; burnt to the color of a hazelnut, with their hair cut short, and its gloss entirely destroyed by the alkali."

After surviving the trails through Death Valley, L. Dow Stephens undressed to bathe at a California ranch and reacted with shocked dismay to his own body: "I found I was nothing but a skeleton. My thighs were not larger than my arm, and the knee joints were like knots on a limb, and on my hip bone the skin was calussed as thick as sole leather and just as hard, caused by lying on the hard ground and rocks."

On Lassen's Cutoff of the Overland Trail, the limited government relief troops lifted travelers off mules, "so entirely disabled had [the travelers] become from the effect of scurvy."

No wonder, then, that seekers felt they had already earned

their gold. "Any man who makes a trip by land to California deserves to find a fortune," wrote one. And while many would cling to pleasant memories of their journey, no one wished the experience upon a friend. "I will never recommend the over Land route to get here, for it is Death to many and the next thing to Death to all that come this route," one forty-niner stated in a comment echoed by others.

That left one sea route and three routes that combined sea travel with limited land travel. Two of these—the route through Nicaragua and the one through central Mexico—drew a modest number of individuals and companies. Most gold voyagers in 1849 sailed around Cape Horn while a significant number of the relatively affluent and impatient spilled through Panama on a route that would grow in popularity as the migration continued. The Cape Horn and Panama travelers would "see the elephant" in ways both similar to and vastly different from their overland compatriots, and both voyagers and overlanders would embark upon added journeys in California and beyond its borders.

The Cape Horn and
Panama trails to California

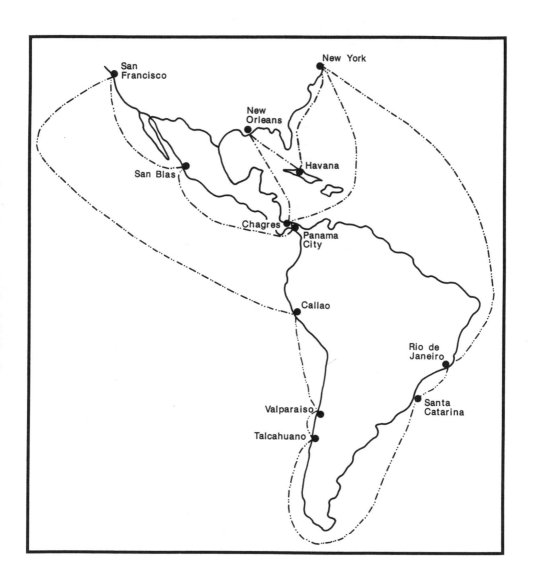

Chapter Three

Sea Travels and California Wanderings

These times, we believe, have harmonized all our minds on one point, viz.: that some other way to California is preferable.

—*L. J. Hall aboard the* Henry Lee, *1849*

If I did not dig more than $2 pr day I would try a new spot, and so it was a new spot all the while.

—*Alonzo Hill from San Francisco, 1854*

While many of the overlanders had launched onto the prairies in hopes of reaching the diggings in three months, Cape Horn travelers departing from northeastern ports knew that their voyage would probably stretch half a year. They faced fifteen thousand or more miles of travel between the East Coast and San Francisco, depending on how far their vessels had to swing into the Atlantic and Pacific. While sleek clipper ships could cut the time by a third, they were in short supply in the initial years of the rush. So were steamships, although the vessels of the recently established Pacific Mail Steamship Company between New York and the West Coast were rapidly pressed into service to transport gold-seekers.

The travelers were often farmers and tradesmen with little or no direct acquaintance with seafaring, despite its New England tradition. One representative group, the Hartford Union Mining and Trading Company, had 122 members, more than a third married, with an average age of twenty-seven. The twenty-three farmers in the group made up the largest occupational category, followed by sixteen joiners and eight machinists. Only two navigators and six seamen were included in the number.

Yet most companies relied on experienced crews, and the Cape Horn route had been used by New England maritimers for forty years. Furthermore, it would prove the safest of all the options, for cholera only brushed Horn travelers, saving its ravages for the overland and Panama routes. Of the thousands going by way of Cape Horn, only about fifty perished en route the first year, some by drowning, others as a result of scurvy and various illnesses.

The vessels carrying the Horn-bound travelers quickly arced out of sight of land, encountered an eight-hundred-mile stretch of "equatorial doldrums," then followed the trade winds to within sight of the Brazilian coast, its mountains ascending dramatically from the water, and sailed into a bay bound by "lofty hills." Here the city of Rio de Janeiro spread invitingly before the voyagers, its whitewashed homes bright against the green vegetation, its churches, monasteries, and convents lending a quaint charm to the hilltops.

Most ships also made a stop at Santa Catarina, an island village of adobe homes and bounteous tropical fruits five hundred miles farther down the coast, then approached "the perpetually storm-lashed waters off Cape Horn," considered "the most daunting stretch of ocean on the planet." Tackling this passage, too, were ships carrying independent European companies— and a few thousand French political dissidents, adventurers, and moral rejects sent California-gold hunting by French authorities on the proceeds of a national lottery.

A few vessels squeezed through the Strait of Magellan rather than "rounding the Horn," but they were assaulted by violent weather and confounded by tricky currents. The Cape passage was preferable—but it could take anywhere from a week to more than a month to accomplish.

Once around the Horn and in the Pacific, the sailing vessels slid into tranquil waters, most making stops at both the mud-daubed village of Talcahuano and the cosmopolitan city of Valparaiso on the Chilean coast. Some ships also weighed anchor at Callao, Peru, "one of the best ports on the Pacific" because even the largest vessels could anchor there safely.

From the South American coastline, the ships made a slow but steady swing to the northwest until they were on about the same latitude as San Francisco Bay yet still far to the west, "sometimes halfway to Hawaii." Here the sails filled with prevailing winds and drew the boats directly eastward to the bay.

In contrast to this elongated journey, the Panama route could be accomplished—with a dose of luck—in eight weeks or less; fortunately for the travelers, in 1846 the United States government had secured a right-of-way agreement with the Republic of Granada, of which Panama was a part.

The earliest gold-seekers on this route paid as much as $450 for a ticket on a Pacific Mail steamer out of New York, the vessels chugging down the eastern seaboard with brief pauses to pick up passengers and mail at Charleston's bar. They proceeded southward along the coast and through the Straits of Florida, many making an interim stop in Havana and steaming through the Gulf of Mexico to New Orleans. There eastern passengers were joined by a mixed group of southerners and midwesterners, including "tall, gaunt Mississippians and Arkansans, Missouri squatters who had pulled up their stakes yet another time, and an ominous number of professed gamblers." The voyagers—on this route, traveling as individuals or in small companies—then were transported into the Caribbean.

As rival passenger steamer companies emerged, this pattern would vary somewhat, but all were bound for the mouth of Panama's Chagres River. Because it was too shallow for most steamers to enter, early gold travelers were "lightered," or transported by canoe, to the village of Chagres, a nondescript muddle of cane huts, two of which soon went by the elegant names of Astor House and Crescent City Hotel. At Chagres, the adventurers rented canopied bungos, native hollowed-log canoes from fifteen to twenty-five feet in length. In one of these, three or four native boatmen would convey parties of two to four gold-seekers up the Chagres River.

At the village, forty-niners got a taste of the transportation problems to come while bartering for the limited number of bungos. One group secured a boatman's services, departed to make a last-minute food purchase, and returned to find their luggage dumped on the beach and the boatman disappearing with a more lucrative fare. They protested vociferously and waded into the water shouting threats, only to "sink into the mud to their necks."

Once secure in a bungo, the travelers embarked on a three-day, forty-mile trip up the wide, swift Chagres, bordered with dense tropical foliage. Little existed in the way of human habitation along this route—an occasional abandoned plantation or thatched dwelling. Fearing alligators, nervous types remained riveted in the bungos as much as possible, but parties camped on the banks at night.

Sometimes bungo operators, a cheerfully mercenary crew, demanded more money to move on. They had a habit of stripping naked in order to pole their boats in comfort; one group containing women paid them to keep their clothes on, but to no avail. The boatmen entertained their passengers with what had to be incongruous renditions of "O Susanna."

The pilgrims left the serenading river guides and bungos at one of two inland villages, Cruces or Gorgona. There travelers bargained with natives for transport of their goods, shouldered the provisions themselves, or bought or rented mules and horses to carry goods and people over one of two rocky and often mud-mired paths. After twenty or so miles, they would reach sun-drenched Panama City, washed on three sides by the Pacific.

Travelers were struck by the rapid apparent Americanization of the Panama City community as a result of the gold migration, with "American hotels, refectories and grocery stores" in evidence "on every street." But the Panamanian people, the tall stone houses, the decaying religious edifices, the imposing battlements, and the "grass-grown plazas" served as reminders of the Latin American location and roots; the city had existed for more than 150 years when the United States became a republic. Accommodations were rudimentary, one hotel being "a miserable little tent, not more than 12 feet square," and containing only "three cots, one table, two plates, two knives and forks." Innkeepers—mostly American—often had a ruffianly look, as if they had been "steeped in water for a length of time."

Now luck or a through ticket for a steamship became essential, as the number of vessels of any kind available to carry passengers up the Pacific coastline failed to meet the demand, especially since many ship crews who had reached San Francisco had deserted for the goldfields rather than continue providing transportation. As cholera—and yellow fever, malaria, and typhoid—scourged Panama City, gold-seekers jammed the beach, jockeying for a prized berth on a steamer, which could set a straight course for California. The available sailing ships were stymied by winds and currents paralleling the coast; like the Cape Horn ships, they had to swing far westward into the Pacific for most of the voyage, sharing the ocean routes not only with the Cape Horn vessels but with those carrying six thousand or more Australians and—from 1851 on—large numbers of Chinese en route to the goldfields.

In truth, most Panama travelers were happy to board anything that looked as if it might reach San Francisco. Those who endured uncomfortable ocean passages on sailing ships were luckier than the men who tried to set off for California in native bungos, for the latter quickly turned back or simply disappeared.

Meanwhile, those fortunate enough to secure a steamer passage expressed great relief, both at departing and at obtaining this mode of travel. "All were so glad to leave this place, where they had experienced so much anxiety and impatience," wrote one ticket holder. The steamers powered their way up the Mexican coast, passing volcanic islands, beaches lush with "palm, cocoa, banana, and orange trees," and mountain ranges "rising faint and blue through belts of cloud." Acapulco officials temporarily rejected any stopovers there because of the specter of cholera, but many steamers did pause at San Blas, a village of cane huts surrounded by dense jungle, and at the white-walled town of Mazatlán. In these locations, Mexicans and the Americans who had taken the central and southern Mexican routes scrambled for passage to San Francisco as well.

The steamer passengers' first view of the California coastline was not heartening, as it "appeared to be a mass of nearly naked rock, nourishing only a few cacti and some stunted shrubs." But such stark scenery alternated with lush valley and "blooming plain," the northwestern trade winds billowed in to cool temperatures, and eventually the bay of San Francisco opened before the travelers, its harbor a thicket of ship masts.

Like their overland counterparts, the Cape Horn and Panama travelers had found much both intimidating and novel in their journey. They had been exposed in sometimes frightening ways to the natural elements, including the full power of ocean squalls. Early in his voyage, one gold-seeker found himself clinging to the bulwarks, "not yet having learned to walk the deck at a very acute angle," as a storm rocked the vessel and dipped him completely under a cascade of seawater. A few weeks later, a sudden squall hit his ship at midnight, hurling sleepers from their berths, extinguishing the dim lighting, and creating a cacophony of sound: "chains rattling and falling—sails madly flapping, yard-arms snapping and masts breaking."

These buffetings were nothing compared with the extended gales that assaulted ships in the volatile climate surrounding Cape Horn and the Strait of Magellan. Storms raged so terribly that ships were thrust backward, that "one minute the end of bowsprit & flying gibbon would be under water, & the next perhaps fifty or seventy five feet above." Gales hurled dishes from tables, battered trunks, and smashed bottles of wine, brandy, and syrup "to the damage of books, beds, clothes and carpets." One vessel encountered a storm so great that "it appeared as if the ocean were stirred up from its lowest depths." The succession of such tempests around the cape or through the strait forced passengers on some ships to lash themselves to their berths in an attempt to gain some rest.

Panama travelers, too, experienced sea storms, but not of this magnitude. Instead, their memorable wet trials occurred in the jungles. Boating up the Chagres or bedded down in a village hut for the night, they would hear the rain sweeping in "like the trampling of myriad feet on the leaves" and would witness jagged streaks of lightning cleaving the sky. They could feel the thunder vibrating in the earth itself. The India rubber blankets most carried proved a godsend in the drenching rains; one man noted, "If I am ever on a committee to award premiums for valuable inventions, Mr. Goodyear will be at the head of my list."

As the travelers huddled under their rubber blankets, the

trails to Panama City—particularly the Gorgona route—turned to black bogs. The trails were not ideal under the best of circumstances. The Spanish had used slave labor to create the cobblestone Cruces-to-Panama road, but it was still judged "about as bad a road as I ever walked upon." The thoroughfare had deteriorated, the cobblestones shifting enough to make walking difficult, especially for those burdened with seventy-five-pound packs—and steep drop-offs often loomed a step away.

Horses followed the hoofmarks of their predecessors, deepening the indentions until most of the road was "worn into holes three feet deep and filled with water and soft mud." This mixture would jet upward with the movement of the horses' hooves, "coating the rider from head to foot." Rider and animal alike grew bone-weary with what should have been a short, fairly easy journey. "We pitied the poor horses, but ourselves more," wrote one traveler. A forty-niner's horse sank into mud that briefly consumed all but its head and tail; after this experience, it resisted both coaxing and beating and collapsed two miles from Panama City.

Conditions on this road failed to improve as the rush continued. An 1853 traveler found it "a narrow trace" worn "so deep that the . . . level was far above our heads, and the track in the earth so narrow that we could touch each bank with our hands as we sat on our mules."

Extreme temperatures also added to the discomfort of those on both the Panama and Cape Horn sea voyages. The energy-sapping equatorial heat melted tar and laid a coat of mildew over everything on ship. But those who set sail for Cape Horn early in the year often ran into snow and hail as they neared the cape; the combination of cold and wet produced chapped hands, dripping noses, and dreams of a warm fire. Men huddled over galley stoves to thaw the ice from their "whiskers."

A small number of wives and of single women braved the heat and cold and other challenges, although many of the first ships had not a single female aboard. On the Panama voyage, considered the "clean fingernails" route, the presence of women was more apparent from the first. Journalist Taylor reported "a lady from Maine," a "French lady," and other women fighting the muddy trails in Panama in 1849; 14 of the 364 passengers taking the *California* from Panama to San Francisco that year were female. (The number of women and families multiplied so that in 1851 one bachelor was complaining of the "frightful" din kept up by the babies and children aboard.)

Women and men alike suffered from the boredom that threatened to smother sea travelers, particularly when vessels lay becalmed and befogged. "The monotony of the last few days is enough to make one despair," wrote a Chilean on board a steamer. With mess crews and sailors to do the work, voyagers were generally free to retreat to their berths, roll up in a quilt, and try to pass the hours in limited physical comfort. In this way, their experience differed sharply from that of the ever-active overlanders, although on some ships, passengers helped with the serving of meals and, as with overlanders, men were responsible for their own wash. A forty-niner on a shipboard washing day noted, "Many of the poor wights were wondering what their wives and mothers would say, to see them bending to the operation with blistered fingers and bleeding knuckles—and, we trow, with little success." One vowed that "if he lived to get home again, he would do nothing on these great days of purification but help his gentle friends in the labors thereof."

In their leisure time, passengers looked to items near at hand for amusement, others rapidly following suit when one hit upon a new way to pass the time. They fashioned drinking cups from coconut shells, crafted knives from "cast-steel and rose-wood," converted dimes into finger rings, and—as they neared San Francisco—devised ways in which to carry and guard the gold they would find. Men displayed boxes with secret locks, special belts, and false-bottomed trunks, and they hand-manufactured quantities of buckskin bags.

Such sedentary pursuits ill prepared men for an active life in the diggings. Added to this drawback was the fact that many voyagers gained weight on the trip, even though they chorused bitterly about the food on the Pacific voyage. In the South and Central American ports and villages, they could enjoy greater variety and abundance than did their overland counterparts, but on the seas day after day and month after month the voyagers faced only salted and dried food—salt pork and salt beef—stone-like sea biscuits, and the barely potable water grown rank and mucous from sitting in wooden casks. One forty-niner complained of worm-filled biscuits and bread made with "musty flour," the good "being saved for sale in San Francisco." Another would recall beating biscuits on an oak deck to remove the worms and bugs. In their ocean isolation, despite having a food supply at hand, the sea travelers were less fortunate than the overland travelers, who had at least the possibility of fresh game, edible plants, and Indian-gathered or -cultivated offerings on long stretches of their journey.

Another discomfort affecting the voyagers—and their weight—was overcrowding; there was no room to exercise. In the words of one ballad, "We lived like hogs penned up to fat, our vessel was so small." Few captains could resist the temptation to squeeze as many fares as possible on board, berthing passengers in lifeboats; even relatively uncrowded ships would seem claustrophobically packed after a couple of months at sea.

The real crowding nightmares were reserved for the Panama City–San Francisco route.* As noted before, tickets for this leg of the journey had been oversold back on the East Coast; added to that problem was the fact that many travelers had chosen to wait and purchase tickets in Panama City, where speculators commanded as much as $1,000 for a $250 first-class steamer passage.

After the gold-seekers began flooding into the city, every foot of space on the first steamer that chugged into the harbor was claimed. Men "bid against each other for nothing more than a place to stand on deck," filling the vessel to "cramnation." On the steamer *California* in 1849, there were two passengers per berth, plus others "on each side of the machinery, the upper and lower forward decks, [and] the long steerage extending from the bows aft on both sides of the engine." One man found the berths such "holes" that he spent a night on deck draped over and around a post and iron bar. "Got up, neck, back and arms nearly broke rubbing on the bars, ropes, etc., etc.," he wrote.

Sailing ships proved even worse at providing space; the own-

---

* The Panama route became faster, cheaper, and more popular in the 1850s, with small steamers on the Chagres and other transportation improvements, chief among them the railroad that in 1855 shortened the Isthmus of Panama passage to less than half a day. Meanwhile, the Nicaraguan route also became popular with Cornelius Vanderbilt's establishing of transportation services.

ers of a schooner "appear[ed] to have taken the *exact dimensions* of the passengers, and filled the vessel accordingly."

Those who still failed to gain a spot on board often ran out of money, and pathetic auctions of their goods on Panama streets became a means of returning home, the quest unfulfilled. For those who were not ready to relinquish the dream, the close, unsanitary quarters in Panama City and on the ships only compounded the continuing threat of cholera. Some swallowed so many quack remedies in hopes of avoiding the disease that "when they really were taken ill, they were already half poisoned with the stuff they had been swallowing." Despite their efforts, as many as one third of the passengers died in the epidemic on some of the Panama–San Francisco vessels.

Ship fever, too, struck even the hardiest. One New Yorker who had survived a lightning bolt, "almost a giant, in stature and strength," could not survive a Panama passage and succumbed to the fever. Those who died were quickly consigned to the sea, their bodies wrapped in canvas and weighted with sand or bricks. On one ship, sufferers "could hear the splash" as fellow passengers who had died of the ravaging and mysterious fever were "tossed overboard with very little ceremony."

Another health threat was scurvy, which debilitated as many as half the passengers on some gold rush ships. And while seasickness seems minor in comparison with the other ailments, some so afflicted wished they were dead. The illness struck one forty-niner almost as soon as his steamer departed New Orleans:

> . . . pretty soon my head began to ache, and it got so exceedingly bad that I thought if someone would knock the top of it off with an iron stanchion I might have some chance of living. . . . Almost instantly the terrible pain left my head and I got very sick in the stomach—such sickness as I never dreamed of.

Vicente Rosales had barely boarded a ship on the South American Pacific coast when he reported everyone aboard thoroughly seasick: "The sides of the ship are covered with dripping vomit, and the cabins and the ladders as well. Everywhere you see green faces and hear the sounds of men retching."

Travelers had to nurse sick companions, a demanding task under the best of circumstances, but particularly trying for one party with a member who drank heavily, gambled away most of his funds, and came down with gonorrhea. "We had him to [wait] on all the way over the Pacific," groused one of his companions.

Whether on the Cape Horn or Panama route, there were frequent grumblings about the pace of the journey. Many Cape Horn passengers contemplated cutting across South America from one of the Atlantic ports, while some Panama passengers jumped ship on the Pacific leg, determined to work their way northward more quickly on land.

Horror stories of the results of such choices abound. A small band of Philadelphians in 1849 decided to leave their ship at Cape Lucas, on the tip of the Baja California peninsula, and make their way northward along the coastline to San Diego. They had to go into the mountain interior to obtain drinking water, and once there, they wandered on a skittering trail across the mountains, following hairpin turns, losing the path altogether in one or another gorge, encountering long waterless stretches and living on "the fruit of the cactus and the leaves of succulent plants." They obtained horses along the way, but the animals gave out halfway to San Diego, forcing the travelers to walk the rest of the journey.

When the steamer *Sarah Sands* became stranded awaiting a new coal supply on the California coast, some of the passengers decided to go overland to San Francisco—only about fifty miles as the crow flies, more than twice that on the trails. In comparison with the Baja California party's experience, the trek should have been easy. It wasn't. Several of the travelers had procured horses, but "their horses [were] taken from them by persons who claimed them as thieves." One man accidentally shot himself fatally "just before he reached the city," and many grew severely ill, primarily from "exposure and fatigue."

Those remaining on shipboard often voiced dissatisfaction with their sea captain, who, unlike an overland leader, was not a member of the company or companies he transported. John Angell's company fired their first captain in Rio de Janeiro, charging that he had treated them "in a brutal manner," particularly in providing "most miserable food, & but little of that." Passengers on the gold rush bark *Bonne Adele* tried to do the same at Valparaiso, but the more obstreperous were abruptly left behind by the irate captain.

One forty-niner on the English sailing vessel *Circassian* out of Panama found the captain "gentlemanly and accommodating" but drunk in his cabin every day from 2:00 P.M. until dark. The first mate, however, "never got top-heavy until the old man had sobered up," so "we got along pretty well."

Food surfaces again and again as a sore point in voyagers' journals and memoirs, perhaps because the quest for it brought out a selfish acquisitiveness even more than did the struggle for room in which to sleep and live. One voyager spotted a minister sliding a dish of peas toward his plate while offering a premeal prayer. A sensitive youth stepped out of character by stealing a freshly baked shortcake and failing to confess when another man was blamed. Journalist Taylor, a generally cheerful and optimistic gold rush observer, showed revulsion over the first and most aggressive passengers on his steamer from Panama City heaping their plates with food, while half of the passengers were lucky to get anything. "I never witnessed so many disgusting exhibitions of the lowest passions of humanity as during the voyage," he concluded.

Extended rivalries between factions developed as well, with intimate living, the tying together of fortunes, and incipient competition placing stresses on even close friends. But men confined together for a long period simply had to practice basic cooperation. Companies remained loosely bound by common background, purpose, and rules of conduct—written or implicit—and when diverse travelers found themselves thrown together, a rough democracy and forbearance usually emerged. After the captain of the first steamer into Panama disappeared rather than bring order to the varying claims of would-be passengers, the travelers waiting on the beach undertook a fair system of ticket distribution by honoring the first names on the passenger list. One 1849 voyager, having wisely bought a ticket for the Panama–San Francisco leg of the journey in New York, found himself with "no berth . . . and neither sheet nor pillow" on the steamer *California* out of Panama because of overcrowding. Still, he concluded, "humanity to the Americans at Panama demanded that not an individual should be refused passage, who could consistently be taken."

Voyagers' sense of community was fed by the hours and days spent together poring over gold maps, the delight and friendly rivalry felt at the appearance of another California-bound ship. Within their own shipboard communities, stolid New Englanders joined for sing-alongs and even formed debating societies, while the frontiersmen aboard tended to pass time convivially in gambling. Some travelers played chess and whist. On one Pacific passage, the voyagers congregated every evening on the poop deck, "some making plans, some singing, others showing off their strength and agility." All talked about the gold and how much they planned to accumulate.

Other pleasures punctuated the journey. Upon crossing the equator a week or so into their journey, travelers would be treated to a parade of "Neptune" and his retinue, sailors painted and wearing "hair of ropes resembling sea weeds." Those crossing the equator for the first time would receive a dose of "medicine" from one of the weirdly garbed crewmen, then have an uncomfortable shave and be "christened" with a dunk into a shallow pool. There would be singing, farcical comedy, and parading enough to inject life and levity into an already dulling voyage.

Some voyagers found a wild glory in the storms; all could appreciate the serene periods. "We are moving through the water in a most delightful manner . . . who does not delight in beholding a vessel with her snowy white canvas spread and swelling in the breeze, gracefully skimming the water," wrote a forty-niner nearing Rio de Janeiro. Of the Callao–Panama City leg of his Cape Horn journey, an 1850 traveler noted "delightful" weather and "sociable" evenings on deck, the ocean "spreading out its smooth surface, unruffled by a ripple around us," the sky clear and bright. And the stupendous, multicolored Pacific sunsets provoked awe in many.

Sea travelers also observed varied marine life—porpoises, pelicans, whales, and especially albatrosses, sometimes as many as a hundred of the birds following a ship at a time. Companies would capture the albatrosses and other large seabirds, measure their wingspans, and then, curiosity satisfied, set them at liberty. Yet some travelers blasted away at albatrosses and any sea creature that presented itself, passengers on one ship greeting a shoal of whales with a hail of lead.

Voyagers reveled in the exotic sights and sounds of the seaports and the wonders of the uncultivated landscapes. A woman from Maine felt one could go "mad with delight" at the Chagres passage, "the birds singing monkeys screeching the Americans laughing and joking the natives grunting as they pushed us along through the rapids." The people who inhabited the ports and villages were usually of interest only inasmuch as they could supply transportation, food, drink (grog shops were popular), and temporary shelter. Some travelers also sought out sex, with venereal disease a common result. Unlike the overlanders dealing with threats or perceived threats from ladrones and Indians, the voyagers felt no need to meet Latin Americans with any trepidation. The only wariness in gold adventurers' accounts related to prices. "They take advantage of foreigners," wrote a Rio visitor in 1850 on hearing that two young argonauts had been charged twenty-four dollars for a dinner.

While this was clearly an expensive repast, prices were usually a mere fraction of those in California. American travelers were generally afforded reasonable access to whatever resources the port community had. Some responded by playing the ugly American, with one observer noting that the Panamanians treated the Americans "with more respect than the conduct of many deserves."

The seafaring gold travelers, like their overland counterparts, clung to an ethnocentric confidence in American superiority. This attitude was fed by some of the rude, unfamiliar living conditions encountered, by a perception of "anarchy and the insecurity of life and property" in Latin America, and by an impatient sense of Yankee enterprise, as evidenced by an American's 1851 observation that if Panama City "could be Yankee-ized, there might be some hopes [for] it; but as it is, it is deplorable."

Whatever the situation travelers found themselves in, always there was the gold to raise spirits and lure them onward. At Rio in March 1849, a young adventurer wrote to his parents, "I heard a captain of a whaling ship say he was at San Francisco and said we had not heard the half about the quantities of gold." When California-bound ships met vessels returning from the golden land in '49, they received shouted assurances that the gold was "as plenty as ever."

Like their overland counterparts, early voyagers had only a dim idea of what lay ahead, but the news that they were nearing San Francisco Bay threw them into a flurry of packing with "retorts, crucibles, gold tests, pick-axes, and shovels, and tin pans" being assembled and "laid on the *top*" of individual piles, each man "determined to be the first off for the mines." Some, however, began to feel the weight of their enterprise as they neared their destination. Appearing suddenly "fearful and anxious," they experienced the truth of Robert Louis Stevenson's insight: "To travel hopefully is a better thing than to arrive." Bayard Taylor found *Panama* passengers switching from exhilaration to near "despondency" on approaching the bay. He attributed this feeling, in part, to their having been enjoying a holiday from the cares of the "laborious life," in part to "the uncertainty of their venture in a region where all the ordinary rules of trade and enterprise would be at fault." On arrival in California, it was time for gold-seekers to test their hopes, to justify the time, effort, and money already spent. For most, that meant more exhausting trails.

Despite the words of one argonaut anthem, prospectors could not "gather the gold on California's shore." The treasure lay far inland, the diggings not easily accessible from coastal points of entry or even from the scattered inland supply centers to which voyagers and overlanders at first gravitated.

Marshall's Coloma Valley strike that inaugurated the rush occurred more than a hundred miles northeast of San Francisco, and as men followed news of "better diggings" the goldfields soon spread along the isolated western edge of the Sierra. The Sacramento River did provide an avenue from the port of San Francisco to newly born Sacramento, about eighty miles inland on the approach to the northern diggings, but with steamers in short supply in the early days, the charge of two hundred dollars for a passage was far beyond the average argonaut's means.

Companies or individuals used all their remaining funds to get themselves and their provisions to a promising site, finding that the diggings reportedly the richest were usually the most remote, accessible only to those with "a good capital."

Treacherous roads and trails became a further aspect of "seeing the elephant" and stimulated much wry humor. The story was told of the fellow traveling in a buggy from Sacramento to the mines, discovering a half mile of decent road in front of a way station, and driving to and fro over it for a full day. To the irritated stationkeeper, he explained that "it was the only good piece of road found since leaving Sacramento" and he "wanted to get the good out of it."

Mud mired supply wagons bound for the diggings and generally made travel miserable. A man who paid fifty cents to lay his blanket in a public tent at a road crossing found the next morning that his blanket bed "had sunk several inches into the wet, soft soil." Forty-niners enjoyed the tale of the traveler who almost lost his horse in a bog in a California road and penned a sign: "This place is not crossible, Not even horsible." He was followed by a burro owner who experienced the same near-loss and added, "This place is not passable, Not even Jackassable."

Another obstacle was snow, which piled thick in the high Sierra even in midsummer. John Steele, a member of a small company moving from the Yuba to the Feather River in January 1851, celebrated the arrival of rain, which he and his fellows hoped "would settle the snow and the frost of the past night" and form a crust on which they could walk. Instead, they kept falling through, "and as the snow was very deep, we would sink to our shoulders." Starting down the incline to the middle fork of the Yuba, Steele's party at first found the presence of snow helpful, but then it "terminated in a vast sheet of ice" on which the party slipped and slid, accumulating "bruises, scratches, and torn clothing." One wag asserted, "The Bible says that the wicked stand in slippery places, but I can't and there ain't one in this crowd who can."

Steele encountered a similar situation on a second journey with a single companion, an old friend named Donnelley. Crossing a tilted plane of ice en route to the middle fork of the Feather River, they slid "from tree to tree," swinging from rocks and branches "down the ledges" until night came on and caught them still poised precariously on the ice.

Frostbite affected many, whether from the effects of dampness—Steele and Donnelley repeatedly waded "to the arms" in fording the ice-laden Feather River—or simply from enduring the cold temperatures for too long. Men were brought into Sacramento Valley ranches with frostbitten feet, with their exposed hands draped with icicles.

Many had become lost. Western travelers in general found mountains "like so many mischievous individuals who would maliciously volunteer all kinds of information to a stranger—one pointing this way, another that way, and a third contradicting the other two by disclosing a curving ravine which leads you nowhere."

Louise Clappe, later known under the nom de plume "Dame Shirley," and her physician husband Fayette twice became lost while traveling in 1851 to the mining camp of Rich Bar on the Feather River, even though Fayette Clappe had traversed the route before. Although the Clappes eventually reached their destination, John Steele and his friend Donnelley had to give up on theirs, in the process almost rupturing a friendship that had survived the trip across the plains and travails in the diggings. Climbing the shadowed side of a tree-covered mountain, flailing through deep snow, the two exploded in disagreement over their course, with Donnelley refusing to consult the pocket compass he carried, instead insisting his direction was right. Steele rejected this course, and suddenly, Donnelley drew his rifle, Steele his pistol.

The two stood as if frozen; then Donnelley cast his rifle into the snow. Steele withdrew his pistol, turned, and strode along the route he had chosen, crying with pain at the severed friendship. After about an hour, Donnelley joined him, acknowledging, "I believe this is the right direction." Donnelley had checked the pocket compass; he had also apparently checked the value of the relationship. "We never mentioned to each other our terrible episode," Steele would recall, "But words could not express my joy in the consciousness that we were still friends."

Steele's story demonstrates the tremendous physical and emotional strains of the hardships on the trails. It is no surprise, then, that at an Onion Valley way station, he and Donnelley entered a scene of increasing bedlam. As a storm howled for days, precluding travel, men drank themselves into delirium tremens, one having to be repeatedly restrained with ropes.

Yet men continued extending their prospecting trails, seduced by the continuing exhilaration of the quest and driven by desperation. In 1851, forty-niner Charles Churchill joined an American prospecting expedition into Mexico on the strength of a Mexican's story that the Apache had driven him from Sonoran diggings yielding six to eight ounces per day. "I feel ashamed of myself for ever going on such a foolish expedition," Churchill, back in California, wrote to his brother in detailing the group's travails.

If Churchill felt the fool, he had plenty of company. Californians continued looking for an excitement to match that of 1849 as the placer gold played out. Then the grand excitement of Fraser River finally came, pulling them through the rugged Pacific Northwest all the way into British Columbia.

Two basic routes beckoned Californians to the imposingly remote Fraser River diggings. The first started with a voyage up the Pacific coast and through Juan de Fuca Strait to the Canadian settlement of Victoria, its surrounding hills in summer 1858 thick with the white canvas tents of the gold-seekers. There men hewed their own small crafts to transport them across the Strait of Georgia to the vicinity of the Fraser River. Hundreds were believed drowned in the Georgia passage; those who survived to reach Bellingham Bay found three inadequate outfitting posts and pushed northward to the Fraser, a torturous watercourse "in comparison to which the rivers of California were gentle brooks."

Struggling up a wild and rocky trail, the travelers found a smattering of civilized comforts at two Hudson's Bay Company outposts, Fort Langley and Fort Hope. Beyond Fort Hope, the trail led past a three-mile stretch of rapids, across the river, onto the heights to avoid the rugged banks, across streams, and over "a hilly road, in places very stony and impassable for loaded horses without a large amount of labor in its improvements." By following this trail, seekers arrived at the main diggings at the fork of the Fraser and Thompson rivers and could proceed up either river, if they wished, along horse trails.

Meanwhile, their fellows embarked from Portland, taking streamers along the Columbia to The Dalles, where service terminated along with settlement. Here large companies formed against a perceived Indian threat and followed varied but roughly parallel routes northward through present-day Washington State, then an extension of Oregon Territory. A popular trail continued along the Columbia's banks to Fort Walla Walla, crossed the Snake River, and reached Priest Rapids, then led through the Grand Coulee Canyon to a frontier fort and along the Okanogan River. Crossing the river near its fork with the Similkameen, gold-seekers forged along the latter into British Columbia, then worked their way northward to another Hudson's Bay outpost, Fort Kamloops on the Thompson River, and followed the Thompson and its tributaries to gold country.

Yet a third way presented itself to some easterners and Canadians—an overland westward trek across Canada. Residents of St. Paul, Minnesota, rallied to promote the town as an outfitting city for such a journey through the Red River and Saskatchewan River valleys.

The Fraser River pilgrimages echoed the previous treks the participants had made to California. All of the problems of camp life resurfaced, including the inevitable, often rancorous splitting of companies, although most groups prudently managed to stay together until the danger of Indian attack lessened. Also resurfacing were all of the attractions of such an adventure. Men exclaimed over the grandeur of the scenery and any natural abundance; former forty-niner John Callbreath did not regret an eighty-mile wrong turn near Fort Walla Walla, "as it gave us a fine view of the beautiful valleys of the Umatilla," and he exulted over such finds as a small creek "literally choked up with beautiful salmon trout."

Callbreath's journey from The Dalles to the diggings took eighty-seven days and covered an estimated 1,550 miles. He was much luckier than the hapless souls on the Red River/Saskatchewan route. Only one 1858 group managed to cross the Rockies, but they abandoned their attempt to reach the Fraser after nine months of travail, including a desolate night spent in a storm-battered forest "almost wishing that some friendly tree would fall and end our sufferings."

But even if a party reached the Fraser—as did Callbreath's—with a brutal northern winter coming on, where, exactly, were they? In the hinterlands of the Hudson's Bay Company empire, immediately restive under a different government that required they pay for licenses to dig gold, and far from any supply ports, cities, towns, and villages.

Most of the adventurers quickly decided prospects were poor and began tumbling back to California. Others pressed northward, uncovering further limited gold deposits, eventually sparking the rush to Cariboo country in which Victoria governor James Douglas harnessed stampede energies by hiring gold-seekers to construct the Lillooet Trail along the Fraser to the new diggings.

Some Fraser River stampeders simply rambled about prospecting, adopting a convoluted and exploratory homeward course. Warren Sadler and two companions prospected their way through Washington Territory and Oregon, arriving back in California more than three years after they had left it. Among the vanguard corps of gold-seekers, they made it a point of pride to blaze "new trails for other persons to follow up . . . never follow[ing] old trails."

Meanwhile, the "Pikes Peak" craze was pulling seekers from the East, the South, and especially the Midwest onto old and new routes to the central Rockies. For many, the trip would be easier than anything experienced by the travelers of 1849, but gold trails were to remain difficult, often dangerous, and sometimes deadly.

Trails to
the Colorado goldfields,
1859

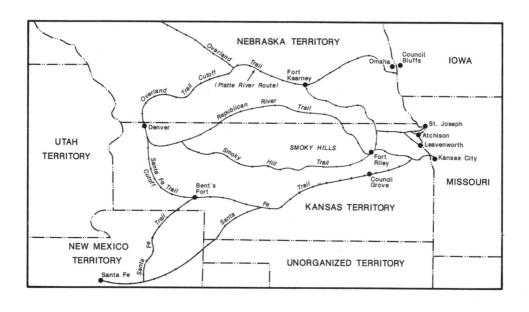

NEBRASKA TERRITORY

IOWA

Council
Bluffs

Omaha

Overland

Trail

Cutoff

Overland

Trail

Fort
Kearney

(Platte River Route)

Republican   River   Trail

St. Joseph

Atchison

Leavenworth

Denver

UTAH
TERRITORY

Smoky   Hill   Trail

SMOKY HILLS

Fort
Riley

Kansas City

Santa Fe
Cutoff

Bent's
Fort

Trail

Council
Grove

MISSOURI

Santa   Fe

KANSAS TERRITORY

Santa Fe Trail

Fe

NEW MEXICO
TERRITORY

Santa

UNORGANIZED TERRITORY

Santa Fe

Chapter Four

One Daunting
Trail After Another

We're "humbugged" if we stay at home,
Or "humbugged" if we leave;
There's few reports that come or go
But what will some deceive

—Lines on fifty-niner G. W. Lucas's wagon cover

W hen the Rocky Mountain excitement exploded, overland travel across prairie and plain had changed little from the days of the forty-niners. There were still no established transportation services, and emigrants continued to plan spring and summer travel to take advantage of grass as livestock feed and to avoid the hazards of cold and snow. A few tried to hasten the journey and lessen its expense by pushing handcarts or simply loading their belongings on their shoulders and backs, but most grouped into wagon companies. These were more loosely arranged than the forty-niner associations. After all, the distance was shorter (some six hundred miles from the Missouri River frontier to Cherry Creek), there were marginally more settlements and way stations, and the main trails—the Overland and Santa Fe—had been traversed by thousands of predecessors.

Still, travel was not easy. "Oh how I did curse the day that I left Pontiac," wrote a Michigan man after slogging through mud from the outfitting town of Leavenworth and making his first night's camp in deep darkness. In particular, difficulties could abound without dependable traveling companions, as Pikes

Peaker W. R. Reed discovered. A short distance west of Fort Kearny on the Overland Trail, he lost his cattle and "broke my shoulder all to d——d smash, running after buffalo." His traveling companions agreed to wait while he returned to the fort to exchange his pony for cattle and have his shoulder set, but when he hurried back onto the trail, they had pushed ahead. Reed had to beg provisions of other travelers before he caught up. "Don't you think I cussed them?" Reed wrote home. He quickly changed companies.

Again, argonauts on the relatively easy Platte River route pushed onto the rolling prairies, most departing from Council Bluffs and the town that had grown up across the river from it, Omaha, and trailing for weeks through the broad, nearly treeless Platte River Valley. The few small settlements that had sprouted along this trail boasted little in the way of population or amenities. One tenuous would-be town, Elkhorn, "consisted of 3 solitary buildings on the open prairie," while the fine-sounding Grand Island City turned out to be "two or three shanties inhabited by old bachelors."

Fort Kearny remained a major stop, its American flag flashing reassuringly through the willows on the river islands as travelers approached. Beyond the fort at the fork of the North and South Platte those bound for the Cherry Creek "mines" turned southward along the latter.* The country was dry and sandy, studded with cacti and wild sage, but eventually the snowcapped Rockies would materialize on the western horizon. From the trading post of Fort St. Vrain on the South Platte in present-day Colorado, it was only a short distance southward along the Platte to the mining region at the junction of the river and Cherry Creek.

Because of the central location of the Colorado diggings between the Overland and Santa Fe trails, the latter was also a popular and fairly easy thoroughfare to Cherry Creek. Santa Fe Trail travelers now launched onto the prairies from Kansas City. Council Grove, 130 miles distant, not only served as the last provisioning settlement of any note but offered the last stands of

* While the word "mines" may lead a reader to envision elaborate underground tunnel systems and industrial development, the gold-seekers used the term for even the simplest surface diggings.

timber travelers would see for some time. Men stopped to cut wood for fuel and wagon repairs, then pushed onto the buffalo range. At Turkey Creek, they could stop at one of two rude turf outposts to buy groceries from a wagon or pay "ten cents per bucket for riley well water." They ferried across the Little Arkansas and reached the big bend of the Arkansas, its water saturated with sand but still cool and pleasant-tasting. On Walnut Creek, they paused at the pole-and-sod way station of one-armed Indian trader Bill Allison, then passed Pawnee Rock and the ruins of Fort Atkinson, making their way over "desolate bluff country" to the Big Timber and luxuriating in the shade of its stubby cottonwood trees after 175 miles without shade. They stopped at Bent's new fort before passing the ruins of the old Bent's fort, which had greeted forty-niners.

At the sight of Pikes Peak to the northwest, the Colorado adventurers left the Santa Fe Trail for a short westerly march, part of it "through sage bushes as high as a man's head," to the junction of the Arkansas and Fountain Creek. There, on the east bank of the stream, some of the immigrants of 1858 established Fountain City, while their fellows followed the creek northward almost to Pikes Peak, then pressed on along the eastern foot of the Front Range to the diggings.

Many Platte River and Santa Fe Trail argonauts, especially those who enjoyed fine weather in fall 1858, wrote home in generally positive terms about their traveling experiences. "We . . . had a glorious time crossing the plains," wrote one. Another pronounced the journey "more a pleasure trip than anything else."

Yet travel took its toll on many, leaving them "hideously hirsute, recklessly ragged, sun-browned, dust-covered . . . sober as judges and slow moving as their own weary oxen." For one thing, the travelers faced many of the same problems forty-niners had encountered. They spent whole days searching for strayed cattle or fording streams. Fifty-niner Romanzo Kingman waded one quicksand-gorged stream eighteen times, helping guide oxen and wagons across. Then his company had to cross the swollen, roiling North Platte. The effort left them "nearly all . . . worn out" and himself "very lame and bottoms of my feet all worn through from the effects of wading the river the few past days."

Rough camping conditions and bad weather also contributed to travelers' discomfort. Frozen and shivering, unable to maintain a night fire because of Indian danger, one winter traveler wrote of covering a draft animal with his own blanket: "Oh heavens my heart's blood almost froze within me, but our teams [are] our lives." Prairie storms still swept upon travelers with unsettling force. One whipped up a whirlwind that destroyed three wagons, scattering pieces across the landscape for miles. Another dumped hail "the size of a quail's egg," leaving marks on the knuckles of the men struggling to hold up tent poles. "Perfect hurricane[s]" dumped streams into tents, soaking their inhabitants, bedclothes, and other belongings.

Although the migration escaped the periodic recurrence of the cholera scourge, accidents continued to take a toll, careless men shooting themselves with depressing regularity. The injured usually joined the hordes of backtrackers along the trails. In 1849, backtrackers had overwhelmingly been those who gave up as soon as they realized the magnitude of their trek. Now, in contrast, the Colorado-bound encountered many who were returning from the diggings in disgust, particularly before knowledge of the richness of John Gregory's and George Jackson's strikes restored the mountains' glow. Even company captains turned around and joined this retreat.

To the extent that companies had formed in response to the Plains Indian threat, most were safe in forging onward with reduced numbers or dividing into smaller units. Bloody clashes between the Plains tribes and United States government troops *had* punctuated the years since 1854, but the Indians were not yet scourging the landscape in retaliation for broken treaties and broken lives. Full hostilities—the "plains aflame"—would come only after the United States had turned on itself and plunged into a civil war. Meanwhile, the natives encountered tended to be friendly and ingratiating, having learned to manipulate the contact as much as possible. One eighteen-year-old gold-seeker reported that "a good looking squaw came to me and said if I would give her my [handkerchief] she would let me lay with her."

Many Indians pressed persistently but nonviolently for food or other items of value. Relinquishing food could be a serious business, though, for some of the Pikes Peakers did not have enough for themselves, having underestimated travel time and thrown their outfits together with a slapdash panache that would haunt them on the road. In addition, most of the Colorado-bound reported some uncomfortable stretches without enough water, grass for the animals, or ready fuel. Few, however, truly suffered from lack of the basic necessities–unless they took the Republican River or Smoky Hill trails.

Those two parallel routes cut more directly westward from the Kansas frontier to Cherry Creek than did the main trails. The two began at the Kansas outfitting towns of Leavenworth and Atchison on the Missouri River and initially followed the Kansas River westward. At Fort Riley the Republican path looped northward along the Republican River and proceeded parallel to the Kansas-Nebraska territorial border, dipped southward into present-day Colorado and joined the Smoky Hill Trail near Big Sandy Creek. The Smoky Hill Trail itself branched onto the river for which it was named, cutting across Kansas Territory in a deceptively straight line on a map.

Promoters assured eager travelers that these routes were well marked and "well supplied with wood, water and camping places" all the way to the diggings, leading one man to complain that "not a word of truth was told him" from the time a travel agent in St. Louis corralled him until he launched upon the Smoky Hill branch. The "trails"—and most vegetation—disappeared west of Fort Riley, giving way to unmarked paths, rugged ravines, dry creek beds, "broken, barren ridges cluttered with rocks," and burning expanses of deep sand. Even the buffalo chips collected for campfires on other trails proved to be difficult to locate. Game was scarce, water supplies disappeared or turned into alkaline water holes, and defiant Indians burned the meager grass necessary for livestock forage.

Travelers on the Republican River route managed to complete their journey—worn, gaunt, and sunburned, but, like many of their counterparts of 1849, generally toughened by the experience. The Smoky Hill route, however, became known as the Starvation Trail.

Horror stories began to filter back to the outfitting towns in late spring 1859. One Smoky Hill company lived on wild onions and prickly pear for nine days, going "three days and nights without a drop of water," a rattlesnake their only meat for two weeks. Others survived on prickly pear and hawk meat, but "scores" sank down along a desolate stretch and never got up again. Unburied bodies and shallow graves became strewn along the route.

The extreme of suffering on this trail is represented in the grisly story of Daniel Blue of Illinois, who set out for the gold-fields with his two brothers in spring 1859. The Blues joined a larger party, which splintered on the Smoky Hill Trail as provisions ran out. The brothers wound up alone save for a dog and a fellow traveler from Chicago. Unable to scare up game in the desolate landscape, they killed and ate the dog. When the Chicago man died of exhaustion, the brothers also fed on his corpse.

The eldest brother began to feel the weakness of approaching death and urged Daniel and the youngest, Charles, to eat his body, that they might eventually be able to return to Illinois and care for his widow and children. Upon his death, the two did as instructed and, gaining energy, pressed westward. But Charles, too, finally succumbed. Daniel, overcome with grief and horror, at first attempted to avoid cannibalizing the body, but he finally consumed the flesh, then gnawed the bones, "breaking them for their marrow," and finally "mangled the skull" and ate Charles's brains.

Arapaho Indians found Daniel Blue, skeletal, almost blind, and unable to walk, his brother's body nearby, the split skull and missing brains confirming his story. Blue survived his ordeal, although, understandably, "his mind never fully recovered."

Ironically, the emaciated, deranged Blue then moved with a speed that long-distance western travelers until that time had only dreamed of, riding into Denver in a stagecoach that had hurtled all the way from the Missouri frontier. Blue's rescuers had delivered him to a brand-new stage station, and in May 1859 he occupied the second coach of the Leavenworth and Pikes Peak Express, a service out of Leavenworth that shortened the plains journey from six weeks to six days.

For $125, an early passenger got a cramped, bone-rattling ride on a hard bench seat, "fifteen inches of seat, with a fat man on one side, a poor widow on the other, a baby in your lap, and bandbox over your head, and three or four more persons immediately in front leaning against your knees." The stage travelers choked on dust, became soaked with rain, or melted with sweat behind the stifling rolled leather curtains. At stage stops, they shared a firmly secured towel, comb, or toothbrush, and swallowed "dry bread or biscuits, rubbery bacon, half-cooked beans, and dishwatery coffee."

Despite such discomforts, the speed was worth the ticket price. But this stage service and the others that quickly sprang up were of little use to most gold immigrants, who needed to carry as many provisions as possible or face exorbitant mining-camp prices on the simplest of items.

Besides, unlike Daniel Blue's, most companies were able to move with some confidence and pleasure across the landscape, like the forty-niners reveling in outdoor living, in hunting opportunities, in new and familiar plants and animals, in the sublime scenery of the Rockies straddling the western horizon. "The first sight we had of them seventy-five miles out," wrote one female traveler, "they looked like silver and gold piled up in the sunlight and I thought, 'Well, we can dig most any place and get the gold.' "

Experiences in the Colorado diggings would make a mockery of the idea of golden plenty, and even those holding the best claims were usually ready to listen to another siren song, to embark on another trail. A man in gold rush Denver convinced 150 to 200 people—men, women, and children—to light out in a caravan for the glittering mines he claimed to have discovered in New Mexico. Their travails included a cattle stampede that badly damaged the wagons, a furnacelike windstorm that burned and blistered their faces and swelled and dried their lips "to such an extent as to render eating almost a physical impossibility," and the laborious raising and lowering of wagons across the peaks of the Sangre de Cristo range. After spending a summer in fruitless search, the group plodded back to Denver, along the way meeting "chilling winds, blasting frosts and dismal rains" that—along with the failure—made the return more difficult than the outward journey.

As gold excitements spilled outward from the Colorado Rockies, the genuine strikes in the Salmon River country in present-day Idaho and in the region just east of the Bitterroots, in present-day Montana, exerted their pull on westerners and easterners alike. These diggings were initially remote even by frontier standards, but such remoteness only whetted the interest of many, especially western prospectors who had found frontier life-style and travel to their liking. For those less inured to hardship, steamer companies quickly offered a way to cover much of the distance. Soon after the first Idaho strikes, water transportation was offered from the West Coast all the way to the Idaho outfitting town of Lewiston with only limited portages. Steamers left Portland following the Columbia River inland to its fork with the Walla Walla, taking the latter and pursuing it to its convergence with the Grande Ronde and on to the Snake, on which Lewiston is located.

On the eighty-mile trek between Lewiston and the first gold rush camp, Orofino, travelers clambered over fallen timbers, descended cavernous ravines, and strained up imposing hills. In the winter, they suffered snow blindness, sometimes forming a weird parade in which the least affected led his companions over the uneven landscape by "carrying the blade of a long-handled miners' shovel under his arm, while the one immediately following him held the end of the handle in his hand and the blade of another shovel under his arm, for the guidance of the one coming after him."

In the summer months, those from the Northeast or upper Midwest could take a train to La Crosse on the Mississippi, book passage for a six-hundred-mile voyage down the river to its juncture with the Missouri River, then secure passage on a steamer up the Missouri as it wended its way more than two thousand miles northwestward to Fort Benton in present-day Montana.*

This second voyage took a month and a half or slightly more—not much longer than a plains crossing, and much easier in terms of daily effort. Still, snags ripped the sturdiest hulls, and

* The Missouri intersected with the Yellowstone near the current North Dakota-Montana border, and entrepreneurs looked for a way to navigate the latter river, which represented a more direct route to the goldfields. But the Yellowstone was even harder to navigate than the Missouri.

sandbars and tricky shallows on the upper Missouri grounded overloaded steamers designed for deep-water travel.

Some steamer captains simply stopped on encountering rapids at Cow Island, depositing voyagers and their freight on the uninhabited strip at the border of "the bad lands," a stretch of "broken bluffs, destitute of timber." This action sparked at least one passengers' "indignation meeting," a concerted effort at demanding redress reminiscent of the "organized protests of the Cape Horners" in 1849. The protestors at this meeting waited two weeks in their hastily constructed brush arbors before another steamer retrieved them—but not their freight—and carried them to within twelve miles of Fort Benton.

Although most Indian bands encountered were friendly, Missouri River travelers also faced a real threat from those Plains Indians aroused to repel invaders; passengers were easy targets for bullets or arrows fired from the shore as the vessel crawled upriver. Then, even if gold-seekers safely rode the steamer all the way to Benton, they were still hundreds of miles east of the Salmon River diggings, across a tortured tangle of mountains.

As the discoveries spread eastward, by 1864 the Montana diggings looped within 120 miles of Fort Benton and could be reached by the Mullan Road, a trail blazed in 1860 to link Benton with Fort Walla Walla, Washington Territory. Even then, most east-to-west argonauts preferred the Overland Trail to Missouri River voyaging. They could now reach the Missouri River frontier via rail, although a man on the Hannibal and St. Joe line in 1864 complained that the ramshackle track "beggar[ed] description." He found "little difference in the depth of the mud in or out of the cars," and almost welcomed a derailment as a chance for uninterrupted sleep.

Arriving at the Missouri outfitting towns, argonauts initially followed the great Platte River Road, then branched onto newly blazed trails—the Lander Road, the Goodale Cutoff. Whole regions jockeyed for identification as alternate trail-supply and outfitting centers, one Minnesota group trying to boost economic development by forging a northern route to the Salmon River goldfields. They covered only twenty miles in the first three days out, as they had to stop and build "several bridges."

Comparable attempts enjoyed partial success at best, even though the United States government tried to help establish and maintain safe, easily traveled roads through its deployment of military force. With the booms in Montana—Bannack, then Virginia City and Helena—work proceeded on the Bozeman Trail, a route forged by former Colorado gold-seeker John Bozeman. The trail was developed from 1863 to 1865 with United States military assistance. But government involvement barely lessened its dangers and might have heightened them by exacerbating Indian anger and fear. This road started at Julesburg, Colorado, on the Platte River route to Denver, briefly joined the Overland Trail along the North Platte to Fort Laramie, then crossed the Wyoming plains and followed the eastern edge of the Bighorn Mountains, eventually meeting the Yellowstone River and shifting westward across southern Montana to the diggings.

Because this route led through prime Sioux, Arapaho, Blackfoot, Crow, and Cheyenne hunting grounds, the Indians adamantly fought the incursion. The United States government responded by establishing trail forts. But even as the road was being mapped out, wary adventurers were testing alternate routes. Old mountain man Jim Bridger in 1864 blazed a trail outside the Indian hunting range on the west side of the Bighorns and had gold-seekers following immediately behind.

The new trails brought great uncertainty. One 1864 Overland Trail party passed both the Bozeman and Bridger cutoffs, as "none of those who had travelled either had been heard from," and road, feed, and water conditions remained unknown.

Whatever route immigrants chose, they encountered more prairie and plains settlement. One 1864 immigrant reported stopping at a small Kansas settlement in his first day's travel to buy "a faucet for our Molasses Keg." A forty-niner might have chuckled in wry delight at the idea that a fellow could purchase anything, much less a faucet, once launched onto the prairies. And he might have guffawed at a traveler's reaction to the information that one stretch of his Kansas passage boasted "no house for fifteen miles"; that seemed "rather 'out of the world.' "

Yet things had not changed as much as it at first appeared. The Kansas farmers were still working to establish themselves and were reluctant to offer any traveler aid beyond a drink of water. At Marysville, with its five hundred residents "the largest place between St. Joe and Denver on the road," one youth asked for the location of the village church and was told that "they did not have such things in this country."

Further, Indian warfare was driving out the Nebraska ranchers. Beyond Fort Kearny, pilgrims found few inhabited outposts; ranches were "deserted and presenting a most gloomy appearance." With wood so scarce, their structures at least served as a handy fuel source, men "taking a rafter off a house or barn or a rail from the fences to replenish their camp fires."

Many travelers formed very limited and casual associations; of his three comrades on the trails to Idaho, a Wisconsin emigrant wrote, "Two of them I had seen but once and was little acquainted with the other. They were bound for Idaho and that was deemed sufficient." But as rumors and evidence of Indian hostilities grew with each westward step, companies, too, grew rather than shrank. One traveler lay awake his first night on the prairies mulling "our danger from guerrillas & Indians" as "everyone we met report[ed] large bands & horrible murders & massacres only 200 miles from our encampment."

On the Bozeman Trail in particular, immigrants feared the worst. A warlike party of 150 Cheyenne and Sioux had met the first wagon train on the trail in 1863, vehemently warning them to stop. Then in 1864 another train fought a six-hour battle with Cheyenne on the Powder River, with four travelers killed and one wounded.

Despite periodic trail discoveries of bloody bodies and hasty graves, incidents of Indian hostilities remained much smaller than popular wisdom would have it. Ironically, some argonauts in this period were safer on the trails than were their families in besieged midwestern frontier settlements, and with nervous, trigger-happy company members, many a stampeder would learn anew "there might be as much danger from frightened individuals in one's own party as from Indians."

The danger was fueled by reports so exaggerated that "the road was said to be strewn with dead bodies and ruins of wagons for miles," and the army troops deployed along the trail could be just as alarmist as the travelers. "They would rush from one end of the train to the other, telling us the Indians were coming from this or that direction," complained one overlander.

Other things besides Indian scares would make a forty-niner or fifty-niner back on the trail in 1864 experience a strong case of déjà vu. "Deliver me from a thunder storm on the plains," wrote one 1864 traveler. A battering storm kept one gold-seeking trio "work[ing] with might and main for an hour holding oil cloth blanket over front of wagon" to keep the conveyance from flooding and the cover from "being blown away by the wind." There were the bedeviling mosquitoes "biting through thick duck trousers and heavy shirt" and "nearly devour[ing]" camp-ers. And there were the personal tensions. One company had gone through two captains before reaching South Pass, and with the resulting squabbling, "it was rather difficult to get [a third one]," the task being accomplished only after "considerable wire-pulling and compromising." Another company exploded in a free-for-all fistfight over rationed water.

Searches for water, fuel, and grass were made worse by the heavy use of the main trails and the aridity of the new ones. Like the forty-niners, travelers of the early 1860s, lulled by false or misleading reports, failed to fill their water casks when they had a chance, and a company on Bridger's Cutoff found only one nonalkaline water source in a 140-mile stretch.

There were, too, the usual problems and duties associated with the animals. One company conducted a frantic two-hour search for their missing cattle, all the while contemplating being "left without a team and with pockets well drained," then dis-covered the animals resting in the bushes "within a stone's throw of the wagon." Soon thereafter, a member reported wearily, "Into the willows the cattle went, and a fine time we had getting them out." Even with animals clearly in place, the guarding of them was an onerous, all-night chore. "This is the 3d night I have herded mules," wrote a traveler, "and [I] have found nei-ther comfort nor convenience in it."

The difficulties of travel within the Idaho and Montana min-ing regions were compounded by the long, brutal winters, even more intimidating than the Sierra storms and cold. Elongated Norwegian snowshoes became de rigueur, with those less adept at this mode of travel following the snow-packed tracks of their fellows. Despite such strategies, an older doctor traveling from Orofino to Lewiston sank into the snow only two miles from a ranch way station, insisting he could go no farther. His companions cajoled for a while; then one announced, "We are not going to leave you here to starve to death and then freeze to death and be eaten alive by the wolves." He threatened to "blow the top" of the prostrate physician's head off if he did not move. From the woods came the howl of wolves, as if on cue, and the shootist fired in their direction, proclaiming, "Old man, the next shot is for you if you don't fetch." The doctor fetched.

Other tired or ill travelers were abandoned, exposing an every-man-for-himself mentality heightened by the demands of the journey. Noting that many cattle were dying along the Lander Road, apparently from eating a poisonous weed, one gold-seeker commented, "None try to discover the cause or invent a remedy so long as 'mine are all right.' It is hard to find a more selfish person than an emigrant."

But, too, there was the camaraderie of the campfire, the familiar buoyant sense of adventure and hope of personal riches, the hunting and sight-seeing. Travelers also now encountered telegraph and stage stations that provided a place to report trouble to and linked them with the wider world. The stations offered an opportunity to mail letters home for twenty-five cents each, although service left much to be desired; one man mailed various letters from the stations, "none of which ever reached those to whom it was addressed."

The stagecoach, of course, provided an alternate mode of travel, albeit still an expensive, impractical—and often frustrating—one. Stage passengers on the Salt Lake City–Virginia City route of more than four hundred miles joked about "walking to Virginia City on the coach" as they "had to push, or at least get out in the boggiest part of the road and wade through gumbo so that the vehicle could move." One man, "tired of paying for the privilege of walking," abandoned a coach and strode all the way into the mining camp ahead of it.

Winter movement remained arduous. The stage run between Virginia City and its successor in the gold rush sweepstakes, Helena, took eighteen hours in the summer; in the winter, when the temperature plunged well below zero, it could take three exhausting days in which passengers fought frostbite and the stage "inched forward behind horses so winded they collapsed on their haunches." Travelers dressed in woolen underwear, thick winter clothes, and woolen socks, then swathed themselves in blankets and buffalo robes. On one such stage ride, nine men crowded in in their bulky winter wrappings, inserting the tenth passenger, "a poor miner with no Arctic gear," in their midst so that he was "insulated from the outside cold like the innermost fish in a sardine can."

On a dry and balmy spring day, stage riding might still be discomfiting. On the same Virginia City–Helena route, drivers took a shortcut over Boulder Hill, with a jouncing ride downward. As the scenery hurtled past, passengers anchored themselves as best they could, trying not to contemplate the precipitous curve the horses had to navigate. On one such journey, so the story goes, the stage occupants did not see both driver and shotgun messenger catapulted from the rattling vehicle. The stage rocketed to the bottom of the hill, its riders "not noticing a jolt of difference between this ride and the usual jarring descent" and ignorant of the loss of the driver until "the runaways, galloping along the valley road, were stopped" as they approached another coach.

As a stage network developed across the upper West, much of it controlled by stagecoach king Ben Holladay of the Overland Stage Line, more and more people availed themselves of the service. But stagecoaches traveled only to established gold camps, where the best claims had already disappeared into someone else's grasp. The growing cadre of long-term prospectors continued to rely on independent movement, walking or riding a horse and packing a mule. Inclined by temperament and calling to get as far as they could from civilization's march, they made and followed rumors of strikes in relatively unexplored areas.

But then, as the old-time argonauts said, "Gold is where you find it." So no one could be too surprised when in 1867, as the transcontinental railroad penetrated the high plains and mountains, gold was discovered in a western area traveled by hundreds of thousands, only fifteen miles from the South Pass of the Rockies and less than two hundred miles north of the proposed path of the Union Pacific railway.

The transcontinental railroad and subsequent lines, like stage service, would prove of limited usefulness for most gold-seekers' far-flung journeys. But the apparent proximity of the "Sweetwater mines" to the railroad's advance helped draw the adventurous, the desperate, the curious. Windswept Cheyenne, born with the rush as a railroad town, became the major outfitting center.

Yet even with a relatively small distance to be covered between railroad town and goldfield, the trails into the rolling plains and the mountain ranges were desolate and dangerous. Scotsman Dave Manson's party of gold-seekers ran out of provisions in a series of arid hills and could locate no game. Manson, who refused to allow a companion to shoot his pet dog for food, grazed along the route "like a cow." Finally managing to shoot two rabbits, he "sat down and ate one of them rabbit's heads up raw." After reaching a cabin, he reported, "I commenced to eat breakfast and I never stopped eating till noon. Lord I was as hollow!"

The Indian threat remained as strong as or stronger than it had been during the Civil War. There were stories such as that about claim-holder Bill Rose, killed by Indians while rushing to assist a wounded teamster. The Indians reportedly removed the sinews of his arms "for tying the steel heads to their arrows" and stripped the sinews of his back for use as bowstrings. "In fact poor Bill seems to have been pretty much used up," concluded one traveler with a whistling-past-the-graveyard callousness.

Although the South Pass gold excitement hiccuped and died as quickly as it had flared, the region would host a larger contingent of gold-seekers during the Black Hills excitement of 1876. Settlements in Iowa, Nebraska, Montana Territory, and Dakota Territory vied for the Black Hills business, but Cheyenne emerged the leader. It was the closest established settlement to the hills—about two hundred miles from the southernmost mining camp of Custer City—and not only had the advantage of being a stop on the Union Pacific cutting straight westward but had Kansas Pacific service from Denver as well. Kansas Pacific promoters encouraged passengers from the East and Midwest to board the train at Kansas City and then stop in Denver to buy mining supplies and "get valuable knowledge by seeing the mines there."

Cheyenne also boasted the limited security provided by nearby forts and "a well-marked and convenient trail" that led northward across the Platte, where first a government ferry, then an iron bridge eliminated wearisome fording. After passing the grassy plains and small streams of present-day eastern Wyoming and running through Fort Laramie, the trail led across the chalk-colored Cheyenne River, "a very pretty Stream about 100 feet wide," to an abrupt geographical boundary—hills "black" with pines etching the sky, a maze of forested slopes and ravines replacing the wide open spaces. Custer City, near the initial strike on French Creek, quickly developed as the mining center of the southern Black Hills, but most gold-seekers pushed onward through the heart of the hill country, to Rapid City and to that most prominent and wicked of the mining camps, Deadwood.

Disembarking from a train in Cheyenne, Jerry Bryan and two friends, all from Illinois, found the outfitting center "chuck full of Black Hillers" and joined in contracting with a freighter to convey them to Custer City for eighteen dollars each. That price included the transport of 150 pounds of provisions per man, the provisions filling the wagons and forcing the "passengers" to walk.

With the Sioux still claiming the Hills by legal right, tales of "Indian Outrages" led the "more timid" to turn around at Cheyenne, some using the excuse of business that suddenly needed to be taken care of at home, others abruptly deciding "that they must wait here to hear from friends." Bryan's small train claimed the relative safety of a large entourage containing "12 wagons about 75 men and two Women." Indians did not assail them, but other difficulties did, chief among them snowstorms that blew down tents and stuffed blankets with snow. A freezing night guard could not rejoice in a crackling fire, as many trains continued to ban the flames that might serve as an invitation to Sioux raiders. Prolonged bad weather caused one man to write, "Only those who have had the experience can understand how great an effect the weather can have upon temper and general morale."

Frequent stage service from Cheyenne to Deadwood soon developed to ease and speed the journey, but the stage sometimes had to wait for days at the Cheyenne River, which periodically rose and raged at an impassable level. And the deep, gluey black mud sucked at wheels, forcing passengers to climb down and scrape the rims and spokes with spades.

Ironically, railroad service directly to mining towns usually materialized, if at all, only when the towns had settled into a decline. Technology could not keep pace with the harum-scarum development that gold rushes precipitated. Besides, prospectors and stampeders were still hurling themselves along the roughest routes over vast distances at the whisper of gold. This number included those who tackled the harsh reaches of the American Southwest in the 1870s. There some found a little gold; none found a lot. The Southwest was stingy in placer gold, and even if there had been huge quartz gold veins streaking the denuded mountains, only a fool would commence a small-scale recovery operation in the midst of arid, Apache-controlled territory.

By the late 1870s, after a series of small, quickly faltering rushes, it should have seemed pretty clear that gold was never going to be the cornerstone of southwestern development. But a passel of stories floating around since the days of the conquistadors, embellished again and again by gold-fevered adventurers, crackpots, and canny businessmen, kept parties wandering sporadically through the deserts. Just about the time prospector Ed Schieffelin discovered Arizona Territory's true bonanza—silver—a party of prospectors departed the silver town of Leadville, Colorado, bound for the Tucson area, where, they were told, gold nuggets were "as plentiful as cockleburs" and "some of the boys were just shoveling up gold dust." They "tramped all the way . . . across the Rockies, down through the San Juan range, past old Santa Fe and Albuquerque to Tucson—a distance of over twelve hundred miles of the hardest, roughest mountain and desert trails" and got lost before reaching the unprofitable "diggins."

A gold party approaching Mexico's Santa Clara diggings in

1889 found the road "impassable" only five miles from their destination and labored a full day with axes, picks, and shovels, to little avail. One reported, "We felt almost hopeless, and finally retired in the wagons to dream of our misfortunes and shiver and freeze."

As the prospectors continued their potholed peregrinations, transportation technology at last connected with a rush. The Colorado Midland Railway led westward from the resort town of Colorado Springs through Ute Pass to the silver town of Leadville and across the Continental Divide to Grand Junction. The line's Florissant station perched only eighteen miles from the Rocky Mountain meadow that suddenly boomed into the gold camp of Cripple Creek in 1891.

As in the Deadwood rush, the railroad itself heavily promoted the new diggings. "A chance of a life-time is worth looking after," proclaimed a Midland ad that concluded, "The ONLY WAY to reach CRIPPLE CREEK is via Florissant and the COLORADO MIDLAND RAILWAY." Passengers switched to a stage at Florissant, endured the long, slow crawl of the vehicle up the hill above Cripple, then held on and sucked in their breaths as the stage rattled fiercely down the incline into the camp, scattering "dogs, chickens, and burros" before jerking to a halt in front of a local hotel.

Cripple Creek's rapid development as an urban industrial center with solid supply and transportation networks separated it firmly from the rush camps of the past. Now a hopeful gold-seeker did not have to band together with others in order to cover the dangerous distance to the diggings. He did not have to haul all his provisions across the landscape or experience first-hand a plains storm with the rain "rushing, pelting, hissing into [his] very bones." He did not have to struggle with broken wagon axles and recalcitrant mules, clamber with wagons and provisions over a series of mountain ridges or ration precious drops of water on the desert or high plains. Nor did he have to fear the power of native Americans now reduced and stashed away on lands nobody wanted.

Historian Frederick Jackson Turner acknowledged in 1893 how much times had changed, declaring the frontier closed and basing that conclusion on 1890 census records reflecting western population density. When gold-hunting meant buying a ticket for a comfortable rail journey and a short stage ride and arriving at a well-supplied urban center resembling industrial towns in Anywhere, U.S.A., gold adventuring was stripped of much of its romantic frontier appeal. And the gold in Cripple had turned out to be overwhelmingly quartz, not the coveted "poor man's gold."

Old prospectors and young would-be adventurers sat by their firesides and dreamed of what had been, of a time when the getting there was as much of an adventure as the hunting, a time when every seeker could feel with every step he took that he almost had a fortune in placer gold in his grasp. And then thousands of Americans from all over the country were swept into a journey just as grand, just as dangerous, just as chimerical and intoxicating as the great California rush.

Landmarks on

the trails to the Yukon,

1898

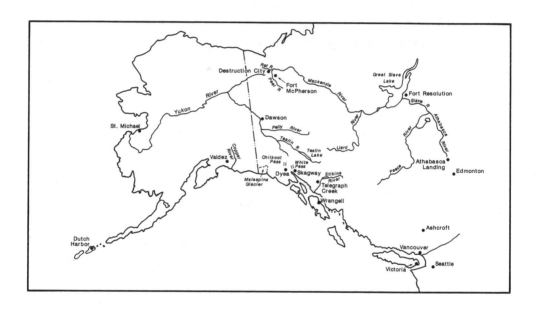

Chapter Five

Bound
for the Far North

For the Cheechako, or newcomer, unfamiliar with the
country and its conditions, there were many severe and
trying tests for his courage.

—Yukon stampeder Marshall Bond

You've read of the trail of '98, but its woe no man may tell;
It was all of a piece and a whole yard wide, and the name
of the brand was "Hell."

—Robert Service

Ιn 1897, Klondike fever inflamed the restless and the desperate
in cities and towns, on farms and ranches throughout the United
States in a frenzy reminiscent of the great California rush almost
fifty years before. The parallels between the two rushes are numerous and striking, among them the international scope of the
response and the high level of interest exhibited by the American populace. Crowds jammed docks and train stations as the
adventurers departed, and everywhere echoed the exuberant
command "Bring me a stocking full of gold!"

The greatest parallel, however, lay in the formidable nature
of the journey itself: more than two thousand miles separated
the new goldfields from Seattle, the closest American city, and
the farther one went, the more rough and remote the landscape
became. Again, the challenge of simply getting there loomed

larger than any other. In a faint but telling echo of the forty-niner who professed to have "an older head on my shoulders by about 1,000 years than when I left the states," an Indiana emigrant wrote, "I am a few days older than when I left . . . and a great deal wiser."

The rush flared, faltered, and faded in the States within the space of roughly a year, failing to approximate the durability of the California excitement. Yet because of its relative brevity, it shared the raw pioneer character of the California rush. In 1849 and the 1850s, technology could not begin to address the transportation needs created by America's rapid westward movement. Now British and American financing and technology could be marshaled fairly quickly to facilitate movement, even in such a challenging environment—but not quickly enough to alter the frontier experiences of most rushers. Thus, massive effort would produce a railroad over one of the toughest stretches of trail in summer 1899, but by then the flood had already turned on itself or eddied elsewhere.

Like the forty-niners, those going to the Klondike sought innovative ways to reach the goldfields, from balloon schemes to "steam sleighs," means that must have seemed foolhardy to anyone with a conception of the territory to be covered. Americans caught up in the contemporary cycling mania bought "Klondike bicycles" to ride to the Yukon. A journalist would find one such conveyance "hanging upon a rock," barely launched upon a particularly precipitous trail, its owner's name and "asylum address" unknown.

The Yukon-bound also resembled the forty-niners and fifty-niners in their outfitting efforts. Alfred McMichael spoke for many when he wrote home from Seattle, "I don't think you will ever understand the strain and excitement there is in preparing for a journey of this kind," especially "in a strange city and among strangers," who seemed bent on "cheating and swindling." Just as placing "California" and "Pikes Peak" before the name of an item in the two previous big rushes had netted merchandisers a tidy profit, so merchants and hawkers now extolled the virtues of "Klondike" boots (with their spikes and armor plate far too cumbersome on the difficult trails), "Klondike" gasoline- or steam-powered sleds, "Klondike" scurvy cures, even an "automatic" gold pan "set on a spindle and operated by clock-work on the gramophone principle."

As in the California rush, items would be discarded along the way, but new challenges differentiated this rush from the mammoth mid-century one.

First, the far northern climate complicated and limited travel. Certainly, those bound for California had journeyed with the nagging knowledge that they must get over the Sierra before winter set in, but that impetus provided only a faint hint of the hazards of the Yukon, where a man might freeze to death if he dared pause on a trail to catch his ice-laden breath, where steamers were suddenly ice-locked in October and stuck until May, where hungry men took an ax to a loaf of frozen bread and still managed only to dent it.

Second, the California argonauts had the assurance of a land of plenty at journey's end. California's fertile valleys and Mexican and Pacific supply networks could sustain a burgeoning population. Despite temporarily skewed supply-demand situations and even bleak periods of destitution or near-destitution, the resources for survival were present both in nature and in commercial centers and trade routes.

By contrast, the Yukon Territory resembled a huge, near-empty icebox occupying a nearly inaccessible attic. Summer brought berries on the mountainsides and plants made immense by the sun that shone twenty hours a day, yet summer was but a blink of the eye in relation to the interminable winters. Goods from outside—foodstuffs and other supplies—had to be conveyed along the Yukon River before it froze or packed with Herculean effort over passes and along frozen waterways. When Yukon stampeders reluctantly abandoned an item, they did so with little promise of replacing it—for any price—at journey's end.

Finally, the argonauts themselves were a new and inexperienced breed. Although one may find romantic references to old California prospectors heading for the Yukon, these men had reached their sixties and seventies and were unlikely candidates for the rigors of Yukon travel. News of grand placer diggings did stir the embers and stimulate talk of following the dream once again, but even the Pikes Peak, Idaho, and Montana veterans were growing gray, and stampedes generally drew young men and middle-aged prospectors enamored of the wandering life.

Hard economic times spurred novice middle-aged prospectors, too, and times were hard in 1897. Cheap transportation costs to outfitting points also fed the stampede; even a fair number of middle-class women joined, both as wives and as independent members of traveling parties. The bulk of the migration remained young men of the middle classes. But that group had become more urbanized, less connected to the natural world, whether inviting or forbidding. Despite a popular-culture emphasis on "manly" athleticism, an army of overwhelmingly urban adventurers, accustomed to treading paved streets and observing nature in the form of Victorian potted plants, ventured onto a plethora of trails and would-be trails into a rugged, virtually uncharted country "almost a million and a half square miles in size—half as big again as the subcontinent of India." No wonder that an observer would find "two men in three" bounced naively into a strange new world and "carried along by the odd man."

Ironically, the trails started out easily because transportation service had developed so extensively. Although some chose the old round-the-Horn water route, trains conveyed most travelers across all or part of the United States' expanse. Ten dollars— even then, a cheap price—bought a Chicago purchaser a seat to the West Coast on a special "gold rush car" plastered with enticing pictures and cluttered with "glass jars full of nuggets and dust," as well as "books, maps, and pamphlets, picks, pans, shovels, hammers, quicksilver, fur, parkas, and heavy boots."

Train travelers expected and got a level of comfort that would have astounded a forty-niner. One gold-seeker assured his mother that the "chair car" on which he sped to San Francisco was comfortable enough, although he was less than enthusiastic about his limited exposure to the "Great American Desert," a "hot, dry dusty place" that billowed clouds of fine white dust into the car. "We stopped at every little mining town on the way, and saw the Indians," he wrote. "It was a horrible day."

Some gold-seekers chose the train service that extended all the way into central Alberta, Canada, succumbing to the blandishments of advocates of an all-Canadian route and disembarking at the village of Edmonton. Most roads, however, initially led to Seattle. Prospectors sporting the de rigueur trail uniform of "red-and-yellow plaid mackinaws" and "overalls tucked into high boots" swarmed over the wharves, their outfits piled nearby. Some spent everything they had in outfitting and found their adventure already at a halt. Meanwhile, better-heeled companies—many backed by "syndicates" in which members contributed a weekly amount to finance their success—still had to decide whether to attack the Yukon from the west or the south and whether to attempt an all-water route, an all-land route, or a combination of the two.

Again, vessels of every size and condition were requisitioned for the gold armada, sailing primarily from San Francisco and Seattle. With the rush caused by the parade of lucky prospectors from the *Excelsior* in San Francisco and the *Portland* in Seattle, even condemned brigs and outmoded schooners were hastily refurbished and set to sea teeming with people, animals, and provisions.

The sea trail considered the easiest to Dawson was the all-water Yukon route. Passengers boarded a steamer, which churned two thousand miles northwestward, through the Pacific and the Aleutian Islands into the Bering Sea, iceberg-clogged even in June. Then the steamers navigated the sea to the bleak trading and transportation outpost of St. Michael on Norton Sound. There smaller steamers were necessary to take the travelers through a stretch of open ocean to the Yukon's delta, then upriver the remaining thirteen hundred miles across the frigid heart of Alaska to Dawson.

Despite the fact that some passengers found themselves building their own steamer transportation from St. Michael in order to proceed, this was the "clean-fingernails" route akin to the Panama Trail to San Francisco. Only the relatively affluent took it, few of them would-be prospectors. It wasn't just the cost and the distance, but the timing. The St. Michael route stayed open only in the summer, when the Yukon "was free of ice but still had enough water to float steamers in the shallows." Impatient gold-seekers would see the summer months, valued because so favorable to mining activity, slipping away in the Bering Sea and on the Yukon itself.

In 1897, the trip turned out to be much longer than travelers anticipated. Klondike fever had hit the West Coast only in July, and the slow-moving steamers failed to outrace the fall icing-over of the Yukon. An estimated twenty-five hundred people were stranded between St. Michael and Dawson in the winter of 1897–98, about eighteen hundred of them newcomers trying to get to Dawson. Among that number was a gold-struck party who had voyaged to St. Michael under the leadership of Seattle's Mayor W. D.Wood, then discovered that they would have to help build a smaller steamer in order to continue. They did so, but halfway up the Yukon to Dawson, the ice solidified around them, locking their vessel in place until spring. In a scene reminiscent of other gold rush travelers' "indignation meetings," they held a series of angry protest sessions and named their impromptu camp Suckerville. Wood quietly footed it back to St. Michael while the passengers used provisions aboard to subsist at Suckerville through an Alaskan winter. They finally steamed into Dawson almost a year after departing Seattle.

Others steaming northward chose to stop along the southern Alaskan coastline and penetrate the interior from there. Some three thousand people attempted to move inland from Valdez. Carrying totally worthless maps displaying "vague and hungry interiors," they set off across the expanse of ice fields that bordered the site. As some turned back in despair, others strained upward over the blinding, mile-high Valdez Glacier, then through a string of valleys to a tributary of the Copper River, where they built boats and navigated as far as a series of wicked rapids that soured the whole already Herculean endeavor. Only two hundred were reported sticking with the enterprise beyond the rapids and reaching the Copper River, where they found themselves firmly in the middle of nowhere.

Even farther down the coast, near the point where the Alaskan panhandle begins to jut narrowly between British Columbia and the gulf, other gold-seeking adventurers tackled the monstrous fifteen-hundred-square-mile Malaspina Glacier, inching through the coastal forests and ravines, clawing across the treacherous, crevasse-filled ice of the glacier, then thrusting into dense, forbidding mountain forests to reach an inland river they could prospect. Men fell into crevasses, succumbed to scurvy, suffocated in avalanches, simply disappeared.

To the east, thousands of people (28,000 to 30,000 in winter 1897–98 and spring 1898) were streaming toward the glorious gold finds at Dawson along two parallel and popular routes. The Chilkoot Pass and White Pass trails started as water routes through Alaska's Inland Passage, with the gold-seekers packed onto the ill-equipped steamers out of Seattle. They first steamed through a series of straits, advancing through mists into an enchantingly still world of placid blue water and deeply hued forests. As they glided northward, huge, somber mountains and glaciers came into view, and icebergs dotted the water. Near journey's end, the travelers, their "gold delirium . . . growing more and more pronounced," passed through long, narrow Lynn Canal to the nascent gold rush settlement of Skagway, either disembarking there or proceeding three miles around a bend to a second settlement, Dyea."*

Neither place had a pier, so pilgrims forked over a few dollars for the privilege of loading their provisions in a small boat. The boatmen unceremoniously jettisoned everything into the shallows and along the shore—"people; horses; dogs; donkeys; mules; cattle; sleds; boats; and boxes, trunks, bags, and sacks filled with goods of all sorts." Argonauts pulled and lugged their belongings to dry spots, then camped beside them, readying for their assault on one of the two passes.

Dyea clustered around a prerush trading post, the gold-seekers felling trees to make camp on the narrow cleared strip between water and dense wood. "You was lucky if you didn't have a stump in the middle of your tent," recalled one traveler. The town would exist for only a little over a year, until rival White Pass boasted a railway. But as the rush crested, Dyea was the most popular gateway to the Klondike.

From the town, one took a deceptively easy wagon road a few miles, then progressed into narrow, gloomy Dyea Canyon and

* At a third location on Lynn Canal, Jack Dalton established an overland road to the old trading post of Fort Selkirk, but the $250 toll fee was enough to render this route unattractive to most.

trudged its two-mile length, inaccessible to wagons and in the summer nearly blocked with "boulders, torn-up trees, and masses of tangled roots." The trail continued "narrow and crowded, lying between two steep mountains" until Sheep Camp was reached, tucked into a valley "about half a mile wide" and within four miles of the Chilkoot summit.

As "the last camping spot below timberline," Sheep Camp quickly became a strange mixture of way station and town. Its Main Street was dotted with saloons and gambling houses and a few primitive stores, hotels, and restaurants to accommodate the transient population that seldom dipped below fifteen hundred. From the location, travelers could contemplate the pass itself, an incline that shot up about one thousand feet in the final half-mile so that a climber could shift to hands and knees "and still seem partially upright."

Sheep Camp owed its continuing vitality to the fact that few people passed through once on their way to the summit. Early in the rush, the Canadian government decreed that no one could go over the pass, the dividing line between American territory and British, without a year's supply of provisions—1,200 to 2,000 pounds of goods such as flour, corned beef, cabbage, baking powder, and candles, all needing careful packing. Outfitting immediately became more expensive and elaborate, especially with Canadian North West Mounted Police stationed on the windy, bone-cold summit to enforce the edict.

If a stampeder had money to spare after all the purchasing and preparing, a Tlingit Indian packer or another gold-seeker-turned-packer could be hired—or a dear-bought pack animal maneuvered to the base of the final ascent. But even the natives, used to the ways of the Anglos, charged a stiff price. At the Scales, a flat rock expanse signaling the final ascent, packers reweighed all goods and hiked their rates to a dollar a pound.

With costs so high, many gold-seekers consumed three months transporting personal provisions from Dyea to the summit, first in back-and-forth trudging along the lower portion of the trail, "caching" the items temporarily in a snowbank or behind a rock. Then they made repeated journeys—often as many as twenty-five—over the boulders between Sheep Camp and the summit.

As in previous rushes, winter snows smoothed the journey in some ways and complicated it in others. Enterprising men soon cut steps out of the ice on the final ascent, charging a toll for their use. A stampeder would toil up the fifteen hundred stairs, deposit provisions on the summit, then go "sliding from side to side and slewing from side to side and, oftimes, almost turning from end to end" back down the icy troughs that quickly developed to the right of the line of toiling climbers. Meanwhile, his summit-deposited supplies might be disappearing under new-fallen snow that could reach a depth of ten feet.

A Christmas present to some travelers that winter of 1897–98 was a roughly constructed tramway on the final ascent; for fifty cents a pound, one could send provisions jerking along above the climbers' heads. A second, more sophisticated tram operation soon replaced the first, and "by spring [1898] freight was being dumped on the summit . . . at the rate of nine tons an hour."

Having achieved the summit with their full retinue of goods, the gold-seekers found a steep descent into a "long, treeless, windswept valley." In the winter, if one had struggled up the slope carrying a sled, he or she could now navigate swiftly yet cautiously downhill—or release a tightly secured sled-load with the cry "Look out below!" One man, trying to maneuver downhill a sled pulled by a dog team, helplessly watched as the sled gained steam, passed the dogs, and "began to pull them down the hill backwards." In summer, jagged rocks and "snake-like roots" made the descent almost as difficult as the ascent. Still, the distance to Lake Lindemann was blessedly short.

At the lake, the weary travelers began building boats for the voyage down the Yukon or portaged around a canyon-bordered watercourse to the next lake, Bennett, where the White Pass Trail travelers were gathering.

The White Pass route, blazed in 1897 in anticipation of gold traffic, was fifteen miles longer than the rival Chilkoot, but boasted a less arduous summit ascent. At White Pass, a heavy sled didn't have to be hoisted but could simply be pulled over. Better yet, pack animals could carry the travelers' loads of provisions—if the pack animals survived to the summit.

The trail began at Skagway, a ramshackle, burgeoning camp

where the Skagway River runs into the bay of the same name. Like the Chilkoot Trail, the White Pass commenced on a good, solid ribbon of flatland, first bordering and crossing the river. After a few miles, stampeders departed the river valley and encountered a series of rugged, rocky hills dotted with spruces and marked by narrow, slick paths, sudden drop-offs, and tricky descents.

The trail was incomplete during the height of the rush, almost disappearing at times or collapsing into a single lane, with progress either way periodically grinding to a complete halt. Rivals for the right-of-way occasionally pulled guns on each other, "the cooler man of the two" bluffing his adversary into a retreat that edged the latter's horses "to the outside of the trail, where they [were] in many cases pushed over the precipice and killed."

As if that were not enough, massive mudholes churned deeper by the hooves of the pack animals and feet of the human travelers resembled the mud pits along the trail to Panama City half a century before. Frequent water crossings wore out even the hardiest, and cooperative attempts to improve the trail quickly gave way to "a general helter-skelter."

People and animals found some respite at the makeshift camp of White Pass City, on the approach to White Pass Summit, and the summit itself posed little problem, being "a steady but not hazardous climb." The descent to Lake Bennett through the "wild and desolate" Tutshi Valley went smoothly enough, as travelers streamed down "gentle slopes, loose gravel, marshes, and sand."

At Lake Bennett, as at Lake Lindemann, most stopped and began scouring the hills for boat lumber, although in the winter a gutsy few instead mushed downriver with sleds and dogs. The number who stayed at Bennett soared in the winter of 1897–98, so that there sprang up on its icy bank "the greatest tent city in the world," a crowded village with a Main Street strewn with "canvas stores, hotels, saloons, doctors' and lawyers' offices, and other places of business."

Out from this business district radiated the boat-building camps; all along Lake Lindemann, as well, and down to Lake Tagish the builders hefted logs and sawed and cursed. The estimated thirty thousand people lining the lakeshores by spring 1898 were busy on "a fleet of more than 7,000 boats." They hacked knotty green spruce from the nearby mountainsides, set up scaffolding, and whipsawed the lumber in a backbreaking, nerve-straining operation in which one worker stood above guiding the saw and the other pulled downward in as straight a line as possible, receiving a shower of sawdust in reward.

Some opted for log rafts or for whatever else they could piece together. "There are more different ways of building boats here," marveled a female traveler. "Some of them are more like boxes than boats." In them, the adventurers launched onto the lakes with their fickle, intermittent winds, weak currents, and shallows. Beyond lay the Yukon proper, dubbed simply "the greatest" by native Indians and a "witch-watercourse," the "coldest and deadliest . . . that I know" by one Yukon resident. British newspaper correspondent Flora Shaw, sent to cover the rush, wrote that the dangers of navigating the lakes and the Yukon were acknowledged by all, but "no one seems to think that he individually is taking any risk." Such an attitude, she noted, only highlighted "the strange disposition of the mining enthusiast," who "without any apparent object to be gained" took "the most desperate chances."

Whole outfits were repeatedly lost, as were vessels and sometimes lives, but it is amazing that the greenhorns, or "cheechakos," navigated their homemade crafts the distance to Dawson as successfully as they did. Their course led down the chain of mountain lakes about fifty miles from Lake Bennett to Miles Canyon, where the rush of water squeezed through hundred-foot-wide canyon walls to one third of its earlier expanse, seething and frothing in a wicked burst of foam over masses of driftwood and tree roots, jagged outcroppings and giant rocks. Shooting from the narrow canyon aperture, the vessels landed in rocky Squaw Rapids, then rocketed on to the Whitehorse Rapids, finally emerging on a relatively calm stretch that after twenty-eight miles led to Lake Laberge.

Both sets of rapids played havoc with boats and rafts, often breaking them to bits. When the first voyagers came through in spring 1898, the pileups and drownings (ten reported within a few days) led the Canadian Mounties to establish checkpoints and ban women and children from tackling the rapids, instruct-ing them to portage the five miles to the end of the Whitehorse Rapids. Some women defied the edict; one reasonably insisted that she didn't care to hike the distance alone, and at least one hiked back and rode the Whitehorse a second time just for fun, belying the grim struggle of the proceedings.

Two more obstacles worth noting awaited the voyagers. The first was the stretch of river beyond Lake Laberge called Thirty-Mile. Its sparkling, bright blue waters revealed the dangerous boulders in midstream, but its "snakelike twists" led so swiftly and ruthlessly onward that many voyagers found themselves swept into shipwreck. Most hung on and found this part of the voyage "quite like taking a long toboggan ride on a water slide."

Having survived the Thirty-Mile—or having dried out and retrieved what they could—travelers drifted downstream on a widening, more tranquil stretch. Where the Teslin River joined in from the southeast, silt slipped into the water, robbing the Yukon of its fine clarity and producing "a continuous hissing sound" as it lapped against the boats.

The hissing river led to the second obstacle, actually two closely placed sets of rapids, the Five Fingers and the Rink. Nei-ther should have been as risky as those already navigated, but the Five Finger Rapids, with "four rock-faced islands split[ting] the Yukon into five channels," looked "dreadful." Many parties portaged part of their goods through dense woods rather than risk them. Those who remained aboard were known to argue about which was safest until they found themselves beyond choice.

Beyond the Five Fingers and the Rink, the Yukon wends placidly another 230 miles, swelling with the influx of the Pelly, White, and Stewart rivers. Gold rush voyagers watched for sand-bars and for the landmark that signaled Dawson—"Moosehead," a large white gash created by an ancient landslide on Dawson's mountain backdrop. Many approached the town with some of the same somber feelings their California predecessors had ex-perienced half a century before: "The responsibilities of the whole trip now begin," wrote one.

*   *   *

Meanwhile, other gold rush voyagers were joining the main flow from the other rivers that fed the Yukon, having followed some highly dubious trails. Two of those paths led northward from British Columbia's Vancouver and Victoria, their merchants eager to snatch a share of the gold rush business. On the Stikine Trail, travelers followed Alaska's narrow southern isthmus, through the crime-ridden settlement of Wrangell and northward along the shallow Stikine River. At the small mining and supply center of Telegraph Creek, the travelers left the river and pushed overland across "frozen swamps, small streams and lakes and mountains" the 160 miles to Teslin Lake, and thence onto the Teslin River by boat to join the Yukon.

To the east, the Ashcroft Trail, named for a settlement through which it passed 125 miles north of Vancouver, plunged through the wild Fraser River and Cariboo areas familiar to rushers of the late 1850s, then along a path blazed by linemen for a stillborn telegraph operation in 1865. It, too, led to the shores of Teslin Lake. Only a few of the fifteen hundred or more travelers who chose the Ashcroft route would reach their destination. They scrambled through muddy bogs, searched despairingly for fodder, watched their horses die at a rate comparable to that of the White Pass Trail, and, like many before them, ascended one tree-encrusted hill only to see a line of others spreading interminably before them.

Farther inland, a few thousand gold pilgrims had set out from Edmonton, Alberta, on trails comparable in difficulty to the glacier trails of the western coast. Gold-seekers took the train to the Alberta village of about seven hundred, its residents so delighted to welcome the trade that they provided a rousing send-off and an occasional "guide." Twelve hundred miles to the northwest, across brutal bogs and marshes, tortuous rivers, thick forests, and the Continental Divide, lay Dawson. Finding trails through this harsh landscape meant adding 500 to 1,300 miles to the distance.

Two basic routes emerged: the primarily overland Peace River Trail and the primarily water Mackenzie River thoroughfare. To get to the Peace, travelers branched onto one of the three exhausting paths, the shortest leading across such territory as "Dog-Eating Prairie," so named because native travelers knew from experience that killing and eating one's dog might well be the only way to survive it. A second led through the foothills of the Rockies, and a third branched slightly northward across Lesser Slave Lake, crossing its seventy-five-mile expanse, and overland to the Peace River.

These routes were marked by moss-covered bogs that almost swallowed horses and by piles of dead timber, which the tired animals had to vault. If travelers reached the Peace with their animals still serviceable—no small feat—they still faced a trip of more than a thousand miles. Near the border of British Columbia and Yukon Territory, they began traveling the Liard River. As with so many gold rush routes, the names of Liard landmarks revealed much about the journey: Hell's Gate, the Grand Canyon, Rapids of the Drowned, Devil's Portage.

Beyond the Liard loomed the Pelly Mountains. Only a few hardy souls still possessed the strength, determination, and luck necessary to penetrate this range to the Pelly River, which would carry them downriver to its junction with the Yukon about two hundred miles from Dawson.

The Mackenzie River route, on the other hand, resembled the Cape Horn option in the California gold rush: relatively safe and reliable, but achingly long. Travelers converged on Athabasca Landing, a Hudson's Bay outpost a hundred miles north of Edmonton, then followed the Athabasca River northward, picked up the Slave River and proceeded to another rude outpost, Fort Resolution on the shores of Great Slave Lake. The lake cut a huge, frosty swath through the western portion of the Laurentian Plateau as it swept into the Northwest Territories.

Argonauts crossed—by boat or dogsled, depending on the time of year—and continued to follow the Mackenzie. Some diverged quickly once well launched on the Mackenzie again, instead choosing to pick up the Liard at its junction with the great river and to follow the Liard all along its loop into British Columbia and Yukon Territory. Others stayed on the Mackenzie's undulating path northwestward between the Mackenzie and Franklin mountains and beyond, almost to the Arctic Ocean. Then, entering the delta that led into the frigid Arctic, they veered onto the Peel and Rat rivers, working their way west and south along those streams and their tributaries, searching out passes through the mountains that complicated the final approach to Dawson.

Obviously, every route was lacking both in ease and speed. Like virtually all their predecessors in the well-publicized rushes, the Klondike-bound soon concluded that they had been sold a bill of goods when it came to the routes themselves; they had quit their jobs, mortgaged their homes, or sold urban businesses only to find themselves "caught like rats in a trap." Of the Ashcroft Trail, American realist author Hamlin Garland wrote, "We had been led into a sort of sack, and the string was tied behind us"—this after following the glowing literature of the Victoria Board of Trade.

Travel was further complicated on the inland trails by seasonal conditions. Early in the spring, those who had adapted to winter travel found themselves caught in a dangerous warp between freeze and thaw. On the Chilkoot, the volatility of the snow in April 1898 led to the largest tragedy of the rush: an avalanche that blanketed ten acres and killed more than eighty people at and around Sheep Camp. On the formerly winter-firm rivers, too, the melting ice and snow spelled trouble. Every day, ice highways dissolved in small but deadly ways, until travelers could not guess which step would send them crashing through the surface. If they didn't drown, they still faced the discomfort and danger of an icy plunge.

With the thaw complete, the threat of drowning still loomed large. Rivers became swollen, rushed with powerful currents, frothed and surged. A crowd of travelers watched in horror as a twenty-three-year-old Seattle man slipped from a slick log crossing into the Skagway River. His fellow travelers rushed to help, but he disappeared immediately, then "rose to the surface fully 200 feet below the point where he first sank, his pack buoying him up, and rolled over and over in the tumbling waters, his face gashed and bleeding from cuts received in striking snags and trees."

Even the seemingly placid summer lakes could prove dangerous. In the summer of 1898, a party taking the Mackenzie River route was crossing Great Slave Lake, a German emigrant standing at the sweep as his wife bailed out water. She glanced up and found him suddenly, silently gone. Turning questioningly to her companions at the oars, she "looked out just in time to see his head disappear." After two days of fruitless searching for the body, the party sailed on.

Long after the lake and river ice had melted and gold-seekers had begun poling, rafting, sailing, and tugging their way along the watercourses, the glacier travelers to the west lived and moved every day on ice and snow. Through the spring and early summer a "rolling sea of ice . . . modeled into hills and hollows" stretched around them, the glare of the snow accompanied by the "blinding brilliancy" of the windswept patches of ice.

The memoirs of Arthur Dietz, one of the few survivors of a New York prospecting company, demonstrate the nightmare of glacier travel. Venturing onto the Malaspina Glacier, Dietz's party quickly experienced severe eye strain, despite their use of colored goggles. The glaring ice seemed to rise up to meet them, their feet seemed to move in confused and ineffectual ways, and their eyes grew raw from constant rubbing. Sleds overturned on the rough ice, and the dogs could not pull all the provisions up the steep inclines without repeated trips.

The sheer grinding monotony of the crossing settled most firmly in Dietz's soul. The party made the effort of cooking every other day or every few days, then scraped the soot from the cooking pot onto the snow, smudging a bit of contrast onto the landscape. "The relief [it] gave to our eyes can hardly be imagined," Dietz would write. "For myself, I got more satisfaction out of seeing those dirty black spots of soot, then anything else on the glacier."

Crevasses could and did swallow a person in one heartstopping, strangely muffled instant. Dietz's brother-in-law, a doctor hauling valuable medicines and other supplies, was the first to go, sledding quietly—dogs, provisions, and all—into a gorgeous, awful oblivion. His companions looked down into "a hopeless blackness" and let down all the rope they had—five hundred feet—then gave up, "burdened with sorrow and completely unnerved."

Without the doctor's soothing cold cream and borax water,

the men rubbed their eyes until they had no eyelashes left. Some dipped into the remaining supply of raw bacon and laid it over their eyes, hoping that the fat would provide relief. But the salt in the bacon only exacerbated their pain.

More men disappeared into the widening fissures. The ice weakened every day, the glacier dripping, then pouring. By August, the glacier expanse behind Dietz and his remaining companions had become unnavigable.

Through the late summer and early fall, everywhere across the expanse of Alaska and the Yukon, water poured. Rains swept the White Pass and Chilkoot routes, and by late August 1897 the White Pass Trail had become "so murky as to resemble a pigpen," with both the gold-seekers and horses "sink[ing] to their knees in the mire in many places."

The next month, the wet weather led to a flood and mudslide at Sheep Camp on the Chilkoot, with three fatalities. "Well, you could not see the people for the mud going back to God's country," recalled one traveler. "People could not get out fast enough. They sold out for anything they could get." The remaining pilgrims on this and other routes shared a common urgency. Unless they were prepared to drive dog teams down or up the Yukon to Dawson, they had to reach or create a haven before winter sent temperatures plunging. Those still on the Yukon in September found ice starting to choke the thoroughfare, making it "an interminable job getting ashore to camp in the swift and congested current and even more dangerous getting back into the stream again in the morning."

From difficult river navigation to the irritations of camp life—including, in the summer, combat with hordes of mosquitoes, gnats, and blackflies—stampeders were continually faced with "plenty that was plaguy and new." The few women on the trails often got stuck with feeding a crowd of relative strangers. As one summed up her unenviable situation, "Cooking for fourteen with no grocery store handy, was enough to give one gray hair." She depended heavily on "dried potatoes, dried onions, and dried eggs," as well as pork grease in place of butter. Another woman either by misfortune or design produced such unappetizing meals that her companions hurled the dishes at her and relieved her of the job. Yet a third, the aggressively independent Belinda Mulrooney, made it clear that she preferred catching fish and shooting game to cooking for her party.

Any food carried along—in addition to other provisions— weighed heavily on a stampeder's back, pack horse, and psyche. Straps of heavy packs cut cruelly into shoulders, so that men carried the imprints long after their journey's end. Emma Kelly, a correspondent for the *Kansas City Star*, was climbing the Chilkoot Pass when she heard a toiling packer ask a fellow returning from the lakes how far it was to the summit.

"How many pounds have you on your back?" was the response.

"One hundred."

"Then it is just one hundred thousand miles to you."

When Kelly asked the man the same question, he observed that she was carrying no pack, smiled, and replied, "About a mile and a half."

The ones who felt the loads most keenly were those most burdened—the pack animals. If many humans found the Yukon trails tough, for the Montana and Wyoming range horses, the mules, burros, ponies, and oxen, they were journeys into Hades. An estimated four thousand horses died on the Edmonton trails. Pack horses fell screaming to their deaths on the Chilkoot. At the deepest level of hell stretched the White Pass Trail.

For that route, the animals' journey began on the same vessels that carried the argonauts. If men were often "packed like pigs," the animals fared even worse. On a typical steamer, the horses were jammed so tightly that they could not lie down, the heads of some "so close to the engines that they were in a state of continual panic, rearing, biting, kicking, and throwing themselves on their halters at the throb of the machinery and the blast of the whistle."

Many forty-niners had exhibited ignorance and inexperience in dealing with animals, but now, almost fifty years later, even greener greenhorns "failed to feed or care for their livestock properly," overloaded the animals, and drove them along with severely unbalanced loads rubbing their backs raw. Little or no feed could be found on the trail, but no one—not even the experienced packers—wanted to displace provisions or payload with feed for the beasts.

Thousands of animals wasted and died on the White Pass, breaking legs on the uneven and slippery terrain, plummeting with a weary misstep into ravines, or sinking in total exhaustion. Dying, writhing beasts clogged the mudholes on the roadway. Stampeders swore that some animals sought death. One man watched a horse "rubbin' its head on the trees as it was goin' along till it caught its . . . halter rope in the tree. Just stepped over the bank and hung itself!"

Those animals that kept going often received not only poor but abusive treatment. While some travelers' actions were vicious in their meanspiritedness, they also demonstrate an irrationality born of frustration and exhaustion. One man navigated White Pass with his dogs, then in a frenzy pushed them into a water hole and "collapsed in the snow in tears." Another built a fire under two worn-out oxen, "slowly roast[ing them] alive" when they still proved unable to move.

Correspondent Kelly at Lake Bennett became incensed at the sight of a packer beating a spent horse that was obviously near death. She got a revolver and shot the animal in the head when the packer moved down the train. The packer returned enraged, cursing, and on the verge of striking her, but the men of Kelly's party backed her action. "The only thing I was censured for by my companions was that I did not shoot the man," she concluded.

The provisions that had cost a horse or mule its life often ended up strewn in a raft wreck or abandoned beside the trails. The contrast between the need to haul as many provisions as possible and the rigors of the trail resulted predictably in jarring extremes of excess and shortage. On the White Pass Trail in August 1897, people tried to give away bacon and sacks of flour, but one wrote, "If you offer a sack to a man he politely refuses it and asks you to visit his tent and take some away." As yet unhampered by the Canadian provision regulations, many abandoned their goods to reach the lakes. There, however, only some forty miles from the place where they desperately tried to give flour away, they found the staple "very scarce and hard to obtain."

Any luxuries, if available, cost the sky. A cup of coffee and a far-from-fresh doughnut sold at the Chilkoot summit for two dollars and fifty cents, "five times the price of a three-course meal in Seattle," and one operation charged stampeders a dollar each simply to stand in its tent out of the cold and wind. Firewood at the windswept summit ran fifty cents a pound. Martha Purdy, traveling to the Klondike with her brother George after her husband backed out of the venture, grew so exhausted from the final climb that George, obviously worried about her, magnanimously declared, "All right. I'll be a sport. Give her a $5 fire."

Many women determined to keep up with their male companions or surpassed them in stamina. Correspondent Kelly, traveling the Chilkoot with a group of men whom she had paid to join for passage to Dawson, grew so stiff and aching that she could barely walk upon reaching Lake Lindemann, but "made no complaint," fearing "I would be thought a nuisance from the start."

Alfred McMichael noted the presence of a number of women on the Chilkoot, some in bloomers, some in skirts, and one in "blue jean overalls, sweater and top boots." He wrote of the last, "She is a hustler too." McMichael was probably referring to the woman's traveling energies and abilities, but women also demonstrated business acumen on the trails. Journalist E. H. Wells found a married couple named Fancheon at Lake Lindemann, the woman "small, bright, pretty," clad in "a complete outfit of male clothing, even to the pantaloons," and brimming with "push and enterprise." She was busy organizing a group to accompany the couple on a skiff she had just purchased. "Mrs. Fancheon is in the Yukon country to get a financial standing in the world, and unless luck is dead against her she will succeed," concluded Wells.

For many, it seemed as if luck was indeed dead against them, especially when trail accidents and illnesses struck. Stampeders came down with dysentery, scurvy, pneumonia, even hepatitis, typhoid, and meningitis. Still, the mortality rate was not high, even on the Edmonton trails where an estimated seventy expired, half of them from scurvy.

The turnaround rate, on the other hand, proved extremely

high, although no higher than that of other rushes. As many as half of the travelers reached a point at which despair or practical considerations drove them backward without reaching the diggings. And during Dawson's provision scare in the early fall of 1897, many who had completed the journey were fighting to depart at the same time that others were fighting to get in. Yet the tenacity with which thousands of stampeders maintained their vision again demonstrates the virulence of gold fever and the desperate conviction that all of the time, effort, and money they had expended had to result in something worthwhile. "Great Scott, but we are earning our grubstakes!" declared one weary argonaut in an echo of the forty-niners' half-prayerful laments. Pride again played a big role in the resolve to move onward; men would winter on the White Pass rather than "turn back and go home to face their friends."

Even after the roughest, most harrowing experiences, Dawson still loomed as the shining mecca that would give meaning to this soul-numbing pilgrimage. A young single woman traveling with relatives found the hordes of men in Sheep Camp on the Chilkoot apparently indifferent to the presence of women and reasoned, "A man can have only one kind of fever at a time, and they surely all had the gold fever," all their thoughts "centered on one thing": getting to Dawson.

How long some had to cling to that goal is illustrated by the story of Emily Craig, twenty-six years old when she and her husband departed from Chicago for the Klondike in August 1897. They and several companions chose the Mackenzie River route via Calgary, employing an "expedition manager" named Lambertus Warmolts, who claimed to know the country. It soon became clear that Warmolts did not, and the company finally arrived at Fort Resolution on Great Slave Lake in October, just as the water froze over for the winter, rendering travel virtually impossible except by dogsled.

It was soon obvious that Warmolts's promised food supplies would not hold out until late June, when the ice broke. Then, he and a companion fell asleep on night watch, and the company's dogs consumed twenty-five or thirty pounds of provisions. Warmolts disappeared with the remaining cash entrusted to him.

The party dissolved in complete frustration, with the Craigs wintering at Fort Resolution and resuming their journey in April. They crossed the lake, built a boat, and launched onto the Mackenzie, enduring icy spills, a fire that burned their tent to the ground, and a "shortcut" that cost them seventeen days. They pulled their boat up the Peel River, stumbling over tree roots and steep banks. Stopping at Fort McPherson, another forlorn cluster of exposed buildings, they moved on to "Destruction City" on the Rat River, "so named as the spot where boats were torn apart, the river no longer being deep enough to float them."

Bereft of money and provisions, the Craigs halted at Destruction City, built a cabin, and passed a second winter on the trail before building another boat. They finally reached the fading gold camp on August 30, 1899—two years after departing Chicago in the first flush of gold rush excitement.

The Craigs had found natives friendly and generous, and, unlike the situation in the American West, "Indian troubles" seem not to have been feared on the Klondike trails, despite some hostility from natives on the Edmonton routes. Instead of Indians, travelers had to look out for a criminal element among themselves. Rushes always drew a certain number of fugitives from justice, and a former Portland police detective on the Chilkoot discovered a number of wanted men among his traveling companions, while journalist Wells observed that on both the White Pass and the Chilkoot "highwaymen and robbers from Montana, New Mexico and Arizona are putting in an appearance."

Certainly, conditions facilitated theft. As stampeders cached part of their supplies, it was only too easy for the dishonest and the desperate to help themselves, despite reports of summary executions of cache thieves. On the Chilkoot, an accused thief shot himself dead when confronted with a stampeders' court. His funeral sermon was delivered by a peripatetic preacher who used the same text E. L. Cleaveland had offered to his New Haven listeners so many years before: "He that maketh haste to be rich shall not be [innocent]." "I thought this odd," wrote one in attendance, "as everyone on the trail, including myself, was on a gold rush."

The Chilkoot, White Pass, and Stikine trails were haunted by a well-oiled criminal network that had its origins in the American frontier West. Its leader, Jefferson Randolph "Soapy" Smith, had perfected his strategies in the western mining boom towns. Although the Mounties saw to it that his operations stopped at the Canadian border, Smith's gang based in Skagway employed a number of ruses to part stampeders from their money along the trails. Carrying heavy-looking packs "stuffed with feathers, hay, or shavings," the con men would beckon fellow travelers to an inviting fire or a specially constructed seat complete with shelf for one's heavy pack, then rob their victim and stage a quick getaway. Some of Smith's men also set up "shell games" on spindly-legged trays along the routes, with others masquerading as pilgrims who had stopped and were now making an excited show of their big winnings.

Even those world-wise and/or lucky enough to escape the web of a Soapy Smith did not escape some very human disappointments and recriminations as "quarreling and split-ups [became] the order of the day" within companies. There are stories of lifelong friends dissolving partnerships and cutting everything in half, even their flour sacks, "each set[ting] off with his . . . broken halves, the flour spilling away from the torn and useless containers."

It is likely that such destructive division of goods has been exaggerated. Though gold-seekers grew exhausted, embittered, agitated, and selfish, they had to become temporarily crazy to sabotage their chances of survival and success. This point is illustrated by the story of two partners who split their goods at Lake Bennett, one taking the tent, one the stove, and halving their boat. When night came, a downpour commenced. "The one in the tent was dry, but he was cold. The one with the stove, he had the stove but he was wet." The next morning the two reconciled, "patched the boat up . . . and went to Dawson."

The nature of the trip undoubtedly fostered a numb indifference to anyone with whom a stampeder did not have a close connection. On the White Pass Trail during one six-day rainstorm, men "begged and pleaded in vain for a cup of coffee, a piece of bread or the shelter of a tent." Two elderly men whose horses had sunk into the mire on the White Pass succumbed to tears as their fellow travelers trudged close by with "not one offer[ing] assistance." On the Chilkoot, a man sprawled within view of the flow of travelers for a whole day "in agony from a broken leg" before a packer hoisted him and carried him to Dyea.

But stampeders did cooperate for their own benefit and that of others. They occasionally banded together to build bridges and improve roads. In the case of the Sheep Camp avalanche, a committee was formed to deal with the disposition of bodies and raised money to pay diggers to look for survivors, closing the trail for three days "at great expense and much hardship." Individual kindnesses were extended—especially to women. When Martha Purdy sank in weariness at Chilkoot Pass, "every man who pass[ed] asked if he could help." Flora Shaw, the matronly-looking, forty-five-year-old newspaper correspondent, reported that men would pause to chat and share well-thumbed pictures of their wives and children while passing her, then wait for her a few hours down the road, ready to help her over a rough spot, even though that delayed their progress and their supper.

If such thoughtfulness alleviated the difficulties and tensions of the trails for many, so did relish for the adventure. A man on the White Pass Trail reassured his family, "I have a horror of business and like only this life and that which California affords. I take kindly to this sort of existence and feel perfectly at home in a tent."

The grand, novel vistas provoked a mixed reaction. One woman judged Lake Lindemann "so beautiful I felt like I'd started life in a new world. Everything dark, dark blue and snow white. I loved it—but I couldn't stifle the terror inside me." For there was something so formidable about the northern terrain that a stampeder taking the time to observe it was more likely to shudder than to rhapsodize. It had to be crossed. It had to be "conquered," and how does one conquer an iceberg, a crevasse-filled glacier, a frigid, snow-packed peak, a roaring, twisting, rapid-dotted watercourse, a desolate stretch with an ominous name like Dog-Eating Prairie? Even when one survives such obstacles, they remain untamed and timelessly threatening.

Yet the Far North continued to draw adventurers from the United States and abroad with the siren song of poor man's gold. When news of the discovery of gold on Nome's beaches in the summer of 1899 filtered into Dawson, almost fifteen hundred people crowded onto the last stern-wheelers out in the fall. Hundreds more braved the winter Yukon Trail, the "constant passing of the sleds" wearing "the surface of the trail down as level as a table" and the route "bloody for miles from the bleeding and limping dogs." Edward Jesson, a youthful gold-hunter who paid $150 for a bicycle in Dawson in February 1900, practiced cycling "inside an eighteen inch sled track without falling down," and then proceeded to ride to Nome, lashing on spruce boughs to replace a broken handlebar and gaining an appreciative audience of startled Indians near Circle City. One commented, "White man he set down walk like hell."

As Jesson cycled northward, stampeders in the States were restlessly counting the days until Nome would be freed from its frigid seasonal quarantine. In May, the familiar scene of crowded ships departing the West Coast to the cheers of excited crowds was enacted once more. Again, Seattle captured the lion's share of gold rush business, with transportation companies fanning the flames. WOULD YOU LIKE TO BE A MILLIONAIRE? asked a Great Northern Railway ad for rail transport to Seattle. "Few men become rich by slow economy," it warned, giving the already mangled and prostrated Puritan work ethic a good swift kick.

Despite the vast distance to be covered, "a 'Nomer' could leave his home in Chicago, New York, or London, and travel all the way to the Nome goldfields by passenger train and steamship," the steamship ticket costing sixty or seventy dollars for a second-class berth. Only in the case of Cripple Creek had a similarly effective public transportation system been in operation; but then, Cripple Creek had been close to burgeoning American population centers.

Nome was close to nothing, not even a tree (the nearest grew a hundred miles away). Getting there involved a voyage across the North Pacific to barren, mountainous Unalaska Island in the Aleutians and from there into the Bering Sea.

The voyagers' biggest problem was covering the last mile or two from their vessels to the golden beaches. With no deep-water harbor, Nome depended on barges that transported passengers and goods from ship to shallows, with stampeders clambering out and wading to shore carrying freight or perhaps, obligingly, another "Nomad." The effort of the whole operation is indicated by the fact that "it cost almost as much to carry a load of freight the last mile through the surf . . . as it had cost to transport it two thousand miles from Seattle."

The gold-seekers slogging through the surf, like every argonaut before them, knew they were taking a gamble, whatever gilt-edged stories of untold riches they chose to believe. In every rush, many also shared the plaint of a man bound for British Columbia's Cariboo in 1862: "If ever I am settled again it will be something more than Gold that will induce me again to forsake all comforts Wife family & everything one holds dear to roam over an unknown region in quest of Gold." But every stampeder who persevered, from California to Nome, was dreaming of a payoff, a brimming pot of gold at rainbow's end. Instead, the diggings offered a bewildering new series of high-stakes gambles that tested the physical, financial, and emotional resources of even the most inveterate risk-taker.

Chapter Six

Life in the Diggings

Let me tell you it is a good deal like work to dig gold those
that have been hard labouring men say they never worked
as hard before.

—*San Francisco resident Jerusha Merrill, 1849*

You have no idea of the hand to mouth sort of style in
which most men in this country are in the habit of living.

—*California mining camp resident Louise Clappe*

W hen gold-seekers first reached the "diggins," thcy entered
into a new environment, a new kind of work, and a rough pattern of living not far removed from that experienced on the
trails. Their survival would depend on hard work, hope, luck,
and adaptability.

Early arrivals briefly enjoyed an unspoiled natural environment—pristine streams, tree-lined banks, picturesque gulches
tucked jewellike into the rugged western and northern landscapes. But most stampeders arrived after the diggings had been
established, the streams muddied, the trees cut, the gulches or
river bars turned into obstacle courses of pits and dirt piles and
tent pegs. Everywhere the earth was being attacked, rearranged,
washed and dumped, so that it appeared to one Swede that "an
army of pigs had rooted" through the California valleys; to observers in Alder Gulch, Montana, as if the ground were "literally

turned inside out." Mining holes proliferated, their edges marked—but not blocked—by the "immense piles of dirt and stones" removed from them. One California laundress navigating among sludge-filled ground sluice holes stumbled into one, fresh laundry and all, and was resurrected only through the efforts of six men with ropes.

As might be expected, accompanying such radical transformations of the landscape was near-constant activity. "So many men, it seems they would be in each other's way," wrote an observer at Alder Gulch. The workers reminded her of "bees around a hive," endlessly "shoveling and wheeling dirt, passing and repassing each other without a hitch." In the Black Hills, the claims along Deadwood Creek and surrounding streams were "scenes of the greatest activity" even at night, and in the Yukon, miners "crowded thickly upon the diggings, on little Klondike tributary pups [subsidiary streams], like prairie dogs atop their burrows."

Milling around campfire and claim was a population amazing in its diversity. In any big stampede, those who had never before been exposed to different cultures marveled at the medley of languages and accents they heard; one's neighbors in the diggings could be southern farmers, Swedish sailors, New England mechanics, German merchants, or African-American laborers.

Among the Anglo-American majority, class distinctions were obscured under miners' garb, dirt, sweat, and a general joy in freedom from conventional social distinctions, so that "it was never safe to presume on a person's character from his dress or appearance." From California to the Klondike, the weathered, bearded, unkempt-looking men the new arrival encountered might turn out to be farm boys from the newcomer's old home county or doctors, lawyers, even ministers. The democratization of appearance was so strong that one Klondiker deemed it almost a "moral disgrace" to be well dressed in Dawson.

In addition to the miner population, a booming camp would also boast a small but growing phalanx of merchants, freighters, prostitutes, launderers, saloonkeepers and innkeepers, gamblers, fortune-tellers, day laborers, and practicing lawyers and doctors. At least one tent saloon would be in evidence, serving as "clubhouse, news center, and business exchange," as post office and even as bank. If the camp had had the time and vigor to develop at all, it also boasted a general store, a mailbox, a blacksmith shop, maybe even a tent restaurant or a rude "hotel."

Many gold sites were smudged into deep, romantically wild crevices between forested mountains. In such locations, the sun penetrated the campsites only for brief periods, but it lit the mountain rims in "golden glory." The immediate environment might be as bare of natural beauty as human effort could make it—"Not a spot of verdure is to be seen on this place," wrote one woman of a California camp. Yet the surrounding green hills offset the disruptions near at hand.

The gold-seekers who reached the camps were frequently debilitated from rough travel yet still intensely eager to dig shining fortunes, especially when shown evidence of such fortunes by lucky prospectors. After viewing the gold gleaned by another miner, a California newcomer found every rock on the path to his tent "had a yellow tinge," even the filthy camp kettle appearing "gilded."

But a man's heart could sink, too, at the odds of capturing a good claim, especially with so much land already worked over or in others' possession. The same California mining area that drew six thousand miners in 1848 was bursting with forty thousand or more in December 1849, leading one pilgrim to complain that "the whole valley has been dug up." Two Idaho newcomers built a makeshift dwelling, then waited "for somebody to tell us of a place where we could dig prospect holes without infringing upon the vested rights of our predecessors." In the Dawson mining district, men speculated "that not more than one in ten, of all who piled into the Klondike [in 1898], got even a chance to stake a promising claim there."

The first prospectors sought out spots where placer gold was most likely to collect—bends or natural blockades in streams, the latter "sandbars and low-lying gravel banks." In accordance with a system that evolved in California and was carried across the West, early arrivals met jointly to create a mining district, to define the size of a claim, and to set up rules for holding and recording the claim. If the press of people became great enough, however, all tended to bedlam. A prospector at Alder Gulch picked a spot and began sinking a shaft, but was "surrounded by stampeders" and "left in disgust at 4 P.M."

The prospects were simply not as unlimited or golden as people had been led to believe. Jerry Bryan's morose diary entry on arrival at Custer, South Dakota, during the Black Hills rush reflects the feelings of many: "Come to camp with our Spirits way down dont like the looks of the country. and I dont like the looks of the men dont believe there is a claim on the creek that will pay wages."

An alternative for newcomers was to buy an existing claim, but chances were good that any claim offered for sale had been salted with gold dust from another location—or with coin filings or soluble gold chlorides, which gave the same bogus results. Even miners "honest as the day" in other matters engaged in this practice under the belief that it was a "fair game."

Good prospects or not, enthusiasm often ebbed as the newly arrived observed the effort involved in gold retrieval. The earliest California prospectors enjoyed relative ease, the labor resembling a game of hide-and-seek in which they skimmed off the surface gold by "picking into rock crevices, searching under large boulders along the rivers, or digging in nearby ravines" using knives, frying pans, buckets. But most stampeders found "work plenty, work of the most trying and laborious kind" as they dug ever deeper, built sluices in which to wash the earth, and even in gold-pocked California often "sifted through a ton of sand and gravel to find an ounce of gold." One Californian declared, "Digging wells or cellars at home would be play alongside of digging into these hills and rocks," while another wrote home that any other man from his neighborhood wanting to come should first be able to dig up a local creek branch with spade and hoe "and wash all of the mud and sand out." An Orofino miner wrote his partner indicating the extent of the work; the two needed to be shoveling snow and getting ready for the spring rains, when they would have a source of water for washing dirt. He estimated the shoveling would take a week, as the snow was eight feet deep and still falling. Then there were "five thousand feet of lumber to be whipsawed, flume and sluice-boxes to be made and reservoirs to be built."

On claims around Nome, argonauts chipped holes in the icy

ground and encountered a "frozen muck or packed mass of unrotted vegetation which, when it thaws, looks and smells like barnyard filth." One company found the walls of its shaft "began to melt and cave in," while the clods of earth scooped from the bottom "were so cold that for a time frost formed on the outside just like a cold piece of iron brought into a warm room in winter."

Even as most placers turned out to be limited or complete chimeras, there existed a subtle social pressure to prove oneself by prospecting and mining. California miner-turned-storekeeper George Davidson would write, "I regret I did not go to selling goods when I first came . . . but I thought if I did and should fail people would say you were too lazy to work." An 1854 "Miners' Commandment" warned that men were not to "grow discouraged and think of going home" to fifty-cent-a-day wages when they might "strike a lead and fifty dollars a day," thus retaining their "manly self-respect."

For many of those who stayed as prospectors and miners, the Puritan work ethic reasserted itself with a vengeance. These stampeders had known, deep down, that success through easy riches stood in contrast to the success through hard labor their culture still taught them to value. Now they resolutely claimed work and perseverance as the keys to success, Californians accepting and judging themselves by the idea that "the sober, hardy, persevering gold-digger . . . never fails." Men carried this belief through subsequent rushes, insisting that "almost everybody . . . who has been industrious and persevering, have made money," that gold "is not picked up loose on every claim but it is here for those who are willing to and able to work for it."

Anyone willing to look realistically at the gold quest would have to conclude that Dame Fortune enjoyed knocking holes in such philosophies. Louise Clappe offered a qualified work-and-succeed creed: "If a person 'work his claim' himself, is economical and industrious, keeps his health, and is satisfied with small gains, he is 'bound' to make money." Yet, she acknowledged, "almost all with whom we are acquainted seem to have *lost*." For one thing, the gold rush environment did not encourage economy, steady industry, health, and satisfaction with small gains. For another, luck often came to the less industrious.

In fact, one California argonaut concluded disgustedly that fortune smiled only on "drunkards, idlers, and fools." Three "drunken, worthless fellows" happened upon a rich California gulch and removed "hundreds with hardly an effort," while "other and more deserving men" labored mightily nearby with meager success. All the way through the rush era, men clung doggedly to the idea that a little luck and a lot of work would bring golden results. They were supported in this belief by the rise of Horatio Alger as a cultural symbol—the lad with both luck and the pluck to achieve his goals.*

Most came to see the rushes as giant lotteries, everything subject to whimsical chance, leading to the gold rush axiom "It is the unexpected that happens." And whether the gold-seeker was a carousing idler or a model of moral industry or both in turn, he was also a gambler living for the big strike, subsisting on intoxicating hope. So men searched, trying to enhance the odds by perusing pocket gold manuals containing regional geologies and gold-washing and quartz-crushing techniques. They listened to self-proclaimed experts among their number, and they developed their own understanding of where gold might be hiding.

Again and again, knowledge touted as scientific was confounded. "Wherever Geology has said that gold *must* be, there, perversely enough, it lies not," wrote Clappe, adding that the opposite was also true. A Baja California stampeder noted with exasperation that Mexican women and children found nuggets and dust while miners working "from strictly scientific and practical principles" failed to make wages. Yet during the whole gold rush period, the "scientific and practical principles" employed by prospectors "reflected no great advancement beyond that of prehistoric ages."

Even experience offered no very reliable guide. A Black Hills miner complained that "d——n fools will dig for gold where an old miner would not expect any" and hit pay dirt. In Alder Gulch, old-timers read the topography and concluded that gold could not be in most of the river bars. It could, as lucky greenhorns found out. Against all odds, a fortunate prospecting party

* Even though Alger books did not appear until 1868, they reflected and reinforced early-nineteenth-century values. See Richard Weiss's *The American Myth of Success: From Horatio Alger to Norman Vincent Peale* (1988).

in Arizona Territory discovered a mesa strewn with large gold nuggets. A rich Alaskan stream was dismissed by one prospector because "the wrong kind of trees grow on the bars, and gold is never found where wild onions or leeks are found in such abundance." Experienced prospectors also ridiculed the idea of gold in the "cow pasture" of Cripple Creek and the "moose pasture" of Bonanza Creek in the Yukon. Then, when Bonanza turned out to be full of rich holes, they laughed at the cheechakos who filed hillside claims rather than creek claims. In each case, of course, the greenhorns profited.

The fact was that placer gold could appear "in virtually any sort of formation, no matter how theoretically unpromising" if "action of some sort at one time deposited it there," and quartz lodes were almost as unpredictable.

Most argonauts exhibited a mixture of hopeful gullibility and cynicism, with the cynicism growing along with their time in the diggings. A Californian wrote,"Lots of men claim to know where gold is—will sell knowledge for a price—all humbug." Yet almost fifty years later, Alaskan stampeding companies were paying as much as six hundred dollars *per member* for a worthless tip, this at a time when the average annual earnings of a nonfarm employee in the United States were less than five hundred dollars.

The idea that Indians possessed gold secrets remained strong, yet men learned to be wary here, too. A Montana prospector in 1862 expressed a rueful disbelief in a Flathead Indian's account of finding "several pieces of yellow metal" in a mountain stream—"what a whopper, and why didn't he pick up all those pieces of yellow metal[?]" Still, there was always the chance that cynicism could keep men from riches. In 1896, shortly after the discovery of Klondike gold by George Carmack and his native companions, two men paused at the mouth of the Klondike, having received word of the strike. But they pushed on to Fortymile, one concluding, "I wouldn't go across the river on that old Siwash's [Carmack's] word."

When men did decide to follow another's instincts or their own, they soon faced not only hard physical labor but the need for knowledge and skill in extracting the grains of gold, coarse or fine. A forty-niner spent a whole day learning to identify and wash gold, panning only about half a dollar's worth in the process. A Colorado prospecting party discovered that "the operation of 'panning' requires much skill for its proper performance," that only experience could aid in the "peculiar motion to be given to the pan." Three fledgling California miners with a wooden "rocker" to sift the gold found themselves unable to "rock and dig at the same time" and grew easily exhausted. Two other Californians erroneously dug mica and presented it to a storekeeper as gold. The merchant took pity on the two, teaching them to pan and recognize gold, and grubstaking them with supplies in return for half of what they could dig.

Any practical mining knowledge at all could prove useful. When young John Steele of Wisconsin arrived in the mining camp of Nevada City, California, in 1850, he had no luck getting a mining job to raise funds for his own gold hunt until he donned a striped ticking shirt, the kind Wisconsin lead miners wore. This won a comment from a stranger: "When you see a boy from Wisconsin wearin' them togs, he'll do in the mines anywhere." On the basis of that assessment, a claim-holder gave Steele a job and reaped the benefits, for although Steele had not worked in the lead mines, he had learned from Wisconsin miners how to secure shaft timbers so they would not collapse.*

Placer miners learned varied means to capture the golden grains and nuggets eroded from veins and washed into riverbeds, where the gold's weight drew most of it to bedrock. They dug into river bars, borders, flats, and old, exposed streambeds; used dams to expose the bottom of active streams; and engaged in "coyoting," digging into hillsides to find hidden streambeds or former streambeds (the term also referred to digging tunnels outward from a shaft). If digging down, the gold-seeker first had to penetrate the "overburden" to bedrock, a few inches or many yards beneath the surface. Whether digging straight down or into a hillside, he would then need to wash and sift the sand, dirt, and gravel to determine whether gold was present and in what quantities.

At its simplest, this involved unearthing soil, then placing it in the miner's all-purpose pan (also used for doing laundry, feed-

* Wisconsin and Illinois mining techniques would significantly influence western mining technology.

ing the pack animal, and making flapjacks) and rotating the pan, "with its top edge barely submerged" so that the soil would sift out, leaving along with the residue—if one proved lucky—the glittering grains, nineteen times heavier than the water, and occasional nuggets of gold.

A second simple method of placering, used when excavating ancient streambeds no longer close to water, was "dry washing"—throwing gold-bearing sand and dirt or gravel pulverized by a mortar onto a sheet and then tossing the contents into the air for the breeze to blow off the chaff.

As one might imagine, it took a long time to wash much soil through either of these methods. However, panning had its place in each placer rush as a means of establishing how rich the gold deposits were. In rushes of any substance, however, the pan soon gave way to devices for handling more volume: the rocker, then the long tom, then elaborate sluices.

The wooden rocker, or "cradle," stood four to six feet long with "an open box or hopper at its upper end and a series of cleats along its sloping bottom." Gold-seekers spaded the dirt into the top and "rocked" the whole apparatus as they poured water through. The rocks and heavy gravel were to remain in the hopper, while the lighter gravel, gold, and sand-choked water rushed downward, the gold collecting behind the cleats (also known as "riffles" or "riffle bars") as the gravel and water passed through. When a number of rockers were in use, the clattering "sounded like the din of a cotton-factory."

From the rocker, it was a natural step to the long tom, an inclined wooden chute from eight to fourteen feet long. As with the rocker, miners threw sand, dirt, and gravel into the upper end, running water through. They didn't rock this fixed contraption, the water doing the work instead, but the flow had to be even. The earth was strained through a piece of perforated sheet iron near the end, then washed through a box with a set of riffle bars, the gold again collecting behind these cleats. (Miners quickly learned to put quicksilver against the cleats to make the gold adhere.)

While a single prospector could get along fine with his pan and could even handle a rocker alone when necessary, the long tom required at least two and preferably three workers: one or two to hoist the earth into the upper end and one to stir it as it reached the sheet iron, picking out rocks, large pebbles, and an infrequent much-prized nugget and speeding the flow into the cleat-lined box. The long tom also required a steady source of water for even flow, and miners rigged up "a connection by either pipe, hose, or 'flume'—an elevated sloping wooden aqueduct." This arrangement could handle greater volume than the pan or rocker, but it proved of limited usefulness in processing large amounts of earth. With all three methods, much gold still wound up washing away.

In each rush of any significance, a partial solution to these problems was introduced with the inevitable switch to sluice mining—and a corresponding increase in the number of workers necessary. At its most basic, sluicing consisted of digging a long trench and using gravel or cobblestones as riffles to capture the gold. But sluicing usually involved constructing a series of interconnecting long wooden troughs to create "an almost indefinite extension of the Tom," sometimes topping a thousand feet. The linked troughs were strung along the ground or elevated on poles. Riffles of cleats, of small, thickly clustered poles, or of auger-drilled boards anchored the bottom end of each sluice box. A nearby water source would be tapped, often by raising water from a stream, in order to provide a flow through the sluice boxes. As the operation got under way, two or three men might be underground digging, others handling the ore car and windlass (a device for hauling ore to the surface), and two sluicing—loading the dirt into the sluice boxes, removing rocks and gravel, and generally seeing that the water did its work in washing the gold into the riffles.

In theory, water washed the gold into the riffles. In actual practice, greenhorns from rush to rush set their sluice boxes at such an angle "that fine gold washed right on down the tailrace," the narrow ditch prepared at the end for the stream of water and earth. Even careful miners could lose a quarter to a half of the precious dust in sluice-washing. Still, this was the most efficient placering method for the majority, who lacked funds and materials to create more technically sophisticated alternatives.

Some argonaut companies experimented with river damming, pooling funds and sweat equity to divert whole rivers to expose what they hoped were rich diggings. The practice enjoyed a vogue in California in 1850 and 1851, groups sinking and losing as much as ten thousand dollars in the dams alone, aside from the expenses of building flumes, digging water channels, and even constructing roads. A few groups hit pay dirt, but river damming was an especially costly and laborious gamble—particularly since the river would inevitably rise and flood the dam.

One placering method for those with capital developed in California shortly after the general failure of the river damming operations when miners rediscovered an ancient Roman mining technique that would basically allow them to wash the hillside away. The incredibly destructive practice of hydraulic placering continued throughout the gold rush era, men with the wherewithal to engineer the necessary dams and massive sluices blasting away with water at hillsides across the West and Far North.

There were three common, crucial elements in all of these mining endeavors. First, location had to be considered. "It takes a rich mine to pay when supplies have to be packed 60 to 80 miles and wood and timber six to eight miles," explained one Yukon prospector. Unless the gold was practically on the surface waiting to be skimmed, its finders had to worry about at least some short-term development, such as viable trails and timbering for shafts.

Related to that consideration was the second: the need to combine effort and resources. Yet groups split over where to dig, how to divide and facilitate the work, and when to move on. From California to Alaska, prospectors complained of nonproductive group members. "Half of them would not work," groused a former member of a California company. An Alaskan prospecting company fizzled out, a member writing, "The boys were all anxious to get to work, until they got something to do."

Usually, groups broke into more convivial remnants, parties of three to six, with members taking turns scouting out new gold locations. These units might splinter further out of necessity— not enough gold found to support even this number—or out of preference for greater autonomy and mobility. Thus emerged the two-man partnership, each aware that "not only your comfort every day but many times your success in the mines depended on the kind of partner you had."

When there *was* work for a group, one or two men with a good claim could hire and pay workers from the day's gold bounty or lease the site. But often, profits proved meager when split. Two Deadwood newcomers, desperate for funds, obtained a job with a claim-holder who halved the day's profits with them. A hard fourteen-hour day netted the two $2.64 each, and they concluded they would starve before they would work for those wages again.

Besides, argonauts had not entered the quest only to hire out to someone else. "I can get 8 Dolrs per day to work in the mines for another man, but I hope I can do better for myself," one wrote in a typical comment. Daily wages did soar to sixteen dollars in some rich California locations, to more than twelve dollars in the Klondike—exceptional pay by eastern standards even in 1898. But such wages were often merely adequate in a gold rush economy, and in this unregulated environment, wages could be hard to collect, with dishonest claim owners paying only under duress or disappearing shortly before wages were due.

Whatever his arrangements with other gold-seekers, the miner also needed water, the third crucial and troublesome element in mining endeavors and a commodity the American West has always lacked in steady supply. Often when good prospect holes were found, water was distant, hard to make use of, and/or limited. One Black Hills group hauled their pay dirt "for considerable distances by wheelbarrows or carts," then moved to another location where they dug long ditches and paid dearly for lumber for sluice boxes, only to have those with prior water rights protest when they tried to divert water from a summer-shrunken creek. In California, prospectors had to wait for those with first rights to finish washing or had to resort to dry washing while first-comers monopolized the water pools.

Some miners and other entrepreneurs turned to the business of supplying claim-holders with water by forming companies, bringing in machinery, damming streams, and building elaborate flumes. In most cases, the expense was so great that, when passed along to the miners, it ate away any gold profits. One Swedish company in California "rinsed the ditch" and found seventeen dollars in gold—but used twenty dollars' worth of water in the rinsing. Whole mining camps folded because of such expenses.

Occasionally, water was present in abundance in the ground itself, saturating gravel and bedrock and creating extra drudgery. Companies were forced to bail or to construct expensive, time-consuming drainage ditches. In the Far North, one extra stroke of the pick could produce a torrent of underground water that might quickly harden to ice.

Often the availability of water for washing depended upon the time of the year. The rainy season swept into the Colorado mountains in the summer, into the California diggings in the fall. Prospectors in the Klondike were surrounded by lively, full-bodied streams in the summer, while prospectors in Alder Gulch, Montana, found the streams contracting to mere trickles in summer heat and wind.

Rain could hamper mining operations rather than help. In California's northern diggings, the skies opened in October or November, floods carrying off miners' cabins, tools, and claim improvements. On Curtis Creek in November 1849, rain "flood-[ed] the streams, obliterating all signs" of one Texan's rich claim. When he returned in the spring, others were working the site.

The scarcity and unpredictability of water sources—and the difficulty of fairly and efficiently allocating those available—contributed to gold-seekers' mobility and sent some back to pan or rocker. "It is a slow method," one man wrote of his experience with a rocker on the American River in the early 1850s, "but it enables a person to use water from a pond or river, where it cannot be raised so as to run into a sluice or long-tom."

The gold he and others were seeking continued to confound any expectations as to its whereabouts. In Bannack, Montana, the grains clung to plant roots, yet in nearby Alder Gulch, they "lay under heavy overburden, twenty feet or more below streambed." On Bonanza and Eldorado creeks in the Klondike, the gold was concentrated within a well-defined area, while at Negro Hill in the Black Hills the good diggings were "scattered over several gulches," not providing "a single focal point about which a rush could center."

Whatever the situation, prospectors relied to a great extent on the same range of tools and methods not only for washing but for removing the earth and securing shafts. If the overburden was thin, they would dig up a large surface area; but when getting to the gold required deep shafts down to and following bedrock, they were faced with the problem of how to get the dirt out. The solution was the windlass, typically a sturdy frame with a rotating log on top with which to lower a barrel on a thick rope into the shaft. Those digging coyote holes into hillsides solved their removal problems by constructing their own rough ore carts to haul the earth. Anyone digging very far into the ground or a hillside also had to worry about timber supports, especially if they attempted to follow the drift, or course of the gold deposit, with more tunnels once inside.

Of course, geography dictated some variation in methods, none more striking than the burning for gold—"burning out a shaft"—practiced in the northern American territories during the winter and year-round in the permafrost of the Klondike, where "numerous pillars of blue smoke jutt[ed] like Corinthian columns from the snow-robed earth to the heavens above." Because the ground was frozen solid a few inches below the surface, men gambling on a likely spot would build a fire over it, thawing seven to fifteen inches, then remove the defrosted dirt and start another fire by throwing in wood and setting it aflame—again and again, through the fifteen to twenty feet of overburden to what they hoped was the placer-gold pay streak. If they passed the long northern winters that way, they would wait for the water from the spring thaw to see if they really had anything after all. And if they did, at least the permafrost made timbering of shafts unnecessary.

Quartz, or lode, miners shared the difficulties of the Klondike placer gold prospector in that they were tackling a highly resistant substance—in this case, solid rock. First, of course, lode prospectors had to locate the vein, either by digging or by finding where it surfaced. Many worked upstream, or up the traces of ancient streams, looking for the origin of the "float," the loose placer gold washed down at some point in the area's history.

Once promising rock was located, a prospector would take a sample to one of the cluttered assay offices that sprang up across the mining frontier. If the assay proved favorable, the quartz vein still had to be drilled and blasted, with development calling for hoisting and pumping machinery, some method of ventilating and lighting the underground tunnels, and ways to haul the ore.

Then, how did one coax gold in any quantity from the extracted rock? Mexicans had partially accomplished the feat by crushing rock manually, then putting it through an arrastra, a circle of stones or blocks over clay with a pole in the circle's center. From the pole jutted one or more wooden arms, at least one fitted with a "drag stone," or heavy weight, reaching to the stones. When a mule was tied to one of the arms and began walking around the circle, the drag stone would trail over the circle surface, pulverizing the ore as a small stream of water facilitated its filtration through a piece of pierced oxhide, or—later—a screen.

Even when dynamite became readily available to facilitate the initial breaking-up process, the arrastra method was slow and crude. The resulting mess had to be further processed to extract the gold, and while the practice of adding mercury to the arrastra floor would aid the separation process, like the placer miner, the quartz entrepreneur might find the take not worth the extraction costs.

The problems and hardships of quartz mining were partially addressed with the introduction to the West of stamp mills—industrial operations with huge "stamping" mechanisms that crushed rock, then reduced it to dust—and with separating pans in which gold and silver amalgam was extracted from waste. Stampeders in areas where placer opportunities were poor or playing out sank their remaining funds into developing a lode claim, helping finance a stamp mill or simply paying for the mill's services in crushing and assessing the ore they found. Only occasionally were such efforts by gold-seekers profitable. For one thing, the ore, even near the surface, might not be worth the effort, especially after transport and assay costs were paid. For another, a prevailing theory that the ore would increase in richness as miners delved deeper seldom proved true. And the mills themselves were inefficiently constructed and run, letting "several millions of dollars [float] off down the creek."

No wonder, then, that most gold-seekers who showed any

interest in quartz mining did so in hopes of selling early. But should they stop after tests on the surface rock proved promising or angle for a better price by "expos[ing] the raw rock beneath"? In doing the latter, they could expose the fact that "many a good outcrop sat upon a meager vein for which nobody would pay a cent."

Whatever their hopes, both placer and quartz miners experienced crushing disappointments, seeing "the billions of their crazy hopes dwindle to millions, the millions to thousands, and the thousands to the price of a meal." "What happened to the fortune I was supposed to amass in a week or two?" asked one forty-niner. "I feel like kicking myself," said a Seattle man working in a Dawson warehouse. "I left a nice comfortable home and a good wife, spent $1,700 to get here with my nephew, and for what?" Five decades of prospectors shared "the sudden wrecking of gorgeous visions."

Their situation was prosaically but poignantly summed up by a Deadwood prospector whose company eagerly awaited the first "cleanup" of the gravel from a claim for which they had dug a ditch, spent precious savings on lumber, constructed flumes, and dealt with water-sharing and leakage problems:

> I do not care to dwell upon our feelings when that clean-up was panned down and the result announced. It was so pitifully small that we realized at once that all our labor and expense, all our hopes and plans, based on belief in the value of the ground, were vain. In my lifetime I have met with disappointments of various kinds, but I can recall none that hurt as deeply as this.

Yet it took only the news of another's fortune to vault many heavenward again. California prospectors J. M. Letts and J. C. Tracy had had little success when they heard that fifteen hundred dollars' worth of gold had been extracted in one hour on a claim at Mormon Bar. Locating on the bar, they went to work with their imaginations fired, Letts wielding a pick with vigor, Tracy making the rocker agitate "as if propelled by the furies." Although the initial results appeared disappointing, they convinced themselves that their spot bore a strong resemblance to the golden one and soldiered on "with renewed energy," Tracy "almost throw[ing] the machine into spasms" as he thought of the fifteen hundred dollars. In this way, the two managed to wash one hundred buckets of dirt before lunch—a prodigious feat. Their yield, panned down, was two dollars' worth of gold.

Ironically, news of rich finds was not necessary to convince many a miner that such finds had been made. Gold-seekers often remained closemouthed about their successes as well as their failures, leading others to conclude that they had something wonderful—or at least worthwhile—whether they did or not. In California, it proved "almost impossible to learn from the miners themselves, unless one happens to be a near acquaintance, the amount of their gains." Prospectors across the West and Far North suspected others of withholding valuable strike information.

The successful had good reason to play down their fortune. The chaotic character of the mining frontier and the lack of facilities for the storage of gold made theft and even murder strong possibilities. Thousands of dollars' worth of gold dust lay in brush shanties and unlocked cabins from California to the Klondike.

Then, if successful prospectors were working a whole new area, they wanted to keep as much as they could for themselves and whatever friends they chose to share the good news with. Seldom did this secrecy work for long, for there were tip-offs. With exaggerated casualness, prospectors would buy a noticeably large amount of mining supplies and provisions in the nearest town, starting a storm of speculation. Or, as happened on Eldorado Creek in the Yukon, suspicious prospectors would find the creek water muddy, a sign that digging was going on despite the discoverers' statements to the contrary. Often excitement seemed to telegraph itself from men trying to appear normal, as in the case of the "too tired and too glad" Alder Gulch prospector.

Still, some carried off stupendous secrecies. Perhaps the greatest was accomplished by Winfield Scott Stratton, the Colorado Springs carpenter who became Cripple Creek's first millionaire. Growing disgusted with his poor-yielding Independence claim on Battle Mountain in 1893, he gave a mining syndicate a thirty-day option on it. Immediately thereafter, retrieving equipment from one of his four shafts, he stumbled across a colossal vein of gold-bearing quartz. For nearly the full month, he kept the information to himself as the syndicate crew went methodically through the other three shafts. They planned to explore the fourth shaft as well, but the syndicate representative had already given it up as a bad job and came to Stratton before the deadline asking him to take the option back. The carpenter, a prospector for almost twenty years before locating this pay streak, quietly agreed but did not trust himself to reach for the option agreement proffered, instead nonchalantly instructing the representative to throw it in the fire. The Independence contained so much gold that Stratton decided "to limit his net to $2,000 a day," and the mine would yield a million dollars annually for him for a number of years.

Stratton kept another secret, too. Two Irishmen, Jimmy Burns and Jimmy Doyle, struck a rich vein on their Battle Mountain claim, the Portland, and began working it under secrecy of darkness. They were aware that claims to veins running through various properties, as most did, were subject to drawn-out litigation under the law of apex. In other words, if the holders of other claims on the mountain heard of the Portland's wealth, they would be fighting to establish ownership of the vein, too, by trying to prove that it "apexed," or surfaced, on their property. When the Irishmen and their partner John Harnan let Stratton in on the secret, he banded together with them to continue the secret operation and to have a top-notch lawyer ready to go when the other Battle Mountain holders got the news. Lawsuits were eventually filed, but to no avail; Stratton, Burns, Doyle, and Harnan gained virtual command of Battle Mountain's immense gold stores.

Most men could only hope someday to hug such secrets—or any worthwhile gold secrets at all. Dame Fortune was particularly cruel to those who stopped just short of success. In the California dry diggings, where the gold was deeply buried, some had dug "two or three days and given up in despair," while those following "working the same holes" quickly realized thousands of dollars. A Cornishman prospected the Cripple Creek area a few years before the rush, following Bob Womack's float trail. He dug a tunnel but stopped "two feet short of a vein that later produced $3,000,000."

Men could not be faulted for stopping too soon. Lode gold in particular came in many puzzling formations. A group of Colorado prospectors in 1874 also came close to Cripple Creek gold but halted when they found only a gray substance one noted was "white iron," more likely sylvanite, "Cripple's characteristic gold ore."

Even in the placer regions, seekers often failed to recognize a good thing when they saw it. One California prospecting party on reaching the Tuolumne River attacked a grassy ledge and found gold flakes on the grass roots, but they were too naïve to realize how unusual that was. They departed with some dust and their tools, ignoring or unaware of the recognized mining practice of leaving one's tools to mark a claim. Only after they had weighed their yield at the local store the next day did the value of their discovery manifest itself—and by then an observant fellow prospector had taken possession of the ledge.

Sometimes seekers knew they were on to a good thing but saw it slip through their fingers. Granville Stuart, his brother James, and a partner located a very promising California claim on a Saturday. Granville was anxious to go back on Sunday to "post" the ground as theirs, but James and the partner wanted to wash clothes instead. "It was by far the most costly washing that ever any three men indulged in," mourned Granville, for another party found the spot on Sunday and claimed it, extracting "over twenty-five thousand dollars in beautiful coarse gold."

Even if a man was working a good claim, he could thwart himself in various ways. In addition to setting sluices at the wrong angle, greenhorns placed their riffles in an uneven line, didn't regulate the water flow appropriately, and failed to stir regularly the sand collecting against the riffles of the rocker, long tom, or sluice. The sand would build up, and the gold would flow "right over [the riffles] with the muddy water."

Generally, how much diggers took from placer claims remains more difficult to pin down than do the profits or losses of the quartz mines. Both placering and quartz mining could be ephemeral, but because the latter required trips to the assay office and some industrial investment and development, the effort tended to last at least a few months and to be fairly well documented. Placer miners, on the other hand, might work an area anywhere from a day to months and did so informally, involving only a small number of men even when elaborate sluicing or hydraulicking was used. They dug, sometimes paying a modest claim recording fee. They paid for any provisions and services with their dust, finding it an enchanting but transitory medium, one that in use "tended to deteriorate" rather alarmingly. Further, storekeepers and other goods-and-service providers made sure the miners got in trade no more—and often less—than the value of their dust. The providers used small sheet-iron or tin blowers "to separate sand from dust," often reducing one or two ounces to "several penny-weights." Many were not above substituting low-grade dust for high-grade.

Miners pocketed what dust they could hold on to after paying the high cost of living and avoiding or succumbing to the gaming tables. They could leave the remote western areas without declaring their wealth to anyone. When they were pressed to do so—as departing prospectors were by ship personnel demanding freight charges on the gold leaving San Francisco—many simply lied, having hidden dust in "luggage, boots, or specially constructed money belts."

Despite the difficulties of accounting created by such subterfuges, it is clear that California's mother lode country yielded far more riches than any subsequent region, with estimates of over $500 million in both placer and lode gold from 1848 to 1857. The Yukon diggings ran a respectable second, producing close to $300 million in their glory days. (Nevada's Comstock Lode—some gold, but mostly silver—also ran close to $300 million in its first fourteen years of development.) The other giant rush, to "Pikes Peak," led to a relatively meager yield of $25 million in Colorado's first ten years of gold production, while the gold production of Montana and Cripple Creek actually surpassed the Yukon rush figure, but not until the twentieth century.*

All of this would have been academic to the average goldseeker, intent as he was on personal gain. What mattered was

* For specific figures, see T.H. Watkins, *Gold and Silver in the West*, pp. 60, 80, 92, 100, 156.

that early in the California rush some prospectors were collecting ten ounces a day at $15 to $16 an ounce, that in the Klondike men washed single pans of gold worth $150 or $200.

Averages were vastly lower, but averages didn't matter much; most gold-seekers were looking to be more fortunate than the rest, to hold, as lucky Colorado prospector Abe Lee hyperbolized, "all of California" in their hands in one grand strike.

So men kept listening and looking and moving, usually relying on the news of new diggings rather than seeking out whole new gold areas, for most "had rather go where gold had been and was still being found, and risk the chances of obtaining a claim, than go abroad seeking uncertainty." Still, they were ready to treat as a certainty the most tenuous of stories, especially in the winter months when digging slowed or stopped altogether. One man who succumbed to the lure of a mad and torturous winter ministampede in Alaska vowed, "It would take all the men in Ambler City with a great big hawser to pull me away from my warm cabin and grub again this winter."

Long-term prospectors learned to make use of the stampede mentality. In Alaska, old-timers would "locate some claims on different streams, let the word get out that good prospects have been found, then [let] others rush in" and do the work "to see what is really there," the old-timers apparently confident of sharing whatever wealth existed.

For those with a fair share of wanderlust undampened by the journey westward and by ups and downs in the diggings, a prospecting life based both on following gold news and on independent searching held much appeal. Warren Sadler, in California in 1849, averaged sixty dollars a day on the Yuba River at a time when a dollar a day was considered good wages in the East, yet he "got uneasy" and moved on.

For those who took to the freedom of the frontier and to the heady game of prospecting, the rhythms of civilization became stultifying. Sadler went back East and tried to remain, but "there was no excitement as I had been used to in the mountains," where he could blaze his own trails when he wished and "continually [learn] something new." He was the true prospector who felt "thoroughly equipped" with "a slab of bacon, a few pounds of flour, a little sugar, coffee, tobacco . . . an old pick and shovel," and perhaps (but not necessarily) a pack mule. With this outfit, reported one California paper, the prospector "scales the mountains, swims rivers, and skims the plains for months, happy as a stuffed goat."

Such a description obscures the uncomfortable realities that wore down less committed prospectors—lugging grub over rough terrain, panning with mosquitoes blackening hands and arms, stretching exhausted bodies in thin blankets on cold and rocky ground, hanging cooking utensils in the trees to keep the wolves away. And even the raw mining camps to which prospectors gravitated offered a life that was a continual exercise in doing without and making do, provisioning and entertaining oneself in the most basic fashion.

Newcomers to any diggings first had to find some kind of shelter, although many western gold-seekers, lulled by mild weather and unsure about the length of their stay in one place, would simply set up camp under an arching tree. Others paid for the privilege of bedding down in a "hotel" or inn. "No beds, no bedclothes except blankets in which thousands have slept before, and all . . . full of body lice," recalled one California argonaut, while a Yukon traveler found his bed "a dirty sheet spread over unplaned boards—only this, and nothing more." An infrequent woman accompanying a man might offer boarding arrangements in a camp, but most newcomers scrambled about collecting anything that could serve as building material for a shelter. Sawmills were scarce (and far more lucrative for their owners than mining), lumber was usually at a premium and necessary for mining operations, and men were too impatient to dig to spend much time on a habitation.

So they burrowed into hillsides and devised a variety of rude "bowers," in California the most popular being "bush cabins" consisting of "a mere strip of canvas or calico suspended as awning from the spreading branches of bush or tree." At Breckenridge Diggings in Colorado, men lodged in scattered "bowers or houses made of the branches or Saplings of the spruce and pine . . . the builders often only occupying them for a single night, when they are taken possession of by the next comer."

A stay for any length of time at one location made a snug

cabin a priority before the rainy season or snow set in. Yet excited, restless argonauts sometimes failed to create even the most basic shelter. When the rains descended at one California placering site, men wrung out their blankets and held them over their heads, then found both themselves and the blankets saturated in mud when they tried to sleep on the ground. Next the rain and snow mixed, the miserable prospectors lying wrapped in their wet, cold blankets with boot-clad feet sticking out and "the water actually running out of the tops of their boots on their feet."

The difference a cabin made is evidenced in the pleasure one Alaska prospector felt in his new home "after six months knocking about, eating and sleeping in a tent, boat, on the sand or under the trees." His primitive cabin seemed a palace, for it provided "[a] place to sit and lie down, a place to put things besides the ground, a roof overhead and walls that keep out the cold and wet." Some cabin residents spoke with pride of "a place for a window," a storage area, a workbench.

But with the jerry-built nature of these structures, it is not surprising that gold-seekers remained exposed to some vagaries of weather. After a Kotzebue Sound prospecting party completed their winter cabin, a heavy rain the first night "pour[ed] in on our beds and our precious supplies." The party covered the roof with tent cloth, which immediately "froze stiff for the winter," rendering them dry for the season but making the cabin a drippy ice cavern the next March. In a similar situation in Idaho in 1862, spring thaw brought water running downhill into a miners' dwelling, creating a current of water the size of a man's arm streaming through the fireplace and requiring a drainage ditch "right through the middle of [the] cabin to carry the water off."

In Alaska and the Klondike in particular, cold assaulted tents and cabins with deadly force. To one post-gold-rush Klondike arrival, "It seemed like an animate thing, creaking insidiously under the crack of the door in a long white streak," coating every nailhead, keyhole, and knob in ice.

In the early days of each gold rush, however, keyholes and knobs seemed to exist only on another planet. So, too, did furniture, window glass, and household furnishings of all kinds.

Stampeders devised their own. "Easy chairs" were constructed of hewn poles and sacks, and windows were improvised by leaving a hole or by removing a log and inserting fruit jars, the gaps filled with clay. In the Yukon, some created windowpanes by "cut[ting] clear pieces of Yukon River ice slightly larger than the window openings and fasten[ing] them in place with wooden buttons." Yukon gold-seekers were equally adept at creating stoves from rocks and mud, with a flat rock for the top. One Alaska company made carpets out of "gunny-sacking stuffed with dry moss" and fashioned from birch branches the candelabra that graced their Christmas table.

In such ways, people improvised all across the mining frontier, leading California mining camp resident Louise Clappe to muse, "How I shall ever be able to content myself to live in a decent, proper, well-behaved house, where toilet tables are toilet tables, and not an ingenious combination of trunk and claret cases, where lanterns are not broken bottles, book cases not candle boxes, and trunks not wash-stands . . . I am sure I do not know." In a classic statement of frontier practicality and mobility, a gold rush Montanan advised his wife on joining him to leave most of the furniture behind, as "people don't use much furniture here anyway, and we may not stop in this country long."

If wanderlust didn't propel people onward, lack of supplies might. Gold-seekers could get by without chairs or window glass, but they couldn't get by without food. "Damn the gold! You can't eat it," became the cry. Long-term prospectors learned to provision carefully and to live off the land when necessary, but each new rush brought too many people catapulting into an inhospitable, agriculturally undeveloped area.

The supply problem was exacerbated by the gold-seekers, primarily young men doing heavy labor outdoors, possessing prodigious appetites. Diets in the diggings consisted of certain staples and depended upon the meager, inconsistent supply networks and the ability to pay wildly fluctuating or simply exorbitant prices—"these humbug prices," as one argonaut put it. The situation often left cooks in a quandary. "I wish I could see you tonight and get some pointers on tomorrow's cooking," one Klondike prospector wrote a female friend. "I should not wonder if even you would be stalled if you had to work where and with what we do. You would know better how to work your way out though."

A forty-niner reported, "It is soon told what we live on, for it is only Flour, Pork and Beef, Beans, Rice, Sugar, Molasses, Vinegar, Coffee and Tea, and dried Apples." Eggs could be had only at eight dollars a dozen, butter at a dollar and fifty cents a pound, potatoes at fifty cents a pound, milk at a dollar per quart, and apples at twenty-five cents apiece. A heavy goldfield reliance on dried apples is indicated by a wry Colorado "Miner's Creed" in which the miner "wonders if there is any other kind of fruit than dried apples, dried apples scalded, and dried apples with the strings in." In the Yukon, the staple diet consisted of "the three B's": bread, beans (or "Klondike strawberries"), and bacon. On the Alaskan coast, a typical gold rush meal consisted of "corn-meal mush, biscuit or flapjacks, hash, bacon, flour gravy, and coffee," together with "a plentiful supply of mosquito sauce" in the summer.

The main staple, flour, could cost two dollars a pound in the California diggings. It was the necessary ingredient for the flapjacks California miners became so expert at concocting that "they could throw them up the chimney, then run around on the outside of the cabin and catch them in the pan." Then there was the sourdough bread so common in the Klondike that old-timers became "sourdoughs."

Gold-seekers applied their ingenuity to food sources. Men searched the hillsides for berries and other edible plants. They produced coffee from ground acorns and brewed tea from mountain mint or cottonwood twigs and spruce needles. Sometimes prospectors lived only on successful animal kills, trying new meats in different ways: porcupine "boiled, roasted, and stewed," moose meat boiled and fried.

Most argonauts in established diggings relied on freighters and hunters to supply their food. For a hefty charge, the former tackled rugged terrain and inclement weather with their wagons, mules, and oxen (and in Alaska and the Yukon, with their dogsleds), bringing plenty of liquor—"high in value in relation to its bulk"—and whatever else they could squeeze in. A freighter's arrival in camp could signal an onion stampede—or a flour or potato one. Meanwhile, hunters ranged the hills, supplying diggings and small towns with meat from deer, antelope, elk, bear, and buffalo.

On California's American River in the winter of 1849–50, with provisions depleted and supply trails treacherous, men went into the mountains with their rifles in search of wild game. Prospectors on the Feather River ate barley intended for pack-mule feed in order to keep themselves alive. One man, after three days without food, reached a Spanish rancho where he entered legend by consuming twenty-seven biscuits "and a corresponding quantity of other eatables, and, of course, drinkables to match."

On the Fraser River in late 1858, men, realizing that there was "no easy mode of conveying the necessaries of life from Victoria or any of the seaport towns," abandoned their claims. In Alaska and the Yukon, prospectors received periodic warnings to leave the country before winter, although some survived by shooting caribou and rationing moldy flour and beans.

People tried to conserve, but the self-discipline under already Spartan conditions, and the naturally enhanced appetite of men living and working outdoors, undercut such efforts. One Alaskan prospecting party bickered over the rationing of their condensed milk until "at last we ripped it all open and let them swim in it while it lasted." Another Alaska prospecting party, seeing the flour disappear, determined "to eat while we have the means, and go without when it is gone."

Food and liquor cravings sometimes led men to make poor trades. A Klondike miner exchanged a claim for a pig. The claim later brought four thousand dollars, leading a Dawsonite to quip, "This little pig went to a high market." Another Yukon miner traded for half a case of rye a claim that would eventually sell for twenty thousand dollars.

Even in the summers, with all supply lines open, prospectors often found it difficult to obtain provisions. Some agriculture efforts did develop, but the gold rush demand was always greater than the supply—so great that miners around Coloma, California, paid a dollar apiece for unripened pears still hanging from the trees; "the purchaser would select his fruit, tag it with his name, and wait hopefully for it to mature." And most gold rush sites remained incredibly remote from farming locations even when the climate was favorable for growing food.

Mining tools and other mining supplies also proved hard to obtain. In May 1848, California merchant Thomas O. Larkin found that "baskets, tin pans shovels, etc. bring any price imaginable at the gold washings." The next month, in San Francisco, he noted that dollar spades and shovels cost ten at the diggings, with offers of as much as fifty dollars for one. In Montana, enterprising men fashioned wagon spokes into pick handles. In the Yukon, Klondike king Charley Anderson "so badly wanted to build a sluicebox that he gave eight hundred dollars for a small keg of bent and blackened nails salvaged from a fire."

Mundane housekeeping items also carried a high price tag. An Indiana argonaut entering the California mining camp of Agua Fria obtained the two needles he needed at a local store but became irate when told that they cost a dollar each. Significantly, he subsided in embarrassment when the storekeeper took his anger as a sign of his newness and ignorance.

Ironically, a dearth of goods occurred as each gold rush site exploded, just when whatever gold there was found its way into miners' pokes. In California in 1848, "there was plenty of gold but very little to buy—a classic supply-demand crisis." Scotsman William Downie's prospecting party struck pay dirt in California and gleaned a bag of gold "so heavy that they frequently argued about who would carry it, yet they were starving." On Eldorado Creek in the Klondike, a young French Canadian reportedly "died from exposure in his little cabin, surrounded by tin oil-cans filled to overflowing with gold-dust and nuggets." In Dawson, "[a] Chicago ham suspended overnight from the outside ridgepole of one's cabin [would] almost certainly disappear, but a sack of dust, in the same situation, would probably be overlooked." In California's initial boom, two thirds of the proceeds from an ounce-a-day claim often went for food, leading one forty-niner to complain, "I would have made money if I could have lived without eating."

Sundays brought a break in routine, but not a restful one. At the Orofino diggings in 1861, a prospector watched "a great army of miners" swell past his cabin door, "almost staggering under the loads of worn implements, which must be repaired by the blacksmith that day for the renewal of hostilities next morning." In addition, "bills are to be settled, provisions and supplies purchased, implements are to be repaired or replaced by new ones, and all the needed preparations for the coming week are to be made." Two years later, a Bannack prospector reported his Sunday routine: a three-mile walk from the diggings to his cabin, a trip to the blacksmith shop, letter reading and writing, clothes washing and a bath, mailing of letters, and the walk back to the diggings with a load of flour "to be ready to commence another week's hard work." Many became uncomfortably aware of the difference between their former dapper appearance and their present state. "I never more shall cut a swell," lamented one song, implying in addition a weakened faith in the gold hunt. Thoughts of home, both of loved ones and comforts, intruded more forcefully, too, on this day.

Escape was sought primarily in the liquor bottle and in the excitement of gambling. The frontier West was fairly saturated in alcohol, with mining camps absolutely sopping with the stuff, men "steep[ing] their souls in drink." Roads and byways were often littered with "empty bottles every few yards." One of the most famous scenes in mining literature is Louise Clappe's depiction of the male residents of Rich Bar, California, in January 1852, engaging in a "perfect Saturnalia," even the steadiest of men succumbing to the weight of isolation, "remorseless, persevering rain," and fraying hopes by going on a three-week bender, at one point all lying "in drunken heaps" around a barroom floor and barking, roaring, and hissing in an inebriated cacophony of sound.

Alcohol performed a number of functions, imbuing the "temporary relationships of the mining camps" with the luster of friendship, "help[ing] to blunt the edge of despair when speculations went awry," and making outlandish hopes and plans seem feasible. An evening in a saloon in Any Mining Camp, U.S.A. was "warranted to inspire the immigrant's mind with energy enough to turn a boulder of granite into a beautiful quartz ledge."

The saloon, along with the general store, was often the only community structure in camp. If it "demoralized and ruined a few," as one Yukon resident wrote, it also "kept most men from going insane."

Waiting at the gaming tables in the mining camp saloon were the nattily attired gamblers with their faro, poker, monte, and other games of chance. "Wherever there is gold, there are gamblers," Bayard Taylor prophetically observed during the California rush. At Nelson Creek in California, a monte banker came to camp, and one forty-niner saw the whole atmosphere change. Men began to drink heavily, eliminating everything else—"work, meals, sleep"—in order to gamble, their hard-dug gold slipping swiftly through their fingers in what became a gold rush pattern.

Even if a camp was too remote, poor, or unpublicized to attract the professional gamblers with their dapper dress and soft hands that could detect the slightest marks on cards, men in the diggings gambled on anything and everything. In the Klondike, "two old-timers bet ten thousand dollars on the accuracy with which they could spit at a crack in the wall."

All leisure was not drinking and gambling, although many observers of mining camp life perceived it as such. Some camps had the near-constant clatter of bowling, or "ten-pin," alleys, and saloons boasted billiard tables, lugged into camps across the West at tremendous trouble and expense. Another favored activity was musicmaking, often accompanied by dancing. Fiddlers received eager mining-camp welcomes. "Oh! joy," wrote one Montana prospector. "It is not often that we have a fiddler and when we do have one, we try to keep him in practice by having a dance every evening." With women absent from the diggings or only sparsely represented, men staged dances as they had on the trails, with some wearing armbands to identify them as female or even occasionally dressing the part.

Sometimes, too, traveling theater troupes made their way to the camps. Miners welcomed the diversion but could be harsh critics, sending a poor performer off a makeshift stage with a volley of pistol shots. In the mining camp of Washington Flats, three brothers comprising a traveling troupe were warned by the proprietor of the only hotel that "this is the toughest camp in California, and if you don't suit the boys they'll throw you in the pit," a forty-square-foot pool of water used for mining. The brothers retreated to a hotel room after their performance, only to hear a group of miners mounting the stairs and kicking at their door. Whisked from the room, the three were taken not to the pit but to the barroom below, where the miners signaled their critical approval with a good meal and the presentation of a tin mustard box of gold nuggets.

The prospectors also made their own diversions. As they had on the trail, they hunted for pleasure and hiked. Back in camp, they organized lying contests and staged mock trials. In one Idaho mining camp during the interminable winter months, the ringing of a cowbell signaled a meeting of the "Hocum Felta" club, which boasted a number of old forty-niners among its members and which was "change[d] at will into a debating society, mock legislature, mock court, spelling school, reading contest or glee club." On the Kobuk River in Alaska in the winter of 1898–99, a company from California organized a literary society to offer lecture topics ranging from "How to Dispose of the City Slums" to "How to Care for a Frostbite."

Holiday celebrations provided an excuse for bounty and merriment in bleak circumstances. They also allowed the stampeders to evoke sweet memories of their life in civilized surroundings and at the same time to poke fun at that distant life. One company entertained another on the Kobuk to a lavish Thanksgiving meal, a member of the host company then seating himself behind a pot with a spruce branch protruding from it. He plucked upon a string stretched taut on a wooden block, thus providing an after-dinner "orchestra" offering "sweet music from behind potted plants."

On an even more basic level, men gathered around their campfires or fireplaces and told stories, played chess, and read old newspapers or books—when they could get the reading material. On the Feather River in California, a gold rush storekeeper kept a collection of novels "by far the largest, the greasiest, and the 'yellowest kivered' of any to be found on the river." In Circle City, Alaska, a few decades later, a trader rented for a stiff fee volumes from his library of two thousand books, including the Bible, the *Encyclopaedia Britannica,* and the works of Darwin and Hume. Successful Yukon prospector Andy Hunker surprised a visitor with his well-thumbed library: six volumes of the work of English historian Edward Gibbon. One

Alaska gold-hunting company on the Kobuk River in 1898 by traveling almost exclusively by water had managed to transport "a whole library of good books," which they traded with other prospectors up and down the river. Two Montana prospectors followed the rumor of a trunk full of books 150 miles "with three big dangerous rivers to cross" from their remote claim. For five dollars each, the two obtained Shakespeare, Byron, an account of Napoleon, and Adam Smith's *Wealth of Nations,* feeling ecstatic with their good fortune.

In Deadwood, South Dakota, adventure books provided a rich fantasy life for at least one gold-seeker, who modestly let slip to the Deadwood federal judge's daughter that he had received a Victoria Cross for taking a cholera-stricken load of ship passengers through from Panama to Havana. Duly impressed, she lent him Stanley's *In Darkest Africa.* Shortly thereafter, another argonaut asked her, "Did you know that that fellow was in Africa with Stanley?" The other miners had become convinced of his claim when they quizzed him, using the book, and he told them "things he couldn't have known if he hadn't been there."

Whether working hard in the diggings or "strain[ing] to find the easiest and best methods of killing time," men lost track of days and hours, especially when walled into deep ravines where they seldom saw the sun or when snowed into winter cabins. One sure daily marker was the odd, plaintive cry of "Oh Joe" that spread in the evenings through California mining camps and eventually across the western mining frontier. Reportedly based upon an incident in which a prospector fell into a hole and called upon his partner for assistance, the call would drift from cabin to cabin through the waning light and darkness.

But in Alaska and the Yukon, even light and darkness proved no sure indicator of time. One Klondike prospector on a three-day summer spree, when told it was eight o'clock, exclaimed, "Hell of a lot of good that does me! Is it eight o'clock in the morning or eight o'clock at night?" The extended winter blackness was even worse; "revelers lost all sense of time . . . and attuned themselves to the ever-present night until they passed out from sheer exhaustion." As a Dawson resident put it, "We had no very clear notion here of the days, the weeks."

PRECIOUS DUST

*   *   *

As the fainthearted (or the wise, depending on how one looks at it) tired of this strange life and filtered out of the goldfields, those remaining reasserted their resolve in various ways. Some cited family reasons. A forty-niner wrote, "I must cheer up and dig and dig till I get plenty of gold and be able to return to my family with the means to make them comfortable." A Pikes Peaker vowed, "I came here to make something for my family and I will do it before I leave the mountains entirely."

Others exhibited a simple independent pride—"You can bet your life I will never come home until I have something more than when I started"—or colored it with a somber personal assessment. Alonzo Hill, working in San Francisco in 1851, wrote to his father, "I can see no object to come Home, I can make nothing there & but little here." Almost fifty years later, an Alaska prospector wrote his family, "If I were to go home 'busted' or almost so . . . What next? Where will I get a job? . . . I have cut loose and learned a new life, and I must stay with it at least another year."

Some were frankly haunted by past failures. Montana argonaut Frank Kirkaldie felt that he had failed to make a decent living in Vermont, Illinois, and Iowa; if he acknowledged defeat in Montana, "it would look like the last term in the decreasing series."

Kirkaldie represented a portion of the stampede population that virtually abandoned the gold quest but still clung to their determination to "make a pile" in some way in the mining areas. As he insisted, "I have come too far—& have spent too much time—deprived myself of the comforts of home etc. in attempting to better my condition—to abandon the enterprise without a persevering struggle." Kirkaldie had a wife and four children with whom he was eager to reunite, but still others had fallen easily into a hedonistic life-style that held them indefinitely. If the gold eluded them, they tried to carouse their troubles away. If it didn't elude them, so much the better; they would "have a hell of a good time with it while it is worth something."

Of course, the gold rush West afforded an escape not only from social constraints but from personal and even national difficulties. During the Civil War, many a seeker agreed with one prospector that digging was "rather a hard way" to make an admittedly modest living, but "better than living in the midst of the troubles at home," especially when family loyalties were divided.

Ultimately, many simply could not relinquish the dream of gold-gathering they had traveled so long and endured so much hardship to achieve. "Sometimes I get desperate and vow to leave the Godforsaken region," John Callbreath wrote to his mother from the Fraser River diggings. "But the thought of losing a good claim, for which I traveled over 1,500 miles deters me from the step, and I again fall back on hope."

Men had only to look around at their more experienced fellows to discern how great the odds were against success. "In 999 cases out of a thousand the men who pursue the varied avocations of western life, do not even acquire a decent competence," wrote a western observer in the 1860s. These men knew that the wiser course would have been to stay home, "practising a wise economy with the modest earnings of their trade or profession," but they continued their western rambling, "partly from habit," partly from pride, partly from hope of a "big strike."

Hope kept the seekers going for a month, a year, a lifetime. Some struck pay dirt, spent it all, and started all over again. Whatever the individual experience, any seeker who stayed on the gold rush frontier for very long witnessed urban centers develop near the diggings, camps turning into towns. He periodically returned to less remote urban outfitting centers or sought temporary or permanent work in the mining and outfitting towns and cities of the West. In other words, he not only moved in the strange new world of the diggings but in the strange new setting of the pulsing, fractious, mud-spattered towns built on gold.

Chapter Seven

"Cities of
the Magic Lantern"

> . . . like the cities of the magic lantern, which a motion of
> the hand can build or annihilate.
>
> —*Bayard Taylor, on gold rush San Francisco after dark*

The towns built on gold rushes often startled western travelers. Dawson was judged "a mad and freakish place," Nome "about the oddest excuse for a town I ever ran across" and "more like the old part of Coney Island than anything else." Gold rush San Francisco was "odd . . . unlike any other place in creation," having been "hatched like chickens by artificial heat."

The towns did hatch and grow in strange ways, barreling along with precipitous, uncontrolled development and offering stampeders rude, tumultuous, and costly living conditions, erratic economic opportunities, and entertainments ranging from the bland to the bizarre. As they did so, many of these urban centers teetered between stability and oblivion—and all struggled with problems associated with frontier identity and rapid change.

Three types of frontier towns boomed with gold or expectations of gold as their lifeblood: the major supply centers far from the diggings themselves, the intermediate centers created on the way to the diggings, and the urban communities that exploded helter-skelter from the mining camps at or near the strikes.

The first type owed its existence to some factor or factors other than precious metal mining. Cheyenne shot into existence as a rowdy railroad town, and San Francisco germinated as "Yerba Buena," a tiny Mexican port with an economy based on the hide-and-tallow trade. Such diverse places would become rush centers by virtue of hosting concentrated argonaut movement and activity—gold-seekers passing through on the way to the diggings for the first time; buying supplies; arranging transportation or deciding on routes; and returning to replenish their supplies, to rest and recreate, to regain their health, to find a job, to figure out what to do next, to escape harsh winters or short provisions in the mining areas. The transformation of such towns into gold rush centers would end with the ebb of the rush itself.

The other towns were birthed in gold and gold hopes. Intermediate centers such as Sacramento and Stockton, Dyea and Skagway were created as primary staging locations for travel to and activity in the diggings. Mining-camps-turned-urban-centers such as Virginia City, Cripple Creek, and Dawson were fulcrums for claim-holders scattered through a mining region.

A town born of or energized by gold fever was often touted as the next great western metropolis. Such grandiose pronouncements were understandable, for boom sites pulsated with the activity of hopeful hordes. In San Francisco in 1853, streets were almost blocked by a "throng of carts, carriages, horsemen, and pedestrians." In Dawson in 1898, the main thoroughfares were "too crowded for comfort, with men, dogs, wagons, pack-trains, and a great bustle and hustle everywhere." A traveler to Nome, citing "a perfect Babel of noise," declared, "The crowding and bustle there was something I have never seen before and never expect to see again."

Often, two booming urban centers in the same area would enter a runoff of sorts for premier regional town status with the loser destined to disappear rapidly. Denver eclipsed rival Auraria; one of two competing town sites at Cripple Creek prevailed in part because of its lower elevation.

Such practical topographical considerations seldom had much bearing, the most important consideration being proximity to the gold. Like the diggings—often alongside them—towns jammed into narrow strips against creeks in claustrophobic ravines, drunkenly mounted high-altitude hillsides, spilled onto arid slices of desert, squatted on "frozen swampland." The whole crazy patchwork of dubious sites culminated with the appearance of Nome on a chill, barren beach of Alaska's Seward Peninsula; as one historian has put it, "It may be possible to imagine a more unlikely setting for a frontier mining town, or for that matter, a town of any kind, but I can't think where. Perhaps there is such a place in Antarctica."

Sometimes, the argonauts themselves founded the towns, almost in the manner of children playing at metropolis-making. A boatload of Klondike prospectors created the short-lived "Derwent" in one day by platting the town, drafting the necessary resolution, and identifying and selecting eighty lots—all before leaving their steamboat and finding a place to put their scheme to work. Landing at a likely site, they all disembarked, passed their resolution, painted "Derwent" on a box cover, and nailed it to a tree. The group then immediately left a rather bemused gold-seeker-turned-town-recorder in sole possession. Even he didn't stay long.

Gold town developers generally were westerners of a different stripe, urban-oriented adventurers—business entrepreneurs —who seized the opportunity for small-scale empire building as they saw a rush develop. One man or a small group would gain ownership of a site by filing a land claim—even a placer claim—or purchasing it for a pittance. They would then throw up a building or two, get some streets platted, proclaim a town, and wait to realize a fortune in town lot sales.

Of course, this was a chancy enterprise. Even in San Francisco, land and town lot speculators lost fortunes, some of the most prominent in 1852–53 going broke by 1855. In the remote mining areas, the risk could be even greater. Still, there were fabulous profits to be made when a site reached its zenith of desirability. In Dawson during its one-year boom, business property zoomed to as much as "five thousand dollars a front foot," and a slick salesman parlayed his profits from selling ten-cent cigars for a dollar fifty cents each into a tidy fortune by investing in town lots.

But the appropriation and high pricing of land created a tangle of contradictory claims—"I have known as many as six claimants for one lot," wrote one San Franciscan. Speculators who had obtained title to city lots by fair means or foul—or claimed dubious title—warred with the immigrants who arrived after the town had been parceled up among the lucky few. As more and more newcomers arrived, they bitterly resented the situation and began "squatting" on land claimed by the early entrepreneurs or took over land set aside for streets and public areas. In Denver in 1860, residents split into two warring camps, the city's first shareholders charged with having taken "at random and without cost vast land tracts which were now priced so high that newcomers couldn't make decent purchases."

Such standoffs could ignite into serious violence. A small nucleus of Sacramento merchants obtained questionable title from John Sutter's son for four fifths of the town's area, then installed themselves as city officials. Incoming stampeders refused to recognize their claims and appropriated city land, contending that " 'Sutter' has no claim or title and one man has as good a right to the Soil as another." Merchants tried to meet this threat by dubbing themselves the law-and-order party, but their rivals formed a "Settlers' Association" to press their own claims. With no state government yet in place, and with a local court decision going against the association, the situation degenerated quickly; in the ensuing controversy the mayor and other town officials were wounded, the sheriff killed.

The episode demonstrates the absence of any clear-cut, effective territorial, state, or federal government control. Merchants, real estate speculators, and other business and civic aspirants usually took the lead in hammering together a local government—everybody else was too busy rushing. But everyone was aware that this government represented certain urban interests and existed—initially, at least—without sanction of any higher government authority.

The struggle over Nome's town lots demonstrates even more clearly the disorder and infighting accompanying gold rush city-building. In this case, it was actually the original locators of the Anvil Creek claims—the first successful prospectors—who attempted to monopolize the town site by claiming a large chunk of coastal property of which Nome was a part. Meanwhile, others designated the town site land by "popular vote." Yet with no federally authorized local government, neither side had any legal right.

Even when given an opportunity to establish a legal local

government, gold town residents often balked, showing antipathy for any government activity except that which immediately aided their own personal quest. After all, few planned to stay long in one place—or did. "Of all the acquaintances I have made here, and they are many, I do not believe one person wishes to remain permanently in the country," an observer wrote from Virginia City in August 1864.

However transient the population, it still increased phenomenally. As with gold production figures, it is often difficult to pin down reliable population figures for remote nineteenth-century sites. Cities developed in a matter of months or weeks, leading Bayard Taylor to marvel of San Francisco that it "seemed to have accomplished in a day the growth of half a century" and of Sacramento, "Can the world match a growth like this?" In January 1848, San Francisco had about 800 inhabitants, 200 of them newly arrived Mormons; by the end of the year, "it would be the great metropolis of the Pacific coast" and by 1851 would boast a population of thirty thousand. Sacramento went from an outpost of four dwellings in April 1849 to an estimated population of ten thousand by the end of the year.

All across the gold frontier, the story was the same. "This place has been built as if with magic hands," wrote one new arrival of Auraria, Colorado, in January 1859. The Pikes Peak rush had barely gotten under way, but already the community consisted of four hundred houses, six hundred "regular" inhabitants, "six good trading houses, two blacksmith forges in full blast, and grog shops without number."

With easily accessible Cripple Creek, people swept in "at the rate of a thousand a month," elevating the population to about 12,500 by the end of 1893. That year, too, saw a dizzying array of services offered:

> twenty-six saloons and gambling houses, four dance halls, twenty-four grocery stores, ten meat markets, nine hotels, nine laundries, three large bathhouses, eleven clothing stores, ten barber shops, nine assay offices, seven bakeries, six bookstores, forty-four lawyers' offices, eight stenography offices, thirty-six mining stock and real estate offices, eleven parlor houses, and twenty-six one-girl cribs.

Skagway grew so fast that its fledgling newspaper identified "old settlers that have grown up with the place" on the basis of three months' residence. Dawson, the "San Francisco of the North," early boasted among its businesses "fully thirty saloons," as well as two barbershops, an "incipient public library," two trading stores, a meat market, two dance halls, two sawmills, "an uncounted number of lawyers, brokers, etc.," and 235 gamblers, the last being a sure sign of a mining region's wealth or potential.

Nome probably topped them all, even the genuine metropolises of San Francisco and Denver, for staggering growth. In June 1899, it existed as the site of three tents and a log cabin. By the next summer, it had become "a city of close to 20,000 people."

In fact, those who passed through the inflating gold rush towns were rendered nearly speechless upon returning weeks or months later. Bayard Taylor after an absence of three weeks found San Francisco "seemed actually to have doubled its number of dwellings"; a person could take a stroll one morning, then find the way blocked by "a house complete with a family inside" on the next. A resident of Cripple Creek, returning to the booming center after an absence of about four weeks, could barely detect in the new urban landscape the settlement she had left.

Because of such rapid growth and the initial weakness of local forms of government, any urban planning was immediately overwhelmed. Central business districts emerged willy-nilly but fairly well defined, and usually a distinct red-light district quickly developed nearby. In Deadwood, it lay across Wall Street in an area called, in an interesting connotation of physical space, "the badlands." In Dawson, the river separated "Lousetown" from the city itself.

Aside from these groupings, however, the gold rush town developed with no particular regard to neat lines on a surveyor's plat. This was especially apparent in the placement of streets. In Helena, where "city blocks [took] every conceivable shape," some streets "ran for several blocks only to dead-end or leap to one side." Deadwood developed into a dizzying hodgepodge of winding thoroughfares, of steep mountain stairways ending nowhere in particular, of crazily angled houses, of streets girdling such precipitous slopes that one ascended into buildings on one side and descended into buildings on the other. In Cripple Creek, it was a common joke that a man had broken his neck plunging off Bennett Avenue, which had a drop of fifteen feet from north side to south side.

In addition, gold town streets proved irritatingly narrow. In Deadwood, any "especially stupid yoke of oxen" could bottle-neck all traffic. The crowding was ameliorated only by the fact that "to make traffic of a sort possible a house every now and then stepped back accommodatingly and left space for a 'turn-around.' " In Nome, the main thoroughfares were the width of alleys, while other streets were barely discernible.

Jockeying for clear passage were wagons carrying hay, lumber, firewood, provisions, and—in quartz mining areas—ore for the stamp mills. Stagecoaches rolled in, goods were often piled along the streets awaiting warehousing, and "a perfect forest of signs" nearly blocked visibility.

Yet the rush of people could almost obliterate the signs. Even if some adventurers were departing on rumors of new strikes, others continued to pour in, responding to the strikes on which the town was built. With everyone fueled by a heady mixture of hope and scheming, the spirit of speculation intensified the pace of business. At the peak of the 1860 Pikes Peak rush, one store-keeper in Denver reportedly left a scrawled message on his door: "Gone to bury my wife, be back in half an hour."

Adding to the constant movement was the fact that most of the residents—again, primarily young, transient males—used their dwellings for little except sleep. They ate at restaurants or with one of the few families in town, paying the hardworking wife for the privilege. They socialized and gathered mining news in the saloons, theaters, and dance halls, sought female companionship in the dance halls and both companionship and sexual release in the bordellos and one-girl cribs. They hurried frequently to the post office, hoping for news from home, took their wash to laundresses or Chinese laundrymen, and roamed about looking for old acquaintances.

In truth, they didn't have much to stay home in. In the early stages of town building, even the men suddenly worth thousands "were living under worse conditions of squalor than any share-cropper." A forty-niner on arriving in San Francisco in September 1849 noted that half the population was living "in tents, and cloth covered houses," while many simply lay "round here on the ground under blankets" at night. When the rain fell and the wind blew, soaked and chilled men held on to their tents to keep them "from blowing down in[to] the bay."

Newcomers took what accommodations they could. Destitute forty-niner Phineas Blunt was lucky to get a Sacramento bartending job that included the privilege of sleeping "on a few boards laid on the ground under the [bar] where constantly I hear the rattle of decanter and toddy stick," an experience that galled this moralistic family man. Minister A. M. Hough and his wife on arriving in Virginia City probably obtained the best the town had to offer in a mud-roofed parsonage "of entirely green logs from the woods, and chinked with the mud nearest at hand," boasting one glass window and two muslin-covered ones, as well as a cowskin carpet, "placed hair side up, and stretched tightly while green, and nailed to the floor."

In Cripple Creek, tents gave way only to unpainted board shacks boasting a wood or coal stove, canvas-lined ceiling, newspaper insulation, and perhaps an attic room. In Skagway, dwellings were "mostly small frame structures of the packing-case pattern with about the same amount of finish." In Dawson, homes consisted of "rough board buildings on stilts," tents "with stoves, cooking utensils, and bundles thrown around," and one-room shacks on raw, stripped lots.

As for public accommodations, one 1851 traveler wrote, "Oh! you can have no conception of what a Cal[ifornia] 'Hotel' is" and noted how "the cloth ceiling with the least breath of air goes capering & 'skating' about, over your head & it is a mere chance if the rats do not too." Denver's first hotel "had neither floors nor ceilings and only a sheeting of canvas overhead to keep the weather out." Canvas also covered the window holes and separated the bare-ground sleeping compartments from the saloon. In Dyea, one party of gold-seekers stayed in a "hotel" of "about 20×40 feet" in size, "one board thick all around with the wind whistling through the walls and floor," the latter "about a foot and a half above a pool of water." Minister Hough and his wife lodged in the best inn in Virginia City before moving to their mud-roofed cabin. The bed they rented had only one sheet; the other was being used as a tablecloth until bedtime.

Despite the crudeness of such arrangements, gold town hotels quickly—and sometimes deservedly—established a reputation for luxury and elegance. Dawson entrepreneur Belinda Mulrooney herself shepherded over the White Pass Trail the "cut-glass chandeliers and silverware, china and linen and brass bedsteads" for her soon-to-be-famous Fair View Hotel. Mulrooney would offer steam-heated rooms, Turkish baths, the finest table settings, and electric lights powered by a yacht anchored in the Yukon River.

Although most gold rush hotels little resembled Mulrooney's, they contained enough touches of style and elegance to seem incredibly fine to men who had been living in bowers or tents in the diggings for weeks or months, some of whom had not even seen a carpet in years. The saloons, too, seemed to flash luxury with their intricately carved wooden bars, their precious mirrors and sparkling glass bottles, their green cloth gaming tables.

At first, of course, the saloons were mere tents graced by kegs and wooden benches—often multipurpose tents at that. In one of the Colorado gold-mining towns, a cluster of three small canvases contained "a saloon (for drinking and gambling), store, boot and shoe shop, barber shop, [and] bakery," as well as milk for sale. But these functions quickly separated, and saloons became semipermanent structures, perhaps with an old wagon sideboard propped up on boxes to constitute a bar, the gambling tables set nearby. Then the saloon took more definite form, the gaming tables in the back or in the next room; freighters bringing brass fixtures, richly patterned curtains, even pianos. The proprietor of Dawson's Northern Saloon had a piano carried piece by piece over the Chilkoot Pass; the status of having the only piano in town probably contributed to his three-thousand-dollar-a-day business.

Gold towns' existence depended in large part on freighting and supply networks, but these varied in effectiveness. Port settlements naturally benefited from their proximity to sea trade. The influx of goods into San Francisco reflected a preindustrial pattern of development, but the supplying of the port city of Nome fifty years later was much more dramatic. Advances in American manufacturing and shipping made city building possible with the arrival of a few well-stocked vessels. On one day in May 1900 alone, four ships out of Seattle bore among their cargo everything necessary to construct and set in operation a newspaper and printing plant, a bank, and a saloon, as well as "the rails and most of the equipment for a seven-mile-long narrow gauge railroad, about 300,000 feet of lumber, 600 tons of coal, and many more tons of general merchandise and mining equipment." Ships also sped "knock-down theatres, gambling halls, saloons, hotels, [and] restaurants" to the new city.

Some of the mining towns tucked into the rugged terrain of the American and Canadian interiors lay near navigable waterways, but the expense of bringing supplies in by water was high. The steamer freight charges from San Francisco to Sacramento were higher than those levied on the same items between New York and San Francisco.

Supplies were usually freighted across the West on ox trains or mule trains. The early Montana towns depended on such diverse and distant places as Salt Lake City, the Missouri frontier, and Walla Walla for their supplies. Deadwood was only two hundred miles from the transcontinental railway, but two hundred miles meant that "even with favorable weather the journey was too long for perishable goods."

The most prized cargo bound for the gold towns was the liquor, initially barrels of pure alcohol. Saloonkeepers across the gold frontier would dilute it with water and produce whiskey, bourbon, rye, and cognac using burnt sugar, plug tobacco, mountain sage tea, cinnamon root, cayenne pepper, and other closely guarded ingredients. Liquor was always welcomed; an eighty-wagon train carrying champagne and whiskey to Denver in 1864 stimulated the *Rocky Mountain News* to enthuse, "That's a train what is a train."

Some supply lines allowed for a variety of edibles; in Bodie, California, restaurateur George Callahan offered "oysters from San Francisco Bay as well as fresh duck from Mono Lake and fresh trout from the Sierras." But even in the ports, choices were often limited; 1849 San Francisco had plenty of "fresh beef, bread, [and] potatoes," but fruit, vegetables, and milk were "classed as luxuries," fresh butter "rarely heard of." In Deadwood, any food supplies were welcomed, but fruit and green vegetables fetched huge prices.

As in the diggings, mining town residents learned to adapt. A Nevada City, California, storekeeper openly manufactured butter from tallow and lard and won plaudits from his customers. In Deadwood, storekeepers took apples frozen in transit and made apple cider. In early-day Denver, one mountain man with a gun supplied most of the town with meat, while in Deadwood residents foraged the hillsides and a Chinese man sold lettuce, cabbage, and occasionally cucumbers from his truck garden. One Dawson resident paid fifty dollars for a sack of "the first potatoes raised in the Northern country," and considered it a deal when compared with prices on imported potatoes.

"Expenses 'go marching along' in this country, whether profits do or not," wrote one Virginia City resident. In gold rush San Francisco, eggs reached six dollars a dozen; in the Yukon, ten or twelve. A Californian wolfed down a pear in the mining town of Marysville, then asked its price—$2.50. In Denver in 1860, peaches sold for $1.00 each; in Dawson, apples fetched the same, while a watermelon was priced at $80.00 and canned oysters cost $1.00 apiece. A miner entering Dawson spotted a chicken in a crate with "75" scrawled on the box's side and decided to buy the fowl for dinner. But the "75" was dollars, not cents.

Spring brought the highest prices, as old supplies dwindled and those holding the remaining stock and the first freighters arriving received top dollar. When prices reached $100 to $150 for a hundred-pound sack of flour in Virginia City in spring 1865, a disciplined force of nearby Nevada City residents marched into town, searched buildings, and appropriated 175 hoarded sacks, then sold the flour at reasonable rates to anyone who needed it and established a $50 upper limit for subsequent sales.

The fall 1897 panic in Dawson that led to the Canadian ruling requiring an outfit of a year's provisions was caused by the ignorance of newcomers expecting to buy food in Dawson "with the nuggets they would easily mine," by the typical trading company practice of "stock[ing] up on liquor and hardware rather than food," and by the stranding of steamboats carrying supplies up the Yukon. In late September, the stores of the two Yukon trading companies, the Alaska Commercial Company and the North American Trading and Transportation Company, had sold everything except supplies held for men who had placed advance orders, and "old and experienced Yukon miners" were "pol[ing] out up the Yukon to Juneau [to] escape from the country."

With freeze-up looming, the Mounties and the trading companies began urging people to follow the old miners' example. In early October, flour in Dawson cost three dollars a pound, and "individual stores of flour, bacon, rice, etc., [were] closely guarded day and night in cabins and caches." There were calls for the American government to dispatch a Dawson relief expedition, as an estimated four fifths of the population was American. Meanwhile, in a further complication, those steamboats attempting to bypass Circle City for the more lucrative Dawson market were waylaid by armed Circle City miners who took what they needed, but magnanimously paid Dawson prices.

In one way or another, enough people made their way back to the Alaskan coast or lived off the land and short supplies to deflect the threat of starvation, although the danger left some with prodigious appetites. "Henceforth I live to eat," proclaimed one Dawson gold-seeker, mentally forsaking millions in currency or gold dust in favor of "pyramids of porterhouses" and "obelisks of oysters." The United States government, by now alarmed at the distressing reports, in a far-fetched scheme authorized a $200,000 relief herd of Lapland reindeer to be driven to Dawson. (Less than a quarter of the 539 reindeer would survive the trip, arriving in Circle City more than a year after the scare had ended.)

Shortages were not limited to edibles. A broken-down horse, which its owner had been unable to give away in Tacoma, brought seventy-five dollars in Skagway. A ten-cent hinge in Dawson cost a dollar-fifty. Paper was so scarce that newspapers were often printed on wrapping paper and purchases went unwrapped—one Dawsonite buying moose steak from a butcher was handed the raw meat on a sharp stick and had an uncomfortable walk home fending off hungry dogs. Lumber remained so difficult to obtain that men hurried out of mining towns to greet the lumber wagons, straddling the loads to claim possession.

Sources of light and heat cost dearly, too. In gold rush California, candles were four dollars a pound. In the Yukon, where the cold and darkness seemed eternal, kerosene in the winter of 1897–98 cost twenty dollars per gallon—at that price, wrote one Dawsonite, it "smells like perfume, don't you know?" He fantasized that upon his return home he would visit the corner drugstore and "order six lamps at once, have them filled with 12-cent coal oil by the grocer's boy and then put the whole half dozen hard at work furiously shedding light for my personal benefit." His assessment of kerosene's value is reinforced by one Klondiker's reaction to a personal loss of eight thousand dollars through theft: "Thank God they didn't take the kerosene."

Balancing supply and demand was a difficult enterprise. Fortunes *were* being made in trade, and great fortunes were begun, as exemplified by Levi Strauss's sale of sturdy jeans to California miners, by meat-packing magnate Philip D. Armour's start of a business empire in his California mining camp butcher shop, by future Pullman-car king George Pullman's success as a hardware store proprietor (and claim owner) in the Colorado rush.

But most supply ventures were likely to crumble before the speculator had profit in hand. For example, some forty-niners jumped into lumber speculation, sinking what money they had into expeditions to the Pacific coast forests and returning to San Francisco to find that too many others had had the same idea, driving lumber costs so far down that the suppliers went bankrupt.

Still others arranged to have manufactured goods and foods shipped from the East. Alonzo Hill in San Francisco in 1850 urged his father to send "as large an invoice of Men's thick boots as you can obtain on 8 or 10 months credit," as well as "as large an invoice of Brogans as you can procure on the same credit" and "as many corn brooms as will be available to you on fair terms." He also asked for ax handles, stools, dried apples, *"good butter,"* rum, and "good old *cheese,"* enclosed in "air tight tin boxes."

One argonaut in late 1850 cast a skeptical eye on such ventures, warning that "you can't bring anything new to California," that a speculator would immediately see "thousands of the very same kind of articles you thought no one else was smart enough to think of but yourselves." The speculations by Hill and his father generally did fail to pay off in such a volatile market, with so much time intervening between order and delivery, and with so many hazards in transportation. By the next year, Alonzo was writing, "You . . . think that 'Tis Safer to Send goods on a high than a low market. It will be very difficult to Determine whether you send on a high or low as the market may be up & down 27 Quadrillions times in 4 months." By the time his father dared to send apples unbidden in 1852, Hill's response was bitter: "They will doubtless be worse than an entire loss as I shall have to pay the freight on them, but nevertheless, the freak of your imagination should be encouraged when the expense is so trifling."

Timing was everything. A Seattle man entering egg-starved Dawson received a dollar and fifty cents per egg for a load of twenty-four hundred, but "within a week so many boats had tied up loaded with eggs" that the cost had plummeted to a quarter apiece. Newspapers from the East fetched a dollar apiece in 1849 San Francisco, where newcomers retrieved papers they had used in packing and enjoyed a profit. A Yukon resident nearly fifty years later pocketed fifteen dollars for "an ancient newspaper soaked in bacon grease," while another purchased a Seattle paper for fifty dollars, then more than recouped his money by charging people a dollar each to hear it read.

As Hill's comment attests, freight costs ate away at profits and made losses bigger. In Virginia City, a prospector-turned-storekeeper paid freight charges of thirty-five cents a pound for a load of "six hundred pairs of heavy leather boots," only to find that "the boots are not what are liked here by the miners." Then, those trying to establish stable trading operations paid heavily for store space and for business licenses, often the only reliable source of income for gold town governments. And merchants' fortunes depended to a great extent upon the fortunes of the gold-seekers around them; many storekeepers granted extensive credit and grubstaked miners in a futile attempt to improve long-term financial prospects for the store and for the town. Meanwhile, they had to figure out as much as a year ahead what would be wanted and needed.

Other entrepreneurial ventures beckoned. A simple saloon or whiskey-supply business could easily be launched, although competition was fierce. In Summit City, California, a fellow "with a sack of flour, a little salt and bacon, a fry pan, and coffee pot" could set up a "high-class restaurant" on a tree stump and attract a hungry crowd. Legend has it that one hungry diner at such an outdoor dining area, guilty over how much he had consumed, "at once went into partnership with the proprietor, contracted for lumber," and in a quarter of an hour had started building a hotel.

Those who knew valuable trades also found plenty of economic opportunity. Barbers set up lucrative businesses under trees. Blacksmiths were swamped with work honing mining tools, shoeing animals, repairing wheels, even producing "the parts needed for the first quartz mills."

Any service cost dearly—so much so that in early San Francisco clothes and linens were sent by clipper ship to Hawaii and even to China for cleaning. In Dawson, "most men wore their shirts until they could no longer stand them and then threw them away," sometimes buying the same clothes back from an entrepreneur who collected the discards, then laundered and resold them. From California to Nome, argonauts learned to treat quarters as the smallest unit of money, for nothing, "with the possible exception of a box of matches and a two-cent postage stamp," sold for less. "The expenses of living in this place have reached a point beyond anything that I ever dreamt of," wrote a "Nomad" in summer 1900.

Even entering a western urban area could be costly. The pilots of small boats meeting ships entering San Francisco Bay charged four dollars per passenger to transport travelers to shore and refused to barter for less. Tollkeepers who had received lucrative franchises from territorial or county governments established themselves on overland trails. Thus, a group entering Virginia City in June 1865 "had some difficulty in getting through the Toll Gate, as the toll was $10 and we had 'nary a red [cent].' "

PRECIOUS DUST

* * *

Newly arrived or returning argonauts found the urban mining towns both depressing and exciting. They were ugly, with their raw, makeshift buildings, their debris, crazy-quilt streets, obliteration of greenery, and—depending on the locale and time of year—omnivorous mud or pervasive dust. In the industrial gold-mining town of Central City, Colorado, all was "odd, grotesque, unusual" with "no feature [that could] be called attractive." Virginia City was judged "a very dull, desolate looking place," and Circle City a "motley collection of sodden dwellings and dripping roofs" sagging along "straggling thoroughfares." Of Dawson, one woman wrote, "On landing, even in our most enthusiastic moments, we could not have said the place was beautiful."

Again, however, timing was crucial. When a town was booming, the vitality of its residents obscured its drabness, the presence or promise of gold gilding its most tawdry outlines. Men finding gold and feeling secure in their claims acted as if their fortune would last forever. They fed others' fever by coming into town with bulging pouches, "treat[ing] all hands," and liberally sprinkling the glittering dust "with the most lordly nonchalance" on bartops, counters, and boudoir tables. In the wake of a strike, gold was "poured out and weighed almost as carelessly as rice or pepper in the States."

One result of such grandiose carelessness—and of the ephemeral nature of gold dust—was the spilling of precious grains. In 1849, San Francisco residents dug up an area in front of the United States Hotel, scooping the dirt in their hands and gently blowing it in a miniaturized version of dry washing. In that way, one boy realized fourteen dollars' worth of gold in one day in grains escaped from miners' pokes.

When banks were established, they would become prime sifting grounds, but in the early days, saloons were the hot spots for gold scavenging. Barkeepers placed velvet cloth under the scales with which they weighed out purchases, capturing the fine specks in the thick pile. (In Virginia City, saloonkeepers talked one woman into cutting hunks from her thick Brussels carpet for this purpose.) Bartenders also grew their fingernails long in order to capture some of the dust they weighed, "pan[ning] their own fingernails at night" for "a neat profit." In a similar technique, waiters dampened their hands "so that some of the gold stuck to their fingers and was transferred to secret pockets in chamois vests." Even the lowly spittoon-cleaner could make a profit from the gold spilled upon the sawdust floors, and children were sometimes allowed to pan the floor sweepings, realizing hundreds of dollars from their "claims."

If such success stories stimulated further gold fever, even more did the gleaming evidence of someone's good fortune. In Virginia City, the latest big nugget unearthed would occupy a place of honor in "a trustworthy saloon" or the local newspaper office, the paper loyally proclaiming it "the biggest yet found in Montana." In Deadwood, the newest gold bars from the fabulous Homestake quartz mine would be placed, "shining and arrogant," in the local bank window twice a month, always creating a renewed tempest of interest.

Gold town newspaper editors dished up plenty of propaganda, all too aware that the town's fortunes rode on public perceptions of the gold available. As the shift from short-term placering to long-term, elaborate placer and quartz mining occurred, promoters of mining stock and claim-sellers did their part, too. Newspapers in California's gold districts printed columns and columns of mining lore and statistics—"statistics of the number of feet owned by various companies, length of tunnels, depth of shafts, learned descriptions impressively bristling with scientific terms geological and chemical, data on stamp mills, hoisting machinery, force pumps, and the like." Jokesters were soon imitating the glowing reports: "The Skedaddle is a splendid looking vein, and I have no doubt that by it a large fortune will soon be taken from the pockets of its owner."

But the prevailing attitude was best summed up by Deadwood resident Estelline Bennett: "The richest among us had made their money in mines and that was something everybody else expected to do tomorrow." It was this hope that seized young Stephen J. Field, a future Supreme Court justice, when he arrived in gold rush San Francisco and found "something exhilarating and exciting in the atmosphere which made everybody cheerful and buoyant." All greeted him with comments on this "glorious country" they had come to, and he soon caught the spirit: "Though I had but a single dollar in my pocket and no business whatever and did not know where I was to get the next meal, I found myself saying to everybody I met, 'It is a glorious country!' "

This optimism was also fostered, as in the mining camps, by the grand vistas, invigorating climate, and the dramatic natural beauty of most gold towns' surroundings, all of which reinforced romantic nineteenth-century apprehensions of wild, abundant beauty at the very time the town-builders and miners were eradicating any vestiges of it near to hand. Of course, some bleak locations already existed. Nome sat on the dull line between endless expanses of water and endless expanses of tundra, the monotony of its scenery relieved only by the enchanting play of the northern lights.

Still, barrenness might have been preferable to the dangerously cluttered, confused conditions in the gold town itself. Not only were the streets narrow and crowded, but dogs, cats, and even pigs roamed freely through the melee. In gold rush Denver, the pigs pounced upon "anything edible left out of doors for a moment" and constantly tried to enter houses.

Streets and alleys were dotted with tree stumps and refuse: in San Francisco with "egg shells, cabbage leaves, potato parings, onion tops, fish bones, and other articles too numerous to mention"; in Meadow Lake, with "empty bottles, old boots and hats, sardine cans, oyster tins, ham bones, wornout kettles, broken picks and shovels."

Climatic changes also affected ease of movement within the towns. In dry summer weather, people and animals on the streets—even in Nome—were smothered in dust. During fall and spring rains and wet winters, water-soaked thoroughfares turned to mud, then to "deep, soupy mire" as iron freight-wagon wheels and hooves tore up the wet earth. A Frenchman exclaimed of San Francisco, "This is not a town, it is a quagmire." In Virginia City, pack animals were reported to have drowned in the streets, and in Deadwood, six- and eight-mule teams could barely pull through the mud "of a rich quality, its adhesive properties rare, its depth unfathomable, its color indefinable, its extent illimitable, and its usefulness unknown." Gold rush Dawson streets in winter and early spring were "piled high with mounds of broken ice and snow, covered in muck and gravel, some of the individual pieces higher than a man," and any thawing turned thoroughfares to "colloid quagmires of deep muddy lanes." Nome was furrowed with "rivers of mud, two feet deep," making gum boots mandatory, and anyone navigating its Front Street "was at times uncertain whether he would arrive at his destination or, suddenly find himself in China."

Seldom did boomtowns boast adequate exterior lighting or sidewalks, and when sidewalks did exist, they were primitive, limited affairs of plank and barrel-stave. A nighttime traveler in 1849 Sacramento would uneasily hop "from one loose barrel-stave to another," occasionally despite his best efforts walking into "a puddle of liquid mud," hitting a gulley, or receiving a blow from "a scaffold-pole or stray beam." The situation was "still worse" on the outskirts of town among the immigrant tents; there "stumps, trunks, and branches of felled trees were distributed over the soil with delightful uncertainty," while the successful navigator of these could still stumble upon the scattered horned heads of slaughtered cattle or become tangled in one of the myriad ropes picketing testy mules and anchoring tents.

Perhaps the worst aspect of gold town conditions was the utter disregard of sanitary practices by the largely transient population. Sacramento was "one great cesspool of mud, offal, garbage, dead animals and that worst of nuisances consequent upon the entire absence of outhouses." A rudimentary hospital in 1852 San Francisco faced a square still used as a dump ground by city residents and "lined with an immense pile of rubbish and filth of the worst description." At least residents were transporting garbage to a dump; in many gold towns, people simply chucked refuse out the window, as did the Virginia City physician who nonchalantly pitched an amputated hand from his office.

One result of these sanitary problems was a heightened infestation of insects, from fleas to mosquitoes. Another was water pollution. In Nome, sewage and garbage were jettisoned into the Snake River, from which the town drew much of its drinking water. In addition, the mayor reported the alleyways between the saloons on Front Street "almost three feet deep with a glacier of urine," and an army surgeon found "pools of stagnant water, slops, and urine" sheathing the streets and alleys "back of the

principal saloons and in the district occupied by the prostitutes."
The same surgeon found the two public privies sitting below the
high-tide mark and people even using a wrecked schooner as an
odiferous public convenience.

The fledgling town governments tried to address such situations by passing sanitation ordinances, hiring officials to enforce them, and improving the condition of the public areas.
Mining town marshals, far from parading grandly as the romantic figures of gunfighting legends, often spent their time simply
trying to get people to clean up the messes that spilled into the
streets. Work crews, both voluntary and those hired on slim city
treasuries, attempted to improve the streets. In San Francisco, a
street inspector used "carloads of chapparal" and sand to fill and
cover gaping holes, but "in a day or two the gulf was as deep as
ever," and residents had to search "the whole length of a block"
for a place to cross. San Franciscans were known to throw cargoes of overstocked goods—sacks of Chilean flour, containers of
tobacco, even cooking stoves—into the abysses. In later gold
towns, residents and city officials would use straw, packing material, old bedding, other refuse—anything to try to create a firm
surface.

Sanitation problems could become absolutely grisly. Hillside
cemeteries tended to deteriorate in wet weather, sliding the
graves toward town. Nome had no hills, but it hunched by an
ocean that washed the coffins out of the cemetery in a storm. In
Skagway, spring rains and floods brought hundreds of deteriorating carcasses of White Pass Trail pack animals rolling into
town in the Skagway River. Floods actually purged a town of its
"accumulated refuse of months," but also destroyed precious
provisions, exacerbated ill health, and created "ocean[s] of mud."

Fire, too, that most dreaded of gold town disasters, had a
purgative quality recognized even by frontier residents. A doctor
who found gold rush San Francisco "a nasty, dirty, slushy, raviney, sand-hilley place" considered it improved after one of the
great conflagrations—in part because building quality improved,
in part because of the general cleansing the flames provided.

Still, the fires that exploded through the towns with frightening regularity posed a major hazard. Gold rush San Francisco
alone had seven major fires in eighteen months, the least de-

structive burning fifty buildings, the most calamitous destroying "perhaps one-fourth of the city." Cripple Creek weathered one fire that consumed many businesses and dwellings only to be hit with a second and equally virulent blaze a few days later.

The frequency of fires can be explained by a number of factors: the prevalence of canvas tents and flimsy wooden structures with cloth on walls and ceilings; the use of crudely installed stoves and chimneys; dry, windy weather conditions; the abundance of "piles of combustibles"; and the use of candles, lanterns, and lamps. Carelessness played a big part—a cigar tossed aside, a candle left to tip over, grease flaming on an unattended stove. More than one marauding blaze began in domestic violence: In Cripple Creek, a bartender and his girlfriend fought, struggled, and knocked over a lit kerosene stove; while in Dawson, a miner and his girlfriend began arguing, one lobbing a lighted lamp at the other. Sometimes, arson was judged the cause, as in Nevada City, California, where residents determined that banished desperadoes had set a brutal March 1851 fire that wiped out food supplies.

Few people lost their lives in boomtown fires, but the flames speedily swallowed whole blocks, creating an unearthly, oddly beautiful sight, as rows of cabins consumed by fire stood eerily "for quite a long time, giving the appearance of golden tinsel houses, glowing with a rich, red color" before crumbling to the ground. Residents turned out at the cry of "Fire!" or at a prearranged series of warning gunshots. They dug ditches, formed bucket brigades, even threw snowballs or roped buildings and wrenched them out of the fire's path.

Boomtowns often had vigorous volunteer fire companies, membership indicating some social standing within the still-fluid community. Yet local governments struggled in this area of concern as elsewhere against a live-for-the-moment mentality. Everyone recognized the fire threat, but "every man was acquiring with such rapidity that all hoped to complete a fortune ere such a disaster should occur," and seldom was fire-fighting equipment high on anyone's list of priorities.

Perhaps the most striking thing about boomtown fires was the rate at which the community rebuilt. Within days, as opportunists panned the ashes for gold, tents packed the stricken area, piano music poured from saloons still under construction, and, weather permitting, goods arrived from the nearest supply centers. Within weeks, a solid line of new and often stronger buildings lent a fresh air of stability to the town. The rapidity of the process bore out one foreigner's general observation: "Even when his house is burning, an American will think only of how to rebuild it."

Many argonauts eddied around the edges of the mad development and reconstruction efforts, still intent on collecting news of gold and finding or following the next strike. These men took menial work they would scorn back home in order to stay near the gold but rejected any enterprise or wage employment that might interfere with their primary quest. Henry Rogers, jobless in Juneau in February 1899 after his Boston company's unraveling, wrote home that he and a partner "could have gotten work if we would guarantee to stay with it and not leave in a couple of months, but of course we could not do that."

Wages were often tempting, though. Carpenters commanded high prices, in San Francisco striking for and receiving sixteen dollars a day in 1849. Even common laborers enjoyed high pay at the peak of a town's surge. In 1849 San Francisco, the laborer's daily wage of up to twelve dollars was greater than a sailor's monthly wage, a situation contributing to the plethora of deserted ships in San Francisco Bay. In Deadwood, the seven dollars a day for shoveling on bedrock enticed "young fellows from the States—many of whom probably had never been paid more than fifty dollars per month." In Nome in summer 1900, laborers received a dollar an hour.

But overcrowding and a virtual mining halt during winter months made jobs and high wages a seasonal proposition. Two wayfarers in San Francisco in spring 1850 "nearly walked the shoes off our feet looking for something to do, but in vain, there are too many seeking employment."

If a mining boom community stabilized around the business of quartz mining, industrial companies needed workers, but they seldom found them among the stampeders. There were two main reasons for this. First, by the time large-scale quartz operations got under way, most stampeders had turned around and headed home, had rushed off to the next strike, or were planning to do so. Second, given the nature of their adventure and the age and temperament of the stampeders, most were the least likely candidates for long-term, monotonous, controlled wage employment in the mines.

Many gold-seekers saw the towns as places to cast off the hard realities of the diggings and the specter of failure in a surge of frenetic leisure activity. The towns accommodated by offering drinking and gambling opportunities galore. From California to Alaska, miners would regularly come into the urban centers and lose their precious dust to the professional gambler—or they would sabotage a successful departure from the goldfields with a stop at a gambling house. In Marysville, California, in 1850, two gold-seekers on their way home lost their whole earnings—fourteen thousand dollars—in gambling and turned back to the mines. A prospector in Nome watched a young man awaiting a steamer lose all he had—three hundred dollars—in a five-minute gambling game, consigning himself to a winter without funds in the raw city with its inflated prices.

Aside from this near-suicidal gambling mania, from hard drinking and the pleasures offered by prostitutes, there were less dissipating diversions. Theaters featured the standard variety performances and the melodramas of the day. Residents often realized that the melodramas paled beside some of the realities of gold town life. In Deadwood, "so much raw real drama trod the Gulch" that "everyday life was melodrama," while in Dawson "there was plenty of the drama or real life" to supersede play-acting. Still, the theaters found plenty of patrons.

Like the early hotels, early gold town theaters claimed more elegance than they possessed. Virginia City's People's Theater "crowed about its parquette and orchestra in its opening announcements, and had its elegant appointments puffed in the paper," but it offered hard wooden benches, muslin-lined walls, one inadequate heating stove, tallow candles for footlights, and a cramped stage with a cambric curtain.

The entertainment itself often proved unsatisfying or decidedly uneven. At the Dramatic Museum Theater in San Francisco, actors changed plays from night to night, their resident playwright writing and feeding scripts to them between scenes as the play progressed. Gold town troupes also had to improvise in stage-setting; in Dawson, *Uncle Tom's Cabin* was presented with the bloodhounds "represented by a single howling malemute puppy drawn across the stage by invisible wires."

Gold rush audiences were generally more sophisticated than one might think, if very free in their opinions. When Hugh McDermott appeared as Richard in *Richard III* in gold rush Sacramento and handed a sword to another player, half the audience signaled its critical evaluation of his performance by rising and calling for the sword to be "plunged into his body."

In some instances, critical judgment might be suspended. A woman billed as a member of "the Royal Theatre, New Zealand" performed in a drama in 1849 San Francisco by racing on stage and striking a dramatic pose, apparently simply "for the purpose of showing the audience that there [was], actually, a female performer." Those in attendance, "to whom the sight of a woman [was] not a frequent occurrence," responded with avid approval.

Nonetheless, many gold towns actually had an impressive range of theatrical offerings. Virginia City hosted "American melodrama, French farce and drama translated for the London stage and pirated to the United States." Dawson residents could enjoy the popular plays of the time, including *Camille*, although one resident found most offerings "of the blood-curdling type."

Often, theater proprietors hedged their bets by including room for dancing, as much a favored activity of urban residents as of mining camp denizens. There were the regular dance halls, too, where a fellow in mud-spattered mining garb could listen to a chanteuse and part with a dollar's worth of gold dust to careen around the floor with a real live girl. In Dawson, this experience was enhanced—if it needed any enhancing—by smooth floors, hanging oil lamps, and high-caliber music, "some of the musicians having played in the best orchestras in America." Saloons, too, used music to lure customers, sometimes stationing whole bands outside their doors.

The "upper crust"—those who early removed themselves from the plebian hodgepodge based on origins, profession, and/or dress and deportment—quickly established dancing schools and benefit balls that allowed contact between men and the few middle-class females in a gold town. In Bannack, balls were such prized affairs that admittance was five dollars' worth of gold and drinkers were barred from the hall.

There were various locally produced amusements. The "Nasty Club" of Summit City, California, created diversion by conducting "mock trials of members on trumped-up charges." Virginia City residents came up with the fad of "plank walking," seeing how long a man could pace an eighteen-foot plank lifted a foot and a half from the floor. People in the gold towns staged footraces and horseraces, dogfights and badger fights, wrestling and target shooting, ball games and impromptu musical evenings. They joined in the late-nineteenth-century American enthusiasm for prizefights not only by turning out to see traveling prizefighters compete but by staging their own contests on dance hall stages. Those with middle-class backgrounds and aspirations formed literary societies and social clubs.

As infighting over political positions developed within the towns, campaigns became another source of entertainment, offering outdoor rallies, torchlight parades, free drinks, and freewheeling invective. Holidays were celebrated with great fanfare, and even funerals provided zestful diversion, with brass bands, uniformed volunteer fire companies, flower-bedecked hearses, and drinks to the departed one's memory.

Between these periodic excitements, the arrival of the latest immigrants and news from the East was a major event, whether it came in the form of a steamship from Panama, a snow-skiing mailman from Downieville in the California Sierra, or a stagecoach from Cheyenne. The stage often provided "the grand excitement of the day" in Deadwood, the six white horses galloping through the gulch, the driver brandishing his whip in "skillful skyward sweeps," and the stage containing news of the wider world.

Such movement slowed in the winters along with the town's activities, residents and miners in from the diggings either crowded together in flimsy dwellings or gone to winter around more "civilized hearth fires." But with the coming of spring, the exciting, entertaining surge of people and of varied activity began again. Nowhere was this seasonal ebb and flow more obvious than in Dawson, where winter freeze-up on the Yukon and spring breakup overshadowed other events. Betting reached a frenzy in the spring as residents predicted when the ice would fracture into huge cakes and begin coursing down the river, "smashing and grinding against each other with the noise of a dozen express trains," sometimes with terrified caribou careening along upon them. The first boat soon followed, "put[ting] summer into the hearts of all of us."

But the last boat of fall was almost immediately taking away more people than the first spring boat conveyed as Dawson, like many other gold towns, slipped into rapid decline. How rapid that decline could be is indicated by the fact that ghost mining towns appeared in California within two years of Marshall's find. Mark Twain, visiting the gold metropolis of Meadow Lake, California, shortly after its precipitous rush, mused, "They have built a handsome town and painted it neatly and planned long wide streets, and got ready for a rush of business, and then— jumped aboard the stagecoaches and deserted it!"

For once the placers and easily mined quartz veins were gone, urban centers born of gold rushes could only maintain vitality in one of two ways: through development of long-term industrial mining for gold and other precious metals or through a transition to regional supply center or seat of county or state government. Thus, Lead, South Dakota, remains a gold town today, the Homestake Mine and other area mines having yielded riches for well over a hundred years, thanks in part to improving methods of gold extraction. On the other hand, Denver's gold hopes proved to be overly sanguine, but the city nonetheless mushroomed as political, economic, and social fulcrum for the central Rockies, drawing only part of its vitality from mineral strikes in the mountains.

Even when towns thrived on gold for an extended period, they lost some of the character of a gold rush community quickly. At the height of the 1849 rush, San Francisco's business community was already exchanging slouch felt hats and flannel shirts for "narrow-brimmed black beavers" and white linen, "though indifferently washed."

As populations shrank and stabilized, coins replaced gold dust as currency, government developed at the local and state or territorial level, and industrial or other long-term employment became the norm. The saloon ceased to be the accepted community center, and the town began to have more brick and stone structures and distinctly drawn business, leisure, and residential sections. Banks were established, "symboliz[ing] a degree of financial stability." The arrival of middle-class women and children in any numbers changed the town's moral and civic tone, stimulating the development of such institutions as schools and churches.

Thus, sooner or later, the days of the pioneer merchant gave way to the prominence of an "*arriviste*" class, well-off newcomers with "different conceptions of urban life and civic responsibility." Such a change was behind the contemptuous comment of one Coloradoan about the transformed silver town of Leadville: "Leadville! No, stranger, this ain't Leadville. It's only some infernal Sunday-school town that ain't been named yet."

The arrival of a railroad brought the most noticeable changes in the crude, curiously vital towns that had clung stubbornly to the freedom, excess, squalor, and sheer distinctiveness of the mining camp. It removed "much of the romance and adventure . . . the sense of remoteness and daring," made life somehow seem drabber, made people crave more luxuries, and "introduced a new population which knows not the ways of the 'old-timer.' "

Using the railroad as a gauge of civilized constraints and pursuits, Deadwood, South Dakota, "was young longer than any town that ever grew and prospered because it was off the railroad longer," only waking as "a surprised town with civic and moral obligations" with the iron horse's arrival fifteen years after the gulch community's founding. Interestingly, Granville Bennett, who as a judge knew Deadwood intimately from its chaotic earliest years, announced with the coming of the tracks, "Well, we'll have to lock our doors now."

It is an odd comment; as Bennett's daughter noted, he had had to present a calm demeanor throughout her childhood because "we were surrounded by so much danger he could not let us be afraid." How does one reconcile the idea reflected in Granville Bennett's comment of a gold rush town as protected environment with the evidence of it as a fragmented, unsettled hodgepodge of selfish strangers? How cohesive, how safe, how democratic was the raw new society the Bennetts lived in before the advent of the railroad, the society that argonauts from California to Nome dreamed and schemed and drank and struggled in?

The answers are mixed, but they tell us something about gold rush society and about the America that spawned it—about ideas of social good and individualism, about concepts of law and order, and about the tensions between idealistic democratic beliefs and realities in this strange but still familiar new world.

William Sidney Mount's 1850 oil painting *California News* conveys some of the pleased excitement aroused by word of gold in California, with the young man in the right foreground already apparently lost in golden reverie. (The Museums at Stony Brook)

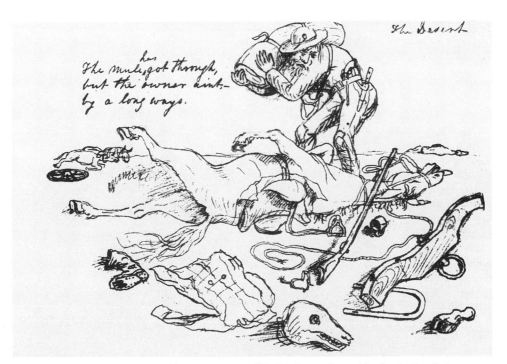

Forty-niner J. Goldsborough Bruff in his journal graphically depicted the ordeal of the California-bound in the western deserts. He wrote, "The mule has got through, but the owner ain't,—by a long ways." (Beinecke Library, Yale)

"Pikes Peak" stampeders had a shorter and generally easier journey than did the forty-niners, but the continuing difficulties of overland travel and disappointments in the diggings would lead many to change "Pikes Peak or Bust" to "Busted, By God!"

(Denver Public Library, Western History Department)

En route to the Yukon River and Dawson, many Klondike gold-seekers struggled with their provisions up the Chilkoot Pass, its final ascent too steep for pack animals. (Special Collections Division, University of Washington Libraries, Hegg 100)

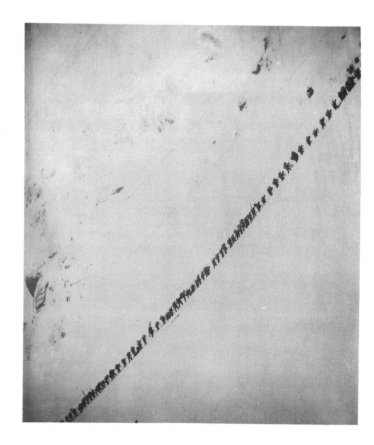

The standard forms of placering included (*clockwise from left foreground*) panning, "rocking the cradle," tunneling, winching ore up from bedrock, sluicing, and working a "long tom."
(The Bancroft Library)

This early California riverbed mining scene provides a hint of the physical effort involved. Although not easily discernible, African Americans are among the workers pictured.
(The Bancroft Library)

Prospectors took their dreams and mining methods across the West. These
"old-timers" in Rockerville, Dakota Territory, feed their sluice
in the late 1880s.
(Library of Congress)

This scene of early placer mining at Cripple Creek includes women and
children, reflecting the site's relative accessibility.
(Denver Public Library, Western History Department)

Chinese miners followed the booms and busts, eking gold from worked-over sites and employing time-consuming techniques rejected by impatient stampeders. William Henry Jackson captured this threesome in Alder Gulch, Montana, about 1870.
(Montana Historical Society, Helena)

Miners' cabins, however rough, were a vast improvement over tents and other temporary accommodations. These Klondike miners even panned for gold within their shelter.
(National Archives of Canada/C5392)

The first photo of Nome, Alaska, reveals the rudimentary beginnings of
gold rush towns. Note the "Stampeders Home Restaurant."
(The Bettman Archive)

Town sites boomed into bustling, disorderly urban centers, as this portrait
of Deadwood, South Dakota, about 1880, attests.
(Nebraska State Historical Society)

Mail from home always drew a crowd; this one waits in Dawson.
(The Bancroft Library)

Prostitutes' lives were usually harsh and chaotic, although this view of Dawson cribs (taken at midnight) conveys a certain stylized order and calm.
(Special Collections Division, University of Washington Libraries,
Hegg 2447)

"It is the only country that I ever was in where a woman received anything like a just compensation for work," wrote one gold rush Californian. Throughout the era, women like Mrs. G. I. Lowe of Dawson seized a variety of opportunities.
(Special Collections Division, University of Washington Libraries, Hegg B-461)

Lucky prospectors could gild their frontier circumstances with imported luxuries; here, Klondike gold discoverer George Washington Carmack joins other affluent Dawsonites in a birthday celebration.
(Yukon Archives/Clayton Betts Collection)

Most prospectors sooner or later faced the "wrecking of gorgeous visions." This Yukon old-timer contemplates the failed quest. (Yukon Archives, Vancouver Public Library Collection, 2176)

Despite crushing disappointments, many refused to consider their gold adventure a devastating failure. These four weathered yet vibrant California argonauts symbolize the resilience, camaraderie, and love of adventure and freedom that would brighten the memories of gold-seekers.
(The Bancroft Library)

Chapter Eight

Self, Society, and the "Battle of Life"

The *Battle of Life* . . . is a reality here; the fallen are trampled into the mud, and are left to the tender mercies of the earth and sky.

—*J.D.B. Stillman, Sacramento, 1849*

In no part of the world have I ever seen help more freely given to the needy, or more ready co-operation in any humane proposition.

—*Bayard Taylor on gold rush California*

W hen Alonzo Hill and the other members of the Bay State and California Mining and Trading Company ended their Cape Horn voyage in September 1849, disembarking in San Francisco, an Irishman was the first to greet them. "Ye have come to a sorrowful country," he announced. "Plenty of Gold, but every man for himself The Devil for us all."

Hill and his companions laughed "heartily," but many gold-seekers—including Hill—would come to embrace such a bitter philosophy from their experience in the gold regions. There American beliefs in frontier self-reliance, an early-nineteenth-century Romantic emphasis on individualism, and the late-nineteenth-century influence of laissez-faire capitalism all would combine with harsh, divisive frontier conditions to encourage atomistic behavior. Yet other participants would insist that gold rushes produced genuine community, members demonstrating cooperation and even generosity beyond that found in more "civilized" regions. The test of communal spirit would often come, as it had on the trails, in dealing with sick, injured, and impoverished stampeders.

Repeatedly, observers attested to the prevalence of the every-man-for-himself mentality. "None caring for his friend, all are anxious to get gold," complained one California newcomer, while another judged miners "sharing nothing in common except the lust for gold, which really only divides them the more." A third concluded, "Their hearts have been left at home." Such complaints echoed in subsequent rushes; a merchant characterized the population at Idaho's Salmon River diggings as "[a] rough unsympathyseing Society, who are all Scrambling for gold."

It was only natural that people who had forged westward to find a fortune to carry home would have little or no interest in developing communal ties beyond those of the temporary prospecting party. Even among those who did form ties on the gold frontier, there was a certain careful anonymity born of freedom from past associations. One California stampeder acknowledged "a free, openhearted way with all here," but characterized relations as limited, "something like a stagecoach or steamboat acquaintance." A forty-niner contended that he knew men for years only by their first names, as "we don't like to appear inquisitive." He had contact with one fellow for more than two years, developing a relationship with him "almost as intimate as brothers" before discovering that the two had been born "within three miles of each other" in Ohio and "both knew each other's family."

Such cultivated anonymity undercut any impetus toward long-term community building. So, too, did lack of social and moral restraint. In the early gold camps and towns, men behaved crudely and boisterously, openly patronizing red-light districts, flaunting their drunkenness, and squiring prostitutes about the public arena without fear of social opprobrium. "There is no restraint," complained one 1858 visitor to the California mining districts, "and no example but that of rudeness, and of a degrading tendency." A forty-niner noted, "California will for the present lower the moral tone of all who come here," while decades later another California boom led one observer to comment, "All combine to excite and ruin."

In each rush, crowds of young men, separated from the associations and common amusements of home, found that "vice seems more alluring here." Many succumbed before they reached any diggings. In Dawson in 1898, newcomers "who never before knew faro or roulette" turned into rash gamblers within a few days. The journey itself "seemed to have sapped their principles, and the whole environment of the place was that of another and a worse world."

While not accepting such a bleak view of gold rush life, one Dawsonite acknowledged, "No organized group was any longer compelling us to anything; there was nothing, outside our own inner compulsion, to live up to." Thus, as Alonzo Hill concluded in California, some found "a stupendous arena for the exercise of their long pent up and most superior Villainy."

The problem of extravagant license was compounded by widespread acceptance of relaxed standards of behavior. A visitor to Nevada City, California, in 1850 finding a bordello the most imposing building noted, "It is looked on as an honorable business and a man goes into that as he would any other speculation."

Even the moralistic and law-abiding often balked at attempts to establish formal institutions of social and moral control. For example, most gold-seekers appear to have been only too glad to escape the strictures of formal religion. They demonstrated a total disregard for the Sabbath, a disregard that seemed to prove New Haven minister Cleaveland's prediction that individuals would find the day's sanctity only in their own hearts. One woman in the Montana goldfields concluded that she had found "a most worshipful community" if one went by the adage "Labor is worship." But, she noted, "of any other kind . . . there is no public manifestation whatever." When Klondike prospectors began successfully working hillsides, or "benches," one minister wistfully remarked that "the only benches not staked are those in my log church." Virginia City's newspaper reported an attempt to organize a Young Men's Christian Association had failed because "the young men of Christian tendencies were not numerous enough to start the organization."

Many would listen occasionally to a preacher, but clung fiercely to their independence from any form of authority. "We think for ourselves on religious subjects," boasted one frontier newspaper. "It is for us . . . to choose between authority and reason; between other men's dicta and our own convictions." In particular, men were bent on rejecting the messages of hellfire and damnation that they felt they had been force-fed. They saw good reason to question such teachings. A man who had suffered through services held by a fire-and-brimstone preacher in Iowa discovered the preacher dealing faro in California and determined "he don't believe in hell-fire and never did, and neither do I."

Some ministers did gain the respect of the gold-seekers. John L. Dyer won plaudits as the "snow-shoe itinerant" in Colorado. Henry Weston "Preacher" Smith impressed Deadwood with his evangelical determination and martyr's death, being killed by Indians while en route to a nearby camp to deliver the gospel. Daniel Sylvester Tuttle was known as the "Bishop of All Outdoors" in Virginia City. Jesuit missionary Father William Judge became the "Saint of Dawson" for providing practical hospital care as well as spiritual direction. When they came into contact with men such as these, stampeders did pour some of their precious dust into the tin cups passed by early church organizers.

Some thought they were flirting with heathenism when they were actually still pounding the straight and narrow path. One youth wrote to his sister from gold rush San Francisco that he attended church only half a day now, as "here we are left to indulge our own peculiar views," and he chose to spend the other half "amid the beauties of nature." When the first church bell rang out over Deadwood from the Congregational church, miners "straightened up from their pans and sluice boxes at the first stroke . . . removed their soft hats, and stood at reverent attention until the last ringing echo died away down the gulch."

But gold-seekers' behavior generally tended to reinforce the lament of a Baptist preacher that he had never encountered "a harder task than to get a man to look through a lump of gold into eternity." Besides, the preacher might be equally affected, even when clinging to his ministerial call; a prospector/minister in Coloma, California, told his congregation one Sunday, "There will be divine service in this house next Sabbath—if, in the meantime, I hear of no new diggin's!"

With preachers themselves out digging for gold or moving on to a new strike, whatever community structure existed often was controlled by a low element, the "daring, unprincipled" gamblers or "any drunken set of blackguards who have money" and thus "can rule those who pretend to decency." The lawless element, whether "roughs" or a smoother-on-the-surface "aristocracy of vice," often seemed to have better luck organizing than did law-abiding citizens, as the examples of the Henry Plummer gang in Montana and the Soapy Smith gang in Alaska will attest.

But despite all these factors, stampeders established or reestablished strong, if limited, communal ties, especially with people from their home regions. By gathering in informal association or in "societies" with others from Texas or New England, from Iowa or Georgia, stampeders "use[d] past regional ties to provide moorings in the chaotic present." Even the fiercely independent, peripatetic prospector who didn't bother to learn last names or the native origins of his western friend reacted with joy to the news that a company from his Ohio hometown was in California's northern diggings: "Nothing less than a double-locked prison would have been able to hold me from going at once to the boys."

In addition, there developed a gold rush code based on straightforward honesty, friendliness without prying, and mutual effort and generosity.

In California, Bayard Taylor found appealing "the general confidence which men were obliged to place, perforce, in each other's honesty." By the time Bodie boomed thirty years later, an identifiable type of "mining-camp man" was recognized as placing his stamp on the social milieu by his free, open ways, by being unconventional yet "liberal-minded, generous to a fault, square-dealing, and devoid of pretense and hypocrisy." This type and his influence was further noted by one Dawsonite, who found "no hypocrisy, and hardly any pretense of being good by those who were not," as well as "a high order to offset what was low."

Further, "men were 'square,' even when they broke some of the Ten Commandments . . . in general, exactly what they appeared to be."

Many commented favorably on the frank, decisive conducting of business. Bayard Taylor considered the California gold rush merchants' "disregard for all the petty arts of money-making" a "refreshing feature of society," while one long-term prospector, on a trip home to New England, judged that "business looked dull and all the men yes all the people seemed to be mean and small in their dealings. In no way frank, and honorable like those in the mountains." In a similar vein, mining camp entrepreneur, Nellie Cashman—who fed, housed, and nursed miners from Tombstone, Arizona, to Dawson—acknowledged "some rascals everywhere," but concluded, "up north, there is a kindly feeling toward humans and a sense of fair play that one doesn't find . . . where men cut each other's business to hack and call it 'competition.' "

Even when they avoided the dissipations of camp and town life, many also rejoiced over their freedom from prying eyes and gossips. A California merchant wrote, "We come and go and nobody wonders and no Mrs. Grundy talks about it. We are free from all fashions and conventionalities of Society." Henry Rogers in Alaska a few decades later wrote home, "Folks up here . . . mind their own business. Seems good not to have a lot of bothering neighbors, wondering what you are doing, and why." In Dawson, where gossip was discouraged, "a man could be good or bad as he pleased, so long as he did not steal or tread on another's toes."

This attitude could lead to callousness and selfishness, of course. In gold rush San Francisco, hundreds refused to fight a fire without a promise of pay, and at a mining camp hotel in southern Oregon, a "doorman" regularly "clubbed away" the men unable to pay $1.50 to $2.00 for a meal before entering.

Yet particularly in the more remote areas, people were expected to pull together and help out. That they did so was attested to by one Yukon stampeder who wrote, "I have seen more downright Good Samaritanism here than I ever knew existed. What is one man's property is the property of all." To fail to live this philosophy put people beyond the pale; in the Black Hills, one man became a pariah when "charged with an act of inhospitality which any prospector considered unpardonable": refusing to give food to a hungry man. By the same token, in the Yukon a fellow who considered himself of sterling moral character was universally despised because in his travels he would use the firewood carefully provided by the last sojourner in a cabin but would fail to replenish it for the next.

Those who lived by the code shared what they had when need presented.* Thus, two hungry California goldfield newcomers were welcomed to a meal at a miners' cabin with the assurance, "If you ever get able to pay us, do so." Fifty years later, Yukon residents found that "nobody was knowingly allowed to go hungry," one noting that a "perfect stranger will come over the trail and walk into the first camp he sees and take dinner." An Alaska prospector reassured his sister that the tradition of largesse had continued even to the big mining operations: "A man never starves in a western mining camp, unless the camp is cut off from supplies from the outside world," as in looking for work "he is supposed to know enough to go into the mess house and sit down at the table, of which ever mine he happens to be at, when meal time arrives"—and "if he don't know enough to, a cook or flunky will always remind him of it."

This spirit of sharing was apparent when a fire swept through the Alder Gulch diggings. "Those who escaped the fire divided with those who lost everything and, by so doing, managed to get along without serious discomfort until supplies arrived." There was something about such situations that bound people closely, that eased the strain, as in Dawson, where "any real deprivations were shared by every one, and so became no deprivations."

In another kind of sharing, prospectors were generous with their friends when it came to staking golden ground, secretly rounding them up to join in the first organized claim locations or hurrying them along to take advantage of a new strike. One group in Alaska, on hearing of the grand Yukon discovery, hogtied a drunken friend and tossed him into a boat "like so much flour in a sack," enabling him to make a valuable claim.

* However, the code did not dictate assistance for anybody joining a new stampede ill prepared. "It is not a question of friendship—it is one of justice," wrote one argonaut. "Those who go must be ready."

The general generosity of spirit cited by those in the gold regions was comically expressed by an Idaho rancher who boarded horses for Orofino miners. A man who learned his horse was missing from the ranch accused the owner of "taking people in." "Take 'em in, of course I take 'em in," he declared. "That's what I built this here cabin for at big expense, to take people in. Don't the Scripter say that if a stranger comes along you must take him in?"

Despite such accusations and misunderstandings, gold rush participants professed themselves pleased at the trustworthy character of many of their fellows. "There is more intelligent and generous good feeling than in any country I ever saw," admitted one goldfield critic, while Bayard Taylor felt that the many professional men among the forty-niners brought an "infusion of intelligence" to the gold centers, as well as "an order and individual security." A quarter century later, a Black Hills argonaut insisted that notoriously wild Deadwood was not "rotten to the core," that the core "was sound and wholesome," made up of "fine men and women who stood for personal and civic virtue against the surge of wickedness."

Reporter James Chisolm found the prospectors in the Sweetwater mining district in 1868 a drinking, gambling, and quarreling crowd, but quite peaceable in contrast with the population of "the roaring hells of railroad towns" from which he had recently escaped. And among them he found "some grave and temperate men of sedate habits and superior intelligence," not book-learned, but their western peregrinations giving them "a deep knowledge of men."

Whatever a gold-seeker's character, however, when resourcefulness, luck, or good health deserted him, he had to hope that the rush community he found himself in, town or camp, would sustain him. Here the idea of community was truly put to the test.

In each rush, men spent everything they had in getting to the goldfields, in prospecting and working claims, in meeting living costs, in speculating, or simply in the mind-numbing escapism of the whiskey decanter and the gaming table. The big rushes produced a backwash of tapped-out humanity flowing into the streets of San Francisco and Sacramento, of Denver and Dead-wood and Dawson.

In Sacramento in late 1849, many were "begging for employment, asking only subsistence." In 1853, William Tecumseh Sherman found amid the "business and bustle" of San Francisco "more poverty than in New York" and received daily pleas for aid. In the early 1850s, many southern Oregon prospectors were able to enjoy "a square meal [only] once or twice a week," subsisting on deer meat the rest of the time. In Deadwood, men who claimed not to have eaten for over a week begged on the streets. In Dawson, men slept indefinitely atop or under saloon billiard tables not only because of crowding but because of impoverishment.

Some became marooned in urban areas without the funds to get home. In Dawson, when a Klondike king, Nels Peterson, offered free passage to Seattle for the first person who spotted one of his two new steamboats arriving, "hundreds who had no other way of quitting the country" hurried to a vantage point above town hoping to win the prize.

Some had trouble escaping even the diggings they had toiled so hard to reach. One would-be prospector, after only two days in the California goldfields, was suffering from dysentery and reduced to "no money, no shoes, no bedding, no provisions" and one threadbare flannel shirt. He carried the blankets for a party of three Philadelphians as a means of leaving the mines.

For almost all, their situation was a tremendous fall from the riches and glory they had secretly or openly anticipated as their due. One Yukon stampeder, a "gentlemanly bank-clerk from Chicago," had been "decidedly uppish, a flyer to the Sun" on the trip to Dawson, but a fellow traveler found him almost immediately departing the Yukon as a ship waiter and too ashamed to speak to his former companion. Another argonaut who had put on considerable airs on the Chilkoot Trail was soon reduced to a job in Dawson "standing before a saloon, playing a cheap wind-instrument with might and main" to attract customers.

Even when a fellow was "getting by" on his own, the knowledge of his present state in contrast with the anticipated one created a palpable wound in his psyche. Henry Rogers, his limited funds melting away in an ill-fated Alaskan sawmill operation that represented his last hope of remaining in the Far North to prospect, visited a tourist steamer to assuage his curiosity about its construction and found himself "viewed with considerable interest by a group of swell looking young ladies & gentlemen." Suddenly, the former Boston shoe clerk, once a member of such well-groomed social circles, was aware of the picture he presented in his "rough clothes & leather knee boots," commenting, "I don't suppose they would recognize me as the same individual if they should see me at home in a dress suit."

Such a statement might be made with frontier pride if a fellow was wresting a comfortable living from the country. But Rogers was not. In fact, his mother was worried enough to write that a friend advised he go to the United States consul to seek food, a suggestion that exasperated Rogers no end: "Doesn't he know that Alaska is part of the U.S. and has no U.S. Consul any more than N.Y. City has."

Various networks of support did develop in most mining camps and towns. Prospectors would take up collections for the destitute, and fraternal orders such as the Odd Fellows and Masons tended to regroup on the frontier and to take care of their own. So did ethnic groups; French argonauts established a welfare society in San Francisco to aid ill and impoverished countrymen, and Jewish associations helped their destitute members in the big rushes. When a core group of middle-class women gathered in a mining town, they often formed a benevolent society. The Salvation Army made its presence felt in Dawson, offering "work for the needy, food for the destitute," and town authorities there offered food in exchange for chopped wood for heating fuel.

With a little help from friends, a fellow could usually scrape together enough for a sea passage home or cover the costs of an overland journey or, later, purchase a cheap railroad ticket to head east. But if a man was ill, his chances of departing or getting home safely lessened, especially since illness and financial need went together as firmly as did rocker and riffle. A physician in Dawson found "a very large amount of charity work to do, because a large number of the newcomers were practically penniless" and "these were the first men . . . to lose their health." Yet even a man who had been doing well in the diggings could join the charity patients because doctor bills and reduced productivity took a devastating toll on his fortunes. As one Colorado argonaut put it, "The expenses of a few weeks sickness here would buy a farm in Minnesota."

Much depended on one's state of health upon arrival in the town, camp, or diggings. Travel conditions were often so debilitating that many "ended their journey . . . utterly prostrated by overexertion, and too often poisoned by unwholesome food, and want of cleanliness." A Cariboo prospector attributed illnesses among almost all his fellows to "the hardships they suffered on the overland journey."

Those who entered the gold towns found their health problems compounded. Although officials liked to tout the towns as healthy places, the concentration of population and the unsanitary conditions naturally compounded or caused illness. From San Francisco to Nome, towns became "center[s] of sickness." The mining camps and the gold regions themselves were considered healthier locales. One Californian wrote, "With all the toil and exposure, there is something invigorating in this mountain air which sharpens the appetite and promotes health, so that some of our party seem never to tire," and another noted that aside from two or three drownings, "I have not heard of a death for months."

Yet it was hard work to stay healthy. Argonauts found in the California camp of Amador "the very dust was alive" with lice, "the clothing and blankets in all the stores literally swarm[ing] with them," so that men had to burn their old clothes and boil clothing and blankets twice weekly. Labor and exposure to the elements, fevered calculations for fortune, and rough living clearly took their toll. One California doctor noted "sickness in some form often overtaking even the most robust, after a few weeks' toil." He had never seen "more broken-down constitutions" and concluded that "few who work in the mines, ever carry home their usual full health."

Certain debilitating illnesses were common from California to the Yukon, chief among them the scurvy that resulted from limited diet. "I would pity the meanest dog in the world that had the scurvy," wrote one stampeder. Without recognizing their need for vitamin C, argonauts realized that fresh fruits and vegetables could quickly remedy even a bad case, but, of course, such items were often unavailable or in short supply. Meanwhile, men grew weak and achy, their skin becoming bruised and "spongy," their gums bleeding, angry rashes spreading over their bodies. One California miner noticed the effects of the disease when he realized that he could not climb a hill without losing his breath or rise easily from a sitting position and that he could dent the skin on his legs with his thumb. By the time the Idaho mining frontier developed in the early 1860s, scurvy had become known as "the winter scourge," and it haunted stampeders all the way to the Far North.

A second common gold frontier illness was dysentery, an intestinal inflammation signaled by abdominal pain and chronic, bloody diarrhea. One California physician reported that among those who passed in 1849 through Happy Valley, the section of San Francisco where argonauts routinely camped, "almost none escaped dysentery," owing in large part to the use of water from "hundreds of little seep-hole wells two and three feet deep." A half century later, a doctor in Dawson estimated that dysentery "in a mild form, afflicts 75 percent of the men landing in Dawson from up Yukon" and attributed the situation to their use of the waters of the Yukon and Klondike. In Nome with its appalling sanitary conditions, the numbers were correspondingly high, with a doctor recording "blood in every public convenience."

Pulmonary diseases—consumption, pneumonia, rheumatism—naturally assailed prospectors laboring knee-deep in freezing water. Fevers were another common scourge. Sometimes they were diagnosed as "intermittent," sometimes—particularly in Colorado, the Black Hills, and the Yukon—as mountain fever. Again, lack of sanitation was recognized as a major factor. A physician in Colorado found a Pleasant Valley mining camp anything but pleasant, "extremely filthy," with "mountain fever prevailing to considerable extent" in the dirty dwellings encircled by mud and water.

In addition to all these sicknesses, there were the epidemics, cholera finally making its way to California in 1850, then again in 1852 and 1854. One outbreak reportedly claimed 5 percent of

San Francisco's citizens. Smallpox devastated Deadwood and other mining towns, and outbreaks of deadly spinal meningitis struck all the way north to Skagway and Dawson. Typhoid fever hit "Nomads" with a vengeance.

"I think there would be less [illness] if Men could work with steady nerves & cool heads as they do at home," mused one Californian. "Excitement makes many sick," wrote another. So did discouragement, infecting the overworked miner with a drifting ennui in which he would fall into bed without eating at close of day, arise to partake of "half raw, soggy pork, heavy beans, and leaden soda-biscuits," then return to bed or "loaf away a night gambling and drinking in some saloon or dance hall."

By engaging in vice, men masked their "sickening depression from the disappointment of too sanguine hopes," yet they further debilitated themselves by gambling away money needed for decent food, clothing, and shelter, by catching venereal disease, and by "[drinking] their meals," liquor being cheaper than food.

Mental illness also surfaced in the goldfields, in part because rushes had lured many already possessing a poor grasp of practical realities, in part because dissolution and despair worked away at even the strong. Some went insane, as did the "promising young physician from New Orleans" who had to be sent home from the California diggings under guard. Lunatic asylums were established in San Francisco and Stockton to house "miners whose oft-disappointed hopes have made them mad," although ever-keen observer Bayard Taylor noted the San Francisco one also contained those deranged by "sudden increases of fortune."

If a stampeder escaped illness, he might still be felled by a mining accident. No safety codes governed placer mining or early efforts at deep-vein mining. In the former, dislodged boulders rolled onto workers (and even smashed into occupied cabins), people fell into the open mining pits, men smothered in collapsed coyote holes. Another common danger lay in attempts to separate the mercury from the gold collected in the riffles. Many would heat the mixture over their cabin fires to burn off the mercury but would in the process breathe the poisonous fumes. Some "smart enough to leave the cabin while the cooking was going on" became fatally ill after eating food contaminated with the mercury.

Lode-mining accidents ranged from falls down shafts to cave-ins. Many involved the use of dynamite. Men would drill a hole, fill it with the explosive sticks, and then blast. But tenderfeet across the West failed to put the primer cap on the first stick inserted. After the explosion, they "would go picking around with picks and hit the bottom stick of unexploded dynamite."

The actual work of mining did not present the only external danger. Violence, often fueled by drink and frustration, flared repeatedly on the gold frontier, most of it of a minor nature but leaving men injured and in serious circumstances. Other health dangers lay in the bitter dry heat of the Southwest (Arizona Territory prospectors endured 115 degrees in the shade, when they could find it) and the bitter cold of the Northwest and Far North. Frostbite lurked as an ever-present threat during the long Yukon winters; men who paused along a trail to shake some feeling into freezing limbs could become frozen in place, like macabre mannequins. Others froze in their cabins, too tired to rise when life-preserving fires went out. And there was the man who entered legend by stumbling drunk into a Yukon winter night. He fell or lay down in the snow, and wasn't found until spring, his body preserved in a block of ice created by a fresh-water spring.

Gold-seekers tried to take safety precautions in their new environments, recognizing that "no man can win a fortune from the mountains without health." Further, given the remoteness of many gold diggings, a fellow could not count on calling a doctor, even if he could afford one. Nor were women, traditionally the nurses by virtue of their gender, often available. Thus, prospectors outside the urban mining centers had to learn how to take care of themselves and to rely on their partners and neighboring miners for aid.

This reliance on laypeople was not unusual in the mid-nineteenth century, when alternative medical practitioners such as herbalists, homeopaths, and hydropaths were successfully promoting the slogan "Every man his own doctor." These practitioners stressed a choice of treatments and mild, nature-centered remedies in response to the standard yet extreme methods of regular "allopathic" doctors—bleeding, blistering, and purging.

Gold rush veterans mixed vinegar with water and cooled the liquid in the shade, drinking it through a day in the diggings to prevent sunstroke. They also used vinegar with potatoes as a scurvy preventative or treatment, slicing the potatoes and coating them with the liquid. In the Far North, they rubbed each other's faces with snow at the first sign of frostbite: a pasty white nose.

In case of injury or illness, prospectors searched out and collected medicinal plants. Soft turpentine from pine trees was used to coat cuts and wounds. Spruce bark tea became a popular antidyspeptic and scurvy treatment, and one Black Hills miner found his mountain fever alleviated "by the use of alternate strong doses of teas made from the wild sage and the bark of the Oregon grape root." Prospectors even experimented with hydropathic treatments, two Californians successfully aiding a man with malarial shivering fits by rolling him naked "in a blanket wet with cold water," covering it with dry blankets, filling him with hot tea, then "unrolling" and drying him off.

Many miners carried with them a small store of standard medicines—morphine, calomel, opium, camphor balls, and quinine. One California argonaut "doctored himself through a bad case of dysentery, using morphine and charcoal." Some prospectors carried the patent medicines that promised a cure for everything and delivered a dose of alcohol or an opium derivative. Some adopted innovative and no doubt useful treatments, such as the five- to ten-mile "scurvy trails" created by stampeders on Alaska's Kobuk River to provide the exercise they needed—along with proper diet—to combat scurvy.

Others were innovatively bizarre or desperate in their methods. One group of California prospectors tried to ward off scurvy by burying themselves "up to their necks in the earth, hoping thereby to soak up healing powers from the soil" and appointed one or two men "to walk about as guards against predators, animal or human." One prospector, incapacitated and nearly mute for six days with a swollen, abscessed jaw, finally "got mad and hit himself a lick on the place, which broke the abscess," enabling him to talk again.

A Swede in California moaned, "To be sick here is ten times worse than elsewhere. . . . Who cares if I am sick or well? No one." But as long as a fellow was a member of a prospecting or mining group, or had a dependable partner, he could usually expect some practical aid and solicitous care during an illness. A much-reproduced "Miners' Commandment" proclaimed, "Neither shalt thou forsake thy brother miner while on the couch of sickness and pain, but shall carefully watch over him."

A member of scurvy-stricken Oliver Goldsmith's California prospecting party probably saved his life by bringing to him bean plants in leaf that had sprouted from a split provision sack along a trail. Physician Howard Kelly in Dawson the winter of 1898–99 noted that "as a rule, the men nursed their partners through these illnesses with as much tenderness as women."

Good partners were valuable not only for their own solicitude but for the fact that they involved other miners. Two members of a California prospecting trio nursing the third enlisted the help of "the men on the bar," who "were very kind and sat up with him nights, and were willing to do anything for him." Such informal nursing networks were not uniformly effective. One Feather River victim of a mining accident and typhoid fever was tended for months by the other miners around him, "some of them, it is true, kind and good," but others "neglectful and careless."

Whatever the level of nursing care, miners showed themselves ready to dig into their pouches and pockets for those in need. In California's Sonora mining area in the winter of 1849–50, a reported seven hundred Mexican miners were "sheltered, doctored, nursed, and maintained by miners' subscription alone." Volunteers circled the camps, raising the funds for medical care or for a sick, destitute man's return home, demonstrating the truth of an Alaska miner's observation that sickness "tends to make the well ones kind and charitable and helpful."

But it remained all too easy for a sufferer's need to go unidentified. A foreigner unable to speak English occupied a California claim near John Steele's, impressing the latter with his industry and his "utter loneliness." One day when Steele passed by the spot the foreigner's tent had occupied, he found in its place a grave; the man had sickened and died alone in the tent, the stench finally drawing others to discover the dead man and to bury him wrapped in his cot. A Sacramento woman reproached herself for allowing the cries of a young sufferer in a tent next to hers to go unheeded. She was "hard-worked, hurried all day, and tired out" and "never thought but some one" was answering his pleas for water. When the cries stopped, she looked in and "found him lying dead with not even a friendly hand to close his eyes." In a similar instance in the Klondike, an old prospector starved to death in his cabin, keeping track of his decline in a journal and leaving his neighbors horrified and vaguely remorseful.

"Seems good not to have a lot of bothering neighbors, wondering what you are doing, and why," Henry Rogers had written, but no bothering neighbors meant that no community of concern could be galvanized in response to human need.

Even when the community responded, the effect of the illness was often terribly disruptive. Ill stampeders had to give up promising claims as their finances and health deteriorated. As company members, they went from being productive to useless and even counterproductive, draining their companions' time, energy, and resources. The two California partners who had help from other miners in nursing the third member of their party still spent weeks without working their claim. The first time they left their recuperating friend alone in order to do some claim work, he crawled from the cabin and suffered a relapse, consigning them once again to constant nursing when they had the opportunity to make forty to fifty dollars a day mining. When the third partner finally recovered, he was very irritable and insisted that the three leave the claim for new diggings. They did so, but it soon became clear that he would not be satisfied in any diggings and the trio split up, the disgruntled man departing for Sacramento to pursue his original trade of carpentering.

The man may have found himself too weak to resume his old trade. Alonzo Hill had told his father, "You may expect very much from me if I have my Health," but after a siege of typhoid fever in 1850, he wrote, "I am not stout enough to take hold of labor as I did previous to my fever." The long-term effects of severe illness are apparent in Hill's letters home; the prefever letters have a certain measured buoyancy missing from the vinegar-tinged later ones. Hill's experience demonstrated the truth of George Swain's admonition to his brother William: "If you get sick there, it will go hard with you."

Recognizing this danger, still-healthy men in gold rush Nome scurried to sell good claims and depart when a typhoid epidemic began to rage. But the gold fever often proved stronger than the worst living conditions and most virulent disease. Reaching Sacramento after a brief and disastrous stay in the diggings, forty-niner Phineas Blunt was suffering so from scurvy that "the moment I bruise the skin on my leg or hand it makes an ugly sore." Jobless and broke, he wrote, "I am not more than half clothed for such [wet] weather, but must get along with what I have, as I have no means of getting money. Cold chills run over me all day and all night I cannot keep warm."

Soon thereafter, Blunt obtained his job tending and sleeping under a bar and was able to see a doctor to obtain medicine. As soon as he had back some measure of health, he returned to the diggings and established a profitable claim with a damming company, but again found the scurvy "getting fast hold" of him, his weight dropping quickly and his leg bones aching so that he found rest impossible.

At least Blunt had been able to consult a doctor while in Sacramento, but a physician could often do little or nothing to change unhealthy conditions. Sacramento physician Beeryman Bryant established a small hospital and took scurvy patients, but was still able to provide them only with meals of fried or boiled pork, gravy from the pork grease, and strong coffee. In the Yukon, Howard Kelly prescribed for a scurvy-racked miner "lime-juice, meat, and fresh vegetables—when he could get them." In Nome during its typhoid epidemic, residents initially ignored a doctor's advice to boil drinking water.

The big rushes drew many physicians; one Colorado stampeder wrote, "Doctors are plenty, probably 3 to 1 from any other business. I think I am acquainted with as many as twenty." But few practiced medicine. Some hung their shingles over crude packing boxes, shanties, and cabins or created hospitals; Bryant put together his Sacramento "home for the sick" by using willow poles and canvas, constructing bunks, and stuffing the canvas bed ticking with dried grass.

On the surface, gold rush physicians appeared to command large sums. "They will hardly look at a man's tongue for less than an ounce of gold," complained one Californian. In Dawson almost fifty years later, it cost seventeen dollars for the most fleeting consultation. But, as with other enterprises, the cost of living shrank such sums. The doctor who charged an ounce of gold dust might have to turn around and pay a patient five or six ounces to freight medical supplies to the next mining camp. Then, too, medical fees fluctuated along with everything else, often plummeting with the arrival of competition, and many physicians found themselves performing charity work almost exclusively. For all these reasons, a trained physician might choose to hide his profession.

Many doctors also became disgusted over the number of quacks practicing in the goldfields, "in their grasping cupidity . . . drain[ing] the poor miner of all his hard-earned dust." A highly trained French physician found a former shoemaker from New York and a Parisian architect both practicing as doctors in California, and he observed with consternation a saloon counter that doubled as a drugstore.

The Frenchman thought this state of affairs the product of an American "predilection for liberty and the free practice of medicine" and an ignorance of "the true doctor's role," with Americans "demand[ing] a quick cure for a disease which they themselves diagnose" and refusing to submit to physical examinations.

Argonauts demonstrated the truth of his statement by looking for miracle medications and cures. Gold town newspapers bulged with such rosy promises as that of a Dawson masseuse: "Scurvy prevented and cured by new method. Lost vitality restored." Even here, where the Canadian government required medical licenses and often clapped unlicensed practitioners in jail, quack doctors proliferated.

Still, stampeders often recognized or deduced medical quack-ery. When Charles Ferguson's home remedies were not working on a sick California mining partner, Ferguson sought out a doctor ten miles away. The fellow listened to his recitation "in profound silence and with closed eyes, as though he was taking a mental review of all the cases in the books from Esculapius or Galen down to that hour," then pompously confirmed Ferguson's treatment as the right course, although he offered to see the patient for one hundred dollars. When Ferguson asked him his fee for the just-ended consultation, the doctor grandly waived it, saying "perhaps he should sometimes have a stubborn case" and want Ferguson's assistance. "That settled it with me," wrote Ferguson. "I knew then he was an imposter." Apparently no real doctor would imply such equality of knowledge, however patronizingly, nor would he reject a fee.

A California prospecting expedition encountered another party containing a young man who had just tangled with a grizzly bear. The animal had ripped off "the ear and scalp on one side of his head with part of the parietal wall of the skull, about two inches square, exposing to view the palpitating but uninjured brain." The man's companions had located a quack doctor who determined to pour "a mixture of salt, red pepper, and vinegar as an antiseptic upon [the patient's] exposed brain," explaining that his mother used this concoction in a move "unexcelled in saving meat from spoiling." Those present objected strenuously and found two doctors who knew what they were doing; the patient survived.

The dissatisfaction many felt was voiced by a California stampeder who complained, "There has been three doctors or things they call doctors working at me for some time. . . . Have now paid out all my gold to the doctors and they leave me worse in health." Doctors themselves were repeatedly frustrated by the same lack of resources that plagued other frontier folk. Physicians tried to carry medical supplies with them, but they still had to be resourceful. A California physician called to a mining camp to examine a young argonaut doubled up with painful cramps immediately asked for mustard, but there was none to be had. He then ordered the man's friends "to fry flapjacks very large and thick" and lay them on the man's abdomen. They followed orders, "bringing them right from the frying pan and so hot they almost blistered the skin," and the treatment proved effective.

The gold rush towns often boasted a range of formal health services: doctors practicing as doctors, rudimentary hospitals, voluntary aid organizations, even limited care for the indigent sick provided by city governments. But all of those services tended to be erratic in both quality and quantity. At their worst, they only deepened the sufferer's plight, as in gold rush San Francisco, where the city government contracted for public health care with a physician who provided abysmal service, cramming untended patients together on pallets in a public ward. When the city revoked the contract, he tried to keep the remaining patients and bill the city, only succeeding with those who were too weak to move, although their existence was so miserable that some finally dragged themselves out of his clutches.

In truth, city coffers could not support even the best-intentioned effort; Father Judge's hospital in Dawson was established with credit from the local lumber mill and trading company and the fund-raising efforts of the limited number of women in the town. Again, a crisis-response pattern was the rule. During epidemics, city officials tried to quarantine sufferers in "pest houses" and impose sanitary regulations, but with little success. When the flow of needy immigrants grew, the governments tried to farm out the indigent sick and the mentally ill to private households, making them dependent on the mercy of strangers. Many still fell through the wide cracks of public assistance.

Besides, the heightened anonymity of the urban areas undercut any good-neighbor impulse and made even those in dire need distrustful of offers of assistance. In addition, men were more likely to be separated from their friends in the towns and cities. Small companies often split up temporarily, the members going on different errands, and never reunited. Longtime companions simply disappeared in the confusing vortex of boomtown life. A Dawson newcomer "spent weeks searching for a missing partner until one day he was asked to act as pallbearer at the funeral of a typhoid victim." A mildly curious glance at the corpse revealed the face of his friend.

Actually, the death rates in the various gold regions were not excessive in comparison with death rates in the East. For example, in the ten months before June 16, 1898, the *Klondike Nugget* reported only twenty-five deaths in Father Judge's hospital. Epidemics made numbers soar, but epidemics could and did ravage the nineteenth-century East as well. The difference lay in who was dying, for in settled areas one might find a preponderance of elderly people and young children. On the gold rush frontier, the deaths were generally "those of vigorous men struck down in their prime."

For those who were present at the death of a friend, the experience seared their being. Shortly after arriving in the California diggings, both L. J. Hall and a friend, John Lyon, were struck with fever and dysentery. They occupied adjoining tents, cheering each other by calling back and forth. Then, on the third day of their mutual illness, Lyon's response became so weak that Hall "crawled to the door of his tent" and found him near death. He summoned two members of their company, one of whom asked, "John, shall I send your gold watch, as you requested me on shipboard, to Bertha, your promised bride?" Hall watched as one tear ran down Lyon's face and he quietly expired. "The men came in from different parts of the creek, saddened by his death," Hall wrote. No one could go to bed thinking of the dead man lying nearby, so in the night they prepared a leaf-strewn grave, wrapped the body in a blanket, and buried it.

Soon thereafter, another member of Hall's party, John Hart, fell ill during a prospecting tramp. Announcing that he would not last the night, he prepared his will by the light of the campfire and called upon his cousin to witness it. The cousin responded by cajoling, "John, it's an awful stormy night to die in—postpone it until morning—a man can die so much handsomer by daylight, when the sun is shining. I would postpone it a few days." Hart still insisted he would die before morning, but a doctor happened along during the night, gave him quinine and whiskey, and apparently revived his health and spirits.

When argonauts were dying, they were known to inject a certain levity into the proceedings. One ailing Deadwood miner asked a visiting minister to fetch his two partners. When they arrived and asked if they could do anything, he reportedly responded that he just wanted to "die like our Lord—between two thieves!"

Men had to depend upon their friends to wrap up their affairs, to provide information and personal effects to their families back home, to cover funeral costs and send any money earned home as well. When Alonzo Hill committed suicide in San Francisco in 1857, after eight fruitless years in California, he left for his partner specific instructions concerning his place of burial and individual relatives who were each to receive a few hundred dollars from his effects, and he made an impassioned plea: "400+ is in my pocket—keep the bal[ance], save what is paid for me in expense O-Wells at this hour I trust you—my friends as mentioned above are poor I do well by you O! be true to my request, and if I can I will let you know where I am."

For some gold-seekers, friends or a whole camp would provide a grand good-bye, complete with stately procession to a hillside cemetery. For others, death meant a quick, anonymous burial. One California prospecting company in a ravine near Downieville came across what appeared to be a cleverly covered prospect hole and dug eagerly, thinking it must be a rich find. Instead, they discovered the corpse of a young man of about twenty, with no identification and a bullet hole in his head.

This was not the only company to disturb a corpse in their search for gold. There are stories of prospectors "dig[ging] for treasure among the dead" in gold camp cemeteries and one delightfully macabre tale of a funeral turned to prospectors' frenzy when a mourner spotted gold in the hole into which the coffin was to be lowered.

In some cases, it was feasible to ship bodies home; of those who died in the port of Skagway during its first eleven years of existence, more than half "were returned to their homes and families in other places." But far inland at Dawson, local burial was still the rule. Like digging for gold, digging graves could only be accomplished a good part of the Yukon year by burning off the soil layer by layer. "Burning and digging and burning some more, over and over," wrote one Dawson resident. "Let me tell, you, Dawson made a lot of us forget the itch for gold."

Here was bitter irony, the work of forever burying golden

dreams the same as the effort of realizing them. The living remembered their comrades, those who had "made a pathway with their dust," and turned again to the struggle to balance self-interest and communal need—including the need for a well-ordered, crime-free environment.

Chapter Nine

"No Law but Miners' Law"

There seemed to be no organized government.

—*Argonaut John Steele of Nevada City, California, 1850*

There are 30 saloons, all orderly, no law but miners' law.

—*Peter Trimble Rowe on Circle City, Alaska, 1896*

In 1897, a Canadian official on a vessel bound for gold rush Dawson found his fellow passengers to be American adventurers eagerly devising a government to impose upon the burgeoning town. He listened in bemused silence, then announced, "Gentlemen, any interference of yours in our Canadian affairs will be quite unnecessary," as the North West Mounted Police would "attend to law and order and everything else."

His statement was met with incredulity. "What the hell" were the North West Mounted Police? When the official explained, the Americans wanted to know, "Who the hell sent them here?" Told that they had been dispatched by the Canadian government in Ottawa, some four thousand miles distant, they demanded "what in hell" the Canadian government had to do with affairs in Dawson. The official had hard work "to convince these saucy fellows that the Klondike diggings actually were in Canada!"

The Americans' response demonstrated an ethnocentric ignorance but also reflected the realities of the previous fifty years of gold-hunting. With argonauts repeatedly surging into western territories far removed from the jurisdiction or the practical extension of established federal or territorial government, they had been compelled to make their own. Given the tenuous nature, fluidity, and isolation of gold camp life, the task had proven a mammoth one, even when the gold frontier did not match "Wild West" images of violence and disorder.

In the republican, romantically independent spirit of America in the mid-nineteenth century, most stampeders had an abiding distrust of a controlling federal government. This entity existed to provide support services to aid in their individual enterprises—by providing troops for protection along the gold rush trails, for example—without hindering their personal freedoms, such as the freedom to glean the gold from public lands without paying a cent to the government for the privilege.

The federal government managed little in the way of assistance but cooperated all too well in leaving the argonauts to exercise any freedoms they cared to claim. With people pouring toward California in 1849, the United States Congress adjourned without setting up any mechanism for a territorial government. Even when California went directly to statehood status in 1850, federal courts were not established until 1851.

Distances were a tremendous problem. In the gold rush West, a camp might be two hundred miles from the nearest town, more than twice that from the nearest government center. The government's task was made even more difficult by the fact that western areas were still geographically ill defined. Great chunks of western land carried the label "unorganized" or had been made unwieldy territories—Oregon Territory originally covered the whole Pacific Northwest, and gold rush Denver was established as a distant outpost of Kansas Territory. As residents of Dakota Territory, Montanans never received an official copy of the territorial laws.

As a result of such isolation, the mining codes set up by the stampeders themselves, provided, for the first eighteen years of the era, "the only laws governing the mines and waters of the public land of the United States." While these mining regulations were readily accepted by the populace and by United States courts when they finally appeared, other attempts to impose structure proved vexing.

For when gold region residents tried to set up their own general governmental structures, they ran into questions of legality that undercut their efforts. Denver residents created their own territory, christened Jefferson, then obtained a city charter from the territorial legislature. But many residents refused to recognize the charter as valid, rightly guessing that the independent creation of the territory would not be approved at the federal level.

In gold rush Deadwood, it was especially easy for anyone so inclined to ignore the provisional government, for all Deadwood residents were flouting federal laws by locating on Sioux land. And if a provisional court determined that a miscreant should be sent to the Dakota Territory prison, it had no authority to enforce such a sentence, as "neither the territorial nor the federal government recognized any rights on the part of miners to be in possession of courts, or even of the land they were occupying."

Even when the stampeders' presence was legal, the territory a reality, problems born of political tensions and continuing geographical isolation sabotaged efforts at order. Montana won territorial status in 1864, but both its territorial assembly that year and the one held in 1866 "were declared illegal by the United States Congress, and all of their acts were voided" because of reported southern secessionist sentiment there. Alaska's Organic Act of 1884 established a very rudimentary territorial government, including "one policeman for nearly a million square miles of land," and it was not until the height of the Nome beach placer rush in 1900 that a more refined act was passed. Even then, it simply divided Alaska's expanse into three giant judicial districts.

In truth, the advent of gold-seekers at first did not seem to warrant any complex administrative and criminal justice apparatus. The number of early stampeders was relatively small and mobile, and no one was yet quite sure which areas would turn out to hold long-term riches and offer economic stability. Then, there was initially a noticeable lack of crime.

The last reason drew much comment. A military official visiting the California diggings in summer 1848 found crime "very infrequent," thefts and robberies almost unheard of, although the entire mining population lived in tents and bush houses and carried with them hefty gold pouches. The next year, men were still leaving "buckskin purses half full of gold-dust" on roadside rocks as they worked at mining "some distance off." Even though the gold urban areas would soon seem to breed crime, in San Francisco in 1849 goods lay piled upon the beaches unguarded and undisturbed. In the mining town of Weaverville in December 1849, an immigrant judged "life & property are safer here than they would be on Boston common, we have no stone house or sentry. . . . It is very Peaceable here in the mines." In Circle City, Alaska, prospectors left their gold dust in unlocked cabins, traded informally in claims, cached food and supplies without fear of their disappearance, and were able to move about with little fear of robbery or murder.

This state of affairs usually lasted only a short time—but then, so did a number of the rushes. In the larger rushes, many residents of the gold regions would complain in a there-goes-the-neighborhood vein that conditions changed with the arrival of foreigners and with a less honorable breed of stampeder, "the bully and the bravo."

The first assertion had little basis, as the Americans often resented foreigners in the diggings and found them convenient scapegoats. The second contains some truth, as inflated stories of easy riches, the chance for unrestrained vice, and improving trails would draw the relatively lazy and crooked to concentrate in the fledgling urban areas. But a California official was closer to explaining the initial peace when he wrote in 1848, "The extent of the country is so great and the gold so abundant that for the present there is room and enough for all."

Social-control factors came into play as well. In the first five years of the California gold rush, stampeders spent more than six million dollars on bowie knives and pistols. Many saw the proliferating weapons as a necessary deterrent to crime and violence or as an efficacious means of settling a dispute where no other means were available. One forty-niner found "good order" maintained in San Francisco in early 1850 by "every man taking care of himself." He saw not "a single police officer," and concluded, "Indeed, there was but little use for them, for had a difficulty occurred—as some did—pistols or knives would have settled the matter long before a peace officer could have gotten there."

Unless under the influence of liquor, many preferred to warn and threaten with firearms rather than shoot them. One Bodie resident declared that he had "more than once . . . seen a whole crowd of men with their guns drawn and not a shot fired." But weapons also had the opposite effect of facilitating frontier violence. A gold rush California doctor noted knife and gunshot wounds aplenty, "especially at the mines, since everyone carries arms and uses them in the smallest quarrel."

The remoteness of the camps and towns meant few escape routes and few modes of transportation available, especially during the winter months or the rainy season. Anyone intent on criminal activity had to be able to remain undetected or to brave an extended, rugged trail out. Thus, one Deadwood resident recalled "no petty crimes," for "pickpockets, burglars, embezzlers, and night-slinking thieves never possess the daring required to operate two hundred slow miles from a railroad."

A related deterrent to conflict and crime was the rapid development of miners' self-government. One pioneer of 1849, perhaps with a bit of romantic idealism, would recall that even the disruptive element "knew that it was safer for law and order to govern; and, with a few desperate exceptions, were willing, to let the lovers of order enjoy their rights and wield their influence."

The "lovers of order" in the diggings wielded their influence in two ways. First, in the area of mining rights, they established mining districts—identifying the area being mined or having potential for mining and setting up basic rules for marking off and recording claims—and created the miners' meetings, or the more formal "miners' associations," which articulated and enforced mining law. And second, they devised the miners' courts which served as a crisis-response method of dealing with crime.

Miners' meetings came into use in the California diggings in late 1848, as first-comers found themselves faced with a new wave of arrivals eager to challenge claims. These meetings—gatherings of the various claimants to delineate territory and resolve conflicts, with majority vote ruling—upheld the rights of whoever could establish prior claim. Taking into consideration the number of seekers a mining area could accommodate, they established the size of the river, bank, bar, or gulch area claimants could work. They also validated the leaving of tools—pick, shovel, or pan—as a means of marking a claim until one chose to abandon it or was able to register it with a recorder, a man entrusted by the community with the claim records.

California miners' meetings were intensely democratic bodies, trying to give every working claimant an equal say. The meetings were usually quite purposeful and decisive, as no one wanted to spend time quibbling when there were fortunes to be unearthed. Thus the decrees carried the authority of general opinion. Forty-niners found miners' rights "well protected," the laws that emerged from miners' meetings "accepted and recognized as binding upon all within the district." When a forty-niner took over work on the claim share of a man gone prospecting, a miners' meeting voted him the man's portion, and he "knew this would hold good, as miners laws stood preeminent to all others."

As the California mining frontier developed, the meetings tackled increasingly complex situations. On the south fork of the American River in 1851, companies who had worked a bar until it was submerged still claimed the bar when other companies exposed it along with the riverbed through exhaustive river-damming and diversion efforts. At the same time, the river diversion flooded some claims, rendering them worthless. In response, claim owners lining the American River and its tributaries set up specific ownership guidelines, of which one argonaut wrote, "I do not think there ever was a set of rules, which, upon trial, more perfectly bore the test than those adopted by that mass convention."

California methods were adopted in each new location across the gold rush frontier. Some rules became standard, such as the awarding of an extra "discovery claim" to the original locater in a new mining area. Some remained flexible; a traveler found that the laws of Colorado mining districts "all differ[ed] more or less, and especially in the matter of holding and working claims." Still, anyone who had been in one mining district could expect to find roughly parallel codes in the next. The biggest variation seemed to be in size of claims. In Deadwood Gulch, prospectors established size at "three hundred feet in length along the path of the gulch, extending . . . from rim to rim on either side." On Nome's beach, claims consisted of a circular space "not much larger than a man could reach with his shovel from the point where he first began to dig."

Nome was unusual in that a small military force stationed nearby played an authoritative role. When the press of disgruntled stampeders became great, the local military commander ruled that "no one could attend any miners' meeting unless he were an actual claim holder with a registered claim in the Nome Mining District."

The decision reflects both the military's degree of involvement and the fact that miners' law remained the rule. Even with their wrangling, Nome creek miners used this power judiciously, rejecting a proposal to eject suspected thieves from the diggings by concluding that "many innocent people might be accused and expelled."

Much tension in the goldfields stemmed from the fact that if latecomers were determined to get a share, beyond the possibility of purchasing or leasing a claim they had only three choices: get the claim size reduced, thus opening up part of the diggings; get the previous claims voided; or jump a claim and hang on tight.

The first strategy was a popular ploy. In the Deadwood three-hundred-foot diggings, Montana prospectors arriving to find all the good claims taken tried to engineer a reduction of claim size to only one hundred feet. Failing to hold sway, they threatened use of arms to effect the change and were dissuaded only by becoming convinced that "the character of the ground—deep bedrock in some parts and meager pay in others" made such smaller claims unprofitable.

Late-arriving Nome beach miners went a step further than the Montana men, attempting to establish a new mining district in place of the one already in existence. They designated a new district headquarters, required a new recording fee, and, of course, voided all previous claims—or tried to. A Nome miners' meeting rejected this usurpation and threatened the challengers' designated mining recorder with violence, thus ending the attempt.

Men looking for a piece of rich diggings also resorted to attempts to void specific claims on technicalities. On Anvil Creek in Alaska, where feeling ran high against the Scandinavians who had locked up the best diggings for themselves and friends, Americans unsuccessfully argued that "it took six stakes for a valid claim" rather than the two used by the original locaters.

The threat of violence in many claim-voiding attempts escalated when stampeders resorted to the last option, claim-jumping. Despite the authority and efficiency of miners' meetings, they were episodic at best; men willing to support their decisions were often scattered, and confusion was created by the continuing influx of stampeders. Thus, claim controversies abounded, and newcomers accomplished flagrant takeovers in the face of whatever authority existed. One Nome beach miner reported his company's claims "covered with beach jumpers," mob law in effect, and the local military unwilling or unable to do anything except provide notices to "vacate," which the jumpers ignored.

Often, such confrontations were clearly of the might-makes-right variety. One company of New Englanders in California in 1849 was accosted by a band of long-haired, buckskin-clad adventurers who announced, "You can give us half or git; we mean business." One of the New Englanders squelched this ultimatum by responding that "the fust man that strikes a pick into these diggins . . . had better fust make his peace with God," as "I'll send his soul to meet his Jesus."

It was not unusual to have successive waves of jumpers jockeying for possession. In particular, as the next chapter will show, Euro-Americans jumped the claims of minorities, assuming them easy victims because of racist attitudes and general antiforeign feeling. Meanwhile, first the miners' meetings and then the federally sanctioned courts repeatedly upheld the earliest claims. The wisest prospectors were those who concluded "it would be well to be in actual possession of the ground at all times" and built dwellings on their claims.

Of course, a prospector could not hold a claim indefinitely without working it, and miners' law required a certain amount of "assessment work" on a periodic basis in order for the claimant to retain digging rights. Since unworked claims represented one of the few opportunities for the Johnny-come-lately, they were avidly pursued and hotly contested when the day came.

Even when the claim situation was relatively calm, criminal activity took hold and grew. The biggest factors remained the sudden influx of a heterogeneous population, the consequent shrinking of economic opportunities, and the lack of social, moral, and especially legal controls. When the numbers dropped, so did crime, as evidenced by the camp of Miner's Delight shortly after the South Pass gold rush had fizzled. All remaining prospectors appeared industrious, "friendly, neighborly, hospitable, honest," leaving their doors unbarred at night and eschewing pistols under pillows. Men knew their neighbors—an impossibility under boom conditions.

Monitoring the boom population and controlling their actions were impossible jobs, but British and Canadian authorities during the Canadian rushes showed some effectiveness. When the Fraser River rush developed in 1858, Victoria's Governor James Douglas, "the sole west coast British authority," acted decisively to claim the gold for the crown and to impose a mining license system. In the Yukon Basin, North West Mounted Police arrived in Fortymile in 1895 and were thus poised to help direct traffic and maintain order when the Klondike boom started. On the two main passes into Canadian territory, the Chilkoot Pass and the White Pass, Mountie superintendent Sam Steele rapidly established rules for entry and travel, watched for known criminals, and maintained a law-and-order presence that discouraged crime beyond Skagway and Dyea.

British and Canadian authorities had two distinct advantages in establishing and maintaining law and order. First, the extreme ruggedness and remoteness of the country precluded much independent movement of lawbreakers. Second, despite the reaction of the voyagers intent on establishing their own government controls, Americans *had* entered another country; they had to adapt to Canadian rules or face a "blue ticket" back to the States. But the contrasts also point up the United States government's inability to address and impose some order upon the stampede movements, in part because of slow, inadequate, piecemeal responses to the mass migrations, in part because of Americans' prickly tendency to circumvent or ignore any governmental controls when they could do so.

As most rushes built, then, confusion, disorder, and crime increased, lending partial credence to the myth of the "wild, wild West." The reality, as with so much else on the frontier, is less clear-cut. Some modern historians contend that crime rates were no higher than—and sometimes lower than—those in the East, but two types of criminal activity, personal physical attacks and robbery/theft, stand out on the gold rush frontier. The first —one-on-one fistfights, knifings, and shootings—reflected the youth, masculine pride, drinking habits, and impetuousness of a large portion of the stampeders.

Some one-on-one violence was stimulated by differences between partners, especially when isolation and "cabin fever" were involved. But partners usually settled their differences by splitting up, and it was public combativeness and the saloon environment that usually fostered personal violence.

A spirit of contentiousness pervaded gold camp and town, one man contending that he had known "one-half of the sovereigns of a mountain village ready to cut the other half's throats over the respective merits of the different styled sewing machines—if nothing else presented."

Tied to this attitude was an edgy, pride-filled jockeying for dominance and a common belief that one had a right, even a mandate—to stand one's ground—"no duty to retreat." Stephen J. Field arrived in gold rush California with "all those notions, in respect to acts of violence, which are instilled into New England youth," including the belief that one should simply "turn away" from a rude person. But he soon learned that "men in California were likely to take very great liberties with a person who acted in such a manner," that "the only way to get along was to hold every man responsible and resent every trespass upon one's rights."

This touchy combativeness cut across occupational and class lines. In gold rush San Francisco, "competing confectioners" laid into one another, as did "lawyers, journalists, politicians, theater managers, actors, sporting men [gamblers], and most anyone else with a grievance," often "in the very midst of town." In Bodie, fighting "was not restricted to any particular social or occupational group."

The most likely individual to become mixed up in violent controversy was a local newspaper editor. For one thing, rivalries between two men with printing presses in the same community resembled Armageddon; in Deadwood, "many and bitter were the controversies in which [the rivals] engaged." Then, frontier newspaper editors were a particularly fiery, imprudent lot when it came to giving their opinions on other public figures, as evidenced by the invective of the Virginia City editor who pronounced one political candidate an "inordinate blackguard" who "erects himself on his slim pins and emits quaint remarks of a smutty nature."

Because combativeness was pervasive and accepted, stampeders sometimes failed to address the violence that resulted. Argonauts on the Feather River showed unfeeling indifference to a young man struggling for his life after being "stabbed very severely in the back during a drunken frolic" without having given "the slightest provocation." Here, as in illness, friends were important to call attention to the victim's need and worth—in the case of violence, to his innocence and respectability. Otherwise, people didn't concern themselves with the results of one-on-one violence, particularly within the confines of a saloon.

Sometimes victims shared that attitude. One man, injured in a drunken argument in Bodie, refused to identify his assailant, explaining, "You see he is drunk and don't know what he is doing. He's a first rate fellow when he's sober but he'd shoot his best friend when he's drunk."

While the "rows" might be overlooked, robberies could not, as they cut to the core of a gold-seeker's quest. Miners' courts dealt with thieves and robbers so summarily that they apparently temporarily squelched further crime. A man who stole three hundred dollars in the California diggings had "both his ears cut off and the letter 'T' branded in his cheek," and subsequent theft in the area was reported of "rare occurrence."

Still, the temptation remained. "After a man had been a year or two in the mines, if he was industrious and temperate, it was supposed that he had gold," and he became a prey. William Hall, "a kind, steady, industrious, upright man" who "crossed the plains from Missouri in the summer of 1850," had sent one summer's earnings home from California by express, "but he probably had his winter's earnings in or about his cabin" when someone approached in the night, shot him in the doorway, and searched the dwelling by candlelight.

Towns were worse. One San Franciscan in 1851 found brigands "so bold that they trip a man in the principal streets in the day time and take his bag of gold." In Auraria during the Pikes Peak rush, "anything and everything movable and of value was likely to turn up missing"—including a whole steam sawmill bound for the diggings. In Nome in 1900, "robbery of every description was in full swing." (Some of these acts obviously were born of need and desperation; the items most often stolen in Bodie were firewood and blankets, both in short supply.)

Leaving with one's gold presented a problem, too. In the Klondike, "if you should overtake a man packing a blanket roll and notice that the straps were dragging hard, then you could safely wager that he carried several hundred ounces of the dust inside." Such clues probably contributed to the fate of three men who disappeared on a trail out of Dawson. "It was pretty sure they had gold dust on them," wrote one Yukon resident. "Springtime, they found the bodies on a little portage in the river near where they was camped—jammed under the ice, and full of bullet holes."

Murders usually related to theft. The most chilling accounts were those of men led by greed to kill their partners. While rare, that did occur; in fact, one of the two discoverers of gold on the Fraser River in 1857 murdered the other while journeying out of the remote region with their find.

Such miscreants were usually dealt with by miners' courts. When a man suspected of robbery or murder or both was apprehended, most of the mining camp population flocked to a meeting place—either out of a sense of duty or out of a combination of curiosity and bloodlust. As with meetings on claim issues, all those concerned—in this case, all camp members, whether prospectors or not—would be asked to participate. They might hear the evidence presented by a volunteer prosecutor and defense attorney and pass judgment as a group, or they might send a delegation to the nearest urban area to obtain a judge and lawyers. More commonly, they would elect a judge from their number and/or empower the traditional twelve-man jury. When miner John Carty was charged with the death of a gambler in a card game dispute, Deadwood residents set up court in the street, "placing a couple of planks on boxes" for the jury bench and setting out a chair for a judge.

This careful arrangement almost came to naught when a "big bully" in the sharply divided crowd tried to incite those present to hang the prisoner. One of the leading mining men hurtled through the mob with his six-shooter in the air, causing "an attempt to stampede," but "the men were packed so closely together that only those on the outskirts found it possible to escape."

Bedlam ensued, with spectators swarming into the court space; the prosecuting attorney "became entangled in the judge's chair" and was "thrown down and trampled somewhat seriously before he could be rescued." Meanwhile, the mining man cowed the "bully" with a blow from his Smith and Wesson.

When things calmed down, "disinterested parties" went over procedural matters and concluded that "nominations for judge and jurors be offered and the assembled men should vote upon them viva voce." At that point, a new arrival announced that in his wagon train had been a man who served as a judge in the municipal court in Chicago, and the barrister was immediately elected judge in the proceedings by "unanimous voice." The crowd was still split between those supporting the miner and those wanting vengeance for a member of the gambling fraternity, but "there was no desire on the part of anyone to delay proceedings or quibble over technicalities," so a jury was quickly seated. Carty received the proverbial slap on the wrist in a conviction on assault and battery, for no penalty was levied for such a mild crime, and a ten-man guard was appointed "to escort the liberated man out of the country."

Many of those convicted were not so lucky. Lacking a criminal justice structure, a jail, or a jailer, the miners' courts relied heavily on the public humiliation of head shaving, ear cropping, whipping, and banishment. A man who shot and wounded another over the latter's wife in Gregory Gulch in 1860 received a head shaving and fifty lashes in front of a crowd of "eight hundred or a thousand men . . . all shouting to put it on harder."

Yet, miners' courts often tempered punishment with mercy.

Because of "his previous popularity and inoffensive conduct," a waiter who stole four hundred dollars in gold dust from his employer at Indian Bar, California, in the early 1850s, received a "very light" sentence of thirty-nine lashes, loss of his claim, and banishment, the last financed by his judges. A horse thief at Gold Creek, Montana, in 1862 swayed the hastily called miners' court with his "age and contrition" and "his sentence was only to refund all of the stolen property that he had and to leave the country within twelve hours"—again, with the financial assistance of the very men doing the sentencing.

The worst punishment, of course, was hanging, and in the Colorado miners' code in 1860, anyone caught stealing more than one hundred dollars was to be hanged. In California, too, thieves faced death; one "harmless, quiet, inoffensive" Swede, found guilty of the crime was dispatched in front of a mixed crowd of solemn spectators and drunken, jeering boors, the executioners performing the rite out of ignorance "in the most awkward manner possible . . . by hauling the writhing body up and down several times in succession."

As an alternative to such ugly spectacles, banishment was just as effective in removing the miscreant from the community. It could also be lethal. One young member of a prospecting expedition out of Deadwood accidentally discharged his weapon while stopped along a trail, killing a fellow gold-seeker. His other companions gave him one hour to disappear—in a region in which the Sioux were picking off travelers with regularity. More deserving of the implications of his sentence was a prospector who "tried to poison his partners with strychnine" on the Fortymile River in Alaska and was summarily banished into the wilderness.

In such ways, those in the gold regions intimidated would-be thieves and murderers and maintained some semblance of law and order. That they did so with some sense of gravity and common purpose is evidenced by the fact that as late as the Klondike rush, when a man received the death sentence for killing his two partners at Valdez, "all in camp had a hand on the [hanging] rope, so each would share in the execution and its responsibility."

The harshness of sentences and their quick execution struck

most observers as necessary and appealingly simple. "I never saw a court of justice with so little humbug about it," wrote one California traveler after viewing a miners' court in action. "The people make good laws and enforce them," wrote a forty-niner who found "Judge lynch is all the law in California." Westerners showed pride in their miners' courts. They would "appeal to law if it is handy," but otherwise would simply "right" a situation immediately. "Thus," a Wyoming observer concluded, "an accomplished rogue would be far more safe in the claws of organized 'justice' than to fall into the hands of these men who 'only do by nature the things contained in the law.' "

Some expressed mixed feelings about the kind of justice the prospectors dispensed. One man witnessing a public whipping in the Colorado diggings wrote, "This is rough business for Sunday but such things must be attended to, but such a day I do not want to witness again."

In truth, miners' justice at times went seriously awry. The wrong men might find their necks in nooses, or the punishment might not fit the crime, even to those administering it. The "harmless, quiet, inoffensive" Swede died in his bungled hanging for stealing six hundred dollars' worth of gold dust, and "almost everybody was surprised at the severity of the sentence; and many, with their hands on the cord, did not believe even then, that it would be carried into effect."

By contrast, in Montana, three members of the corrupt Sheriff Henry Plummer's gang killed an honest deputy in cold blood and were able to walk away from a people's court composed of miners and other goldfield residents. One of the three addressed the court with such eloquence that he was allowed to go free. The others were scheduled to be hanged, but when a letter was read from one to his mother, and the writer began "blubbering and crying," some in the crowd became "tender hearted." There followed three farcical attempts to get a new vote on the execution order. In the confusion, with partisans on both sides crying fraud, the convicted men's sentence was reduced to banishment. "The gallows is still standing and the graves still open," wrote one prospector. "These men belonged to a group of highway men and it was a crime to let them go."

The Montana incident not only reflected misplaced sympathies but revealed how a crowd could vacillate in assuming the duties of judge and jury. A Gregory Gulch hanging in Colorado almost stopped when the prisoner won sympathizers by expounding on his drinking problem, but a last-minute vote went against him, and he did not share the luck of the Montana desperadoes.

Near the end of the rush era, it had become painfully obvious to some that "despite good intentions, the miners could not decide cases impartially." But they were more impartial than the gambling and criminal elements who pushed themselves to prominence as goldfield numbers grew and factions developed. The mining population, recognizing this fact, often withdrew their support from the "people's courts."

If men committed to impartiality tried to hold on to the courts, they faced mobs that sometimes bypassed due process altogether or stood ready to take authority away from them, sometimes making defenders or court members targets of violence. Foreigners in particular were denied due process and given harsh sentences, any Anglo defenders threatened or beaten. When two Latinos on the Feather River were sentenced to be whipped after a series of altercations between Latino and Anglo miners, an observer thought the sentence too harsh, but "so great was the rage and excitement of the crowd" that the adjudicators "could do no less."

A similar incident involved a group of Yaqui Indians who were found in the Sonora area burning the corpses of two white men. Although it was established at the murder trial conducted by miners that the Indians had found the corpses and were simply disposing of them according to tribal custom, two of the jurors "advocated a verdict of guilty on the grounds that if acquitted the frantic people would hang the jury." In this case, justice prevailed, with the jury determining that a volunteer miner militia would uphold an acquittal and proceeding to grant the Indians freedom.

The Carty murder case in Deadwood also offers an example of attempted coercion. There part of the assembled multitude wanted to hang the prisoner before the people's court had had a chance to work. Carty had been brought in by a duly authorized representative of the law, a deputy marshal, who had given the prisoner the choice of a miners' court or a legal court four hundred miles away and who argued forcefully that he had promised Carty a fair trial. That seemed to quell the trouble-makers, but just to be sure, "the court, acting on a motion, made by someone in the crowd," appointed a ten-man militia to guard the prisoner.

While the Deadwood assembly chose to respect the deputy sheriff's call for fair play, if not necessarily his authority, gold rush populations resisted formal law-and-order structures whenever it suited them to do so, clinging to crisis-centered self-government with the advent of local, state, and territorial governments. In 1851, a year after California had become a state, a man dubbed "the Squire" acquired a justice of the peace appointment for the mining camp of Indian Bar on the Feather River. The population there barely tolerated his presence, much less his authority. "The people here wish to have the fun of ruling themselves," wrote one resident. "Miners are as fond of playing at law-making and dispensing, as French novelists are of 'playing at Providence.'" They also grumbled that the Squire had not been elected by popular vote. Thus, when cases came up, the Squire simply became a ceremonial appendage of the miners' courts.

Even with some say in the new legal procedures in the form of a vote, residents didn't trust the new law enforcement structures to be as fair or effective as the ones they had improvised. Miners' courts were known to convene their own tribunals while sheriffs held prisoners for trial and to seize and hang those judged guilty. In the Rogue River Valley, Oregon, when a local elected official ruled in a dispute in favor of a prospector who had ejected his sick partner from cabin and claim, the other prospectors created their own Court of Appeal, which restored the incapacitated man's rights and forced the official's resignation.

Although political interest (and self-interest) could surface among stampeders, most continued suspicious of or indifferent to formal government structures. A Californian commenting on the state's September 1850 entry into the Union countered newspaper reports that "the people here" were eager for statehood: "Humbug! There is not one man in a hundred that cares about it one way or another. All they want is what gold they can get, and the state may go to hell, and they would vamoose for home." In fact, many gold-seekers found "any political ambition" to be "positively suspect," professional politicians to be "anathema."

Further, the stampeders often showed minimal interest in voting and in other citizenship rights and duties. When Californians scheduled a convention to hammer out a territorial government, officials attempted to lure voters in the northern diggings to the polls with fresh beef, but to no avail—"Gold was more powerful, for no one attended. The beef spoiled and the election was defeated."

Gold towns were "noncooperative communities," with residents not only refusing to vote but regularly refusing to pay taxes and to serve on juries. When vigilantism loomed in Bodie in 1880, the *Bodie Daily Free Press* argued, "If good citizens will but half do their duty as witnesses, jurors, etc., there are good courts and officers enough to soon bring about a better state of affairs."

Until then, "inconsequent shootings" and fistfights occurred with regularity, and a criminal class flourished in many gold towns. Robbers in particular haunted the red-light district, the milieu of the urban saloon, gambling hall, and bordello. In Bodie, victims of muggings almost invariably "had spent the evening in a gambling den, saloon, or brothel," were known to have money on them, and were "staggering home drunk late at night when the attack occurred." Three saloon-hopping "Nomads" were slipped knock-out drops and robbed, but they couldn't pinpoint in which place they had received the dose.

Con men did not normally take dust or money by force or stealth, but they took the funds all the same. They operated their games of "French monte" and "pick-i-the-loop," working their destructive magic with cards and thimbles and walnut shells throughout the gold regions, often with impunity. Jefferson "Soapy" Smith, who operated in Denver, in the silver towns of Leadville and Creede, and finally in Skagway, is the best known of these charlatans, but many of his ruses and methods were already flourishing in gold rush California decades before. Californian "Lucky Bill" Thorington, like Smith, marshaled a hired phalanx of "cappers"—ostensibly "miners, laborers, traders, etc., etc."—to step up to his tables and lure the unsuspecting onward with apparent easy gambling successes. In Deadwood, "Ten Die" Brown and his cadre of "sharpers" staged their "sure-thing games."

The bunco men also engaged in outright extortion. In Smith's Skagway, newcomers lugging heavy valises were charged fifty cents for crossing a Smith-controlled dock, a dollar "for storage" if they paused to set the valises down.

In addition, victims were subject to more elaborate charades than the apparent luck of the gaming tables. One member of Brown's Deadwood gang "posed as a hard working prospector, always trying to enlist capital for the development of a valuable mining property," his pockets full of enticing ore samples gleaned from "other mines of value" but "purporting to come from his own." In Skagway, an argonaut fresh off the boat would be conducted by a friendly fellow to an "outfitting office," only to be suddenly and forcibly parted from his carefully hoarded funds.

Many con men were not above murder. They, along with a less visible but active mélange of robbers and murderers, provided a potent threat to the hopes, dreams, and very lives of stampeders. Ultimately, communities had to deal with them.

Just as they dismissed personal violence between apparently willing participants, gold region residents tended to ignore scams and assaults on those unknown to or otherwise considered negligible by the community. Strong response was called forth only by the relatively infrequent attacks on or theft from people whom the community recognized and approved, especially if these individuals were moving outside the saloon milieu and minding their own business. Because a reputation for being unable to prevent robbery, injury, and murder could sabotage a town's or gold region's hopes of growth, citizens worried about the impact of these crimes. So severe was the threat of robbery to the economic interests of the community that Soapy Smith's downfall in Skagway—the growth of community opposition and Smith's fatal altercation with a local surveyor—was precipitated by his gang's lifting of twenty-eight hundred dollars' worth of gold dust from one of the first Klondike prospectors to reach Skagway on his way home in July 1898. A prominent local outfitter articulated the community's sentiment: "If word gets down the river that the first man coming out by way of Skagway was robbed, no one else will come this way."

Another type of robbery regularly aroused whole urban communities and occurred with increasing frequency as the gold frontier developed. Stagecoach robbers relieved passengers of their gold dust and other valuables and in the quartz mining areas hijacked shipments of gold bullion. They often ignored the largest shipments, which were heavily guarded, but for lesser hauls, the risks seemed worth the rewards.

As the urban frontier grew, stagecoach robberies on the trails were to strike even more sharply than other forms of robbery at the economic well-being of the whole area. If neither coach travelers carrying dust and other valuables nor shipments of gold bullion could be safeguarded, then immigrants *and* investors were likely to bypass the vicinity. Wells Fargo detectives arrived in the gold towns hunting highwaymen, and transporters of bullion tried various methods of foiling the outlaws, including armor-plated coaches and locked heavy iron containers.

Another dangerous aspect of gold town life was the random shooting in which men indulged for the hell of it. Towns demonstrated different attitudes toward this activity. The policeman in Bodie who tried to disarm a drunken miner firing at the dawn got shot for his troubles and earned a rebuke concerning laissez-faire from the *Bodie Chronicle,* which pointed out that California had no law prohibiting concealed weapons and Bodie had no law against discharging firearms; thus, "there is no authority vested in any one to compel a citizen to give up his arms, unless he has committed a felony."

Such attitudes continued to undermine any governmental attempts to establish and maintain control. So did the attitude that justice had been better served by miners' courts. "Blood & crime and deeds of Deathless shame are common now, Theft, Burglary, &c.," Alonzo Hill wrote home from California in 1850. "This chance I ascribe to the organization of a Government while People knew that they would suffer Death, or Lynching, or have their ears cropped, They held themselves within bounds of honesty, but they know now if they can have the affair handled by Lawyers & Gamblers which fill most of the high stations that their necks go clear."

In Deadwood, miners' courts simply continued to take precedence over the provisional government court, which was perceived as ineffective. Repeatedly, gold-region officials found that "the only real power lay in the will of the majority, making any form of authority valid only when it was enforced by public sentiment."

Even if the community supported a law, once the concerted power of miners' courts waned, an individual could choose to ignore the law or to challenge it through a lawyer. That forced the often-reluctant community into a defensive position requiring stronger organization and heavy outlays of money—on "legal fees, police, and inspectors"—from meager or nonexistent coffers. If gold town residents recognized that "the law carried a price just as surely as flour, lumber, and nails," they usually weren't willing to pay that price.

Many decried the number of charges dropped and acquittals handed down by the courts. Representative was the *Bodie Chronicle*'s reaction to the acquittal of three men charged with murdering a rival claimant to a ranch property:

> If three men shoot another to death, it is not murder but a pleasant pastime in the mountains—unless it can be proven that the three had conspired to kill their man. Out of all the murders committed in this county in the past two years, not a murderer convicted! What laws; what a country and what a peoples!

The reference to "peoples" indicates that critics realized courts struggled with conviction problems tied to gold rush society itself. Two such problems stand out. The first was the lack of witnesses. In a climate in which minding one's own business was so highly valued and in which people could easily be exposed to retaliatory violence, witnesses often preferred not to come forward. Or they just didn't want to devote the time, as any cases going beyond the local justice courts were taken to distant superior courts (and with frontier travel conditions, ten miles could be distant). Or the witnesses might have moved on by the time the courts, usually with heavy backlogs, reached the case. Or witnesses were bribed into silence or false testimony. Or paid "witnesses" were simply procured for the occasion.

Second, since goldfield residents had shown themselves to be generally unconcerned with any crime that didn't affect them directly, the court reflected their attitude. In Bodie, the justice of the peace "routinely discharged defendants in shooting cases if the shooting occurred during a fight or a threat of a fight and little or no harm was done," with only one case of a wounding through "reckless bravado" going to a higher court.

More disquieting to some than leniency was the arbitrary nature of many judges' rulings. Men gained legal appointment without possessing "a burdensome knowledge of law" and proceeded to "employ [their] own version of horse sense" in the courtroom, jettisoning statutes when it suited them. A Sonoran justice of the peace was challenged on a mule theft decision by a lawyer citing recorded statutes, leading the justice to assert, "I didn't care a damn for his book law . . . I was the law myself." A Colorado mining camp justice responded as piquantly to a lawyer's attempted use of law books in the courtroom: "The court don't go a cent on them air kivered books."

This blunt, no-nonsense approach echoed that of the miners' courts, and if the result actually served justice, the community found it preferable to the machinations of the slick frontier lawyer who continued to be perceived as "making a farce of criminal proceedings and thwarting justice."

Another factor reinforcing suspicion of the legal system was the problem of graft. Again, the citizens of the gold frontier contributed to the situation by "tolerat[ing] considerable corruption." Still, one California gold-seeker was shocked upon gaining an appointment as a San Francisco policeman in 1850 to be told by his captain "that by good management I could make my expenses independent of my regular pay as a policeman." He found the suggestion "not justifiable by honest men," but concluded that "perhaps I ought not to be surprised at anything in California."

Conducive to graft was the fact that gold frontier courts almost invariably "operated on speculation," with local justices of the peace drawing their own pay and that of the officers of the court from the collection of fines. Many justices naturally took advantage of this unregulated situation, raising fines as well as accepting bribes.

In an even better position to benefit from graft than the

justices were the recorders of mining claims, at first appointed by the miners themselves, and the government officials who followed in handling the recording and transfer of claims. Although prospectors tended to take advantage of their early arrival at a rush site by filing claims in the names of distant friends, real and imaginary, the early recorders apparently maintained high standards of honesty. That was in part, perhaps, out of compulsion, as they would unprotected face the unbridled rage of a large segment of the mining community if they did not. Government authorities, however, had more avenues for secret chicanery, if they so chose, and disgruntled miners had their suspicions; even in tightly controlled Dawson, it was commonly rumored "that 'side-door tips' on unrecorded claims could be bought by bribe."

The most blatant example of using one's position for illegal gain had to be that of Arthur H. Noyes, first federal judge of the Nome judicial district. Sent to bring order to the many conflicting claims, Noyes instead "set about stealing the richest mines of the region" by appropriating creek claims for his Alaskan Gold Mining Company within a day of his arrival, physically removing several of the legal owners with the help of "a band of hired bullies." As judge, he would hear no protests on the part of the miners and "even commanded the federal troops under his authority to fend off any attempts the miners might make to take back their claims."

Noyes's action was so high-handed and the cries of the Nome community so loud that he was quickly returned to the States, tried, and convicted; although, as with so many privileged violators of the public trust, he received a light sentence.

On a smaller scale, the corruption tended to run unchecked. Thus a Nome deputy sheriff reportedly "made his living by robbing the drunks he arrested." Honest law enforcers were hamstrung by public indifference or outright resistance, by lack of funds, by the challenges of keeping close tabs on a large, free-floating population. But many of the sheriffs, deputy sheriffs, and policemen showed characteristics of or maintained connections with the very element they were charged to control. "There was often little difference between Bodie's badmen and Bodie's lawmen," writes one historian, in a summation that applies to much of the gold frontier. A Bodie newspaper at the height of the town's boom editorialized that "the police force here is made up almost exclusively of hard cases," three of whom were eventually fired for being in collusion with urban robbers. In Nome, it was rumored that "all the sheriffs and deputy sheriffs had done time in the penitentiary."

The proliferation of such "hard cases" in the police ranks stemmed in part from a community's desire to employ someone who understood the criminal milieu, in part from lack of more honorable takers, and in part from the machinations of corrupt power-brokers to intimidate and control a town.

For with government structures so weak and with so little community cohesiveness, criminals could not only operate virtually unimpeded but could make the gold towns their power bases. Often, they were the only organized group around, as evidenced by the rise of the marauding "Hounds" in San Francisco and by Henry Plummer's "able leadership" of the "Roughs" in Bannack, where "the respectable citizens far outnumbered the desperadoes, but having come from all corners of the earth, they were unacquainted and did not know whom to trust."

Criminal attempts to intimidate and control ranged from the random power play to a pattern of intimidation to long-term consolidation of economic, social, and political power.

In the first category, in one random and particularly chilling incident in California, a band of thirty-five to forty "Sydney Ducks," Australian bullies both feared and despised by the general population of the town of Rough and Ready, accused a young storekeeper from Missouri of stealing a gold nugget that had been placed in one of his business drawers for safekeeping. The Ducks dragged the storekeeper to a hill and whipped him brutally, despite his protestations of innocence and offers to pay the nugget's value. No one dared interfere until a doctor stepped forward with a gun and threatened to "kill the man who attempts to strike him again." The victim died that night; citizens belatedly tried to apprehend the perpetrators, but they had hidden or fled, and the best the community could do was to pronounce banishment.

One community member surmised that the doctor's ultimatum had worked only because the victimizers were tiring of their fun. Sometimes, however, an individual dramatically altered the situation. Two desperadoes rode into Placerville, California, in August 1852 waving revolvers and calling to the residents, "Hunt your holes! Hunt your holes!" While people scattered, a lone miner refused to yield, shooting one of the riders and stimulating the second to disappear quickly. "I'm powerful sorry I had to do it," the miner exclaimed, "but I won't be shot nor run over if I can help it."

Other lone defenders were not so lucky. Dance hall operator Pat Ford, in Washington (later Idaho) Territory's Orofino confronted some of Henry Plummer's highwaymen shooting up the town and was killed in a street gun battle with them. A prospector in the area complained of the "most shameful exhibition of cowardice" on the part of the town residents, "mostly traders and those of the more timid class."

Even when townspeople and miners drove the criminal groups out, they faced a recurring nightmare. In Nevada City, California, in 1851, the citizens expelled a group that had become "so numerous and desperate . . . that they greatly interfered with legitimate business" and made robbery and murder a constant threat. But the malefactors lingered in the area, vowing to burn the town; it burned on March 11, apparently by action of "the banished thugs, and in fulfillment of their threat." Often, citizens would rout loosely organized "roughs," only to have some of them filter back and augment their numbers with new arrivals.

In the next stage of criminal activity, some scoundrels established themselves in long-term positions within a town or region, not building a large enough power base to control it, but enjoying one large enough to ensure impunity from any significant prosecution. Henry Plummer's gang first harassed the Idaho mining camps and towns, then moved to Bannack, where he became sheriff and his hundred or more minions terrorized citizens for sixteen months. Deadwood con man "Ten Die" Brown would "laugh in the face of a man whom he had victimized and invite him to get redress if he could." He "swaggered" through the streets and gambling houses, apparently unconcerned with any penalty the local justice of the peace might impose. Although prospectors celebrated the fact that on the gold rush frontier men could be trusted to be "exactly what they appeared to be," here was the dark side; the scoundrel Brown felt free to make "no pretense whatever of being any better than he was."

The smoother manipulators, however, did give the appearance of being better. These were the ones who consolidated control, becoming, in effect, godfathers. Plummer came close in his key position as Bannack sheriff, but the role of urban godfather required a suave Machiavellian subtlety such as that of dapper gambler William Harrison, who wielded tremendous power in the early days of Denver. He had the criminal element to do his bidding when necessary, in exchange offering them his protection. Because he could get a candidate elected or defeated, "could drive undesirable competition out of town," and could control "the rowdy element," his favor was also eagerly sought by the better class of citizens, thus heightening his control.

Soapy Smith in Skagway pushed to the limits this kind of power. An elegant figure with "flowing black moustaches," he ran the Skagway underworld while distancing himself from it and dominating the business and social structure of the town. He indirectly controlled the local bank, newspaper office, and law enforcement operations, made public gifts to the needy (including a few of the stampeders his gang had hoodwinked), and did favors for prominent citizens.

But for every person who enjoyed Soapy's largesse, there were perhaps thirty or forty who suffered from his gang's criminal activities and the corrupt hold he had on the town. Here, as elsewhere across the urban gold rush frontier, people perceived themselves as without effective legal protection of life and property, locked into a situation in which confusion and injustice predominated because government was weak or corrupt.

In response, vigilantism flourished in the gold regions.*

Again, California led the way with its two vigilante movements San Francisco, one in 1851 and one in 1856.

---

* I am distinguishing the term "vigilantism" from miners' court actions in two ways: Vigilantism surfaced not in the absence or near-absence of government structures but in reaction to their perceived ineffectiveness and unfairness, and it tended to be less democratic, with various groups distinguished by their social, economic, and political interests replacing "people's courts" in appropriating power to themselves.

The 1851 Committee of Vigilance was formed to rein in a group of volunteer "law enforcers" called the "Hounds" or "Regulators" who had "degenerated into a 'plunderbund.'" The five hundred members of the vigilante group, led by the city's businessmen, began patrolling to prevent theft and arson, deported a number of people, and hanged four of the Hounds. The committee worked in direct opposition to duly appointed lawmen trying to stop lynch law. When two of the four Hounds were hanged, a last-minute rescue attempt engineered by the governor and the San Francisco sheriff seemed to deliver them from the noose and grant them the safety of jail, but a squad of vigilantes determined to finish the hanging immediately "swooped down upon the county jail, whisked off the prisoners, and seventeen minutes later helped dispatch them amidst another scene of frenzied excitement."

In 1856, vigilante activity—again, directed by community business leaders—was sparked by two sensational shootings: gambler Charles Cora's fatal wounding of U.S. Marshal W. H. Richardson after Richardson tried to have Cora's lover, a bordello madam, barred from attending a theatrical performance with the gentry; and city supervisor James P. Casey's shooting of newspaper editor James King over allegations about Casey in the paper. The vigilantes this time removed both Cora and Casey from jail and hanged them, hanged two other murderers, and banished twenty-six men.

Again, the vigilantes—at one point marching eight thousand strong—forcefully flouted the structures of state government, ignoring the governor's proclamation announcing their county to be "in a state of insurrection" and exerting such strong community control that the state militia sent to quell them found itself "outnumbered fifty to one." Many militiamen switched allegiance to the vigilantes, who peaceably relinquished control after a three-month reign.

These San Francisco movements echoed through the gold regions at the time they were mounted and reverberated all the way through the rush decades to Alaska. The San Francisco leaders had used their power carefully and had disbanded in a timely fashion, but they had also contributed to a climate of extralegal violence and jettisoned such basic democratic rights as the right to a public trial and legal counsel. Further, with their short-term solutions, the vigilantes relieved the general population of responsibility for good government.

The 1856 San Francisco vigilante action also pointed up another problem—class and occupational bias. The organization did banish politicians "accused of ballot fraud" but did nothing about "corrupt businessmen" involved in massive real estate swindles. When vigilantes became active in Montana Territory, "they kept their hands off any murderer who came from the 'respectable' class," allowing the legal system to work in these defendants' favor.

As vigilante groups formed across the gold frontier, three other significant problems developed. First, vigilante organizations became a means of attacking competing economic or political groups, as had the Sacramento merchants' "law and order" party trying to fend off the lot claims of the "Settlers' Association." Second, whether challenged or not, some groups concentrated power in very few, very secretive hands. In Denver, vigilante power gradually passed into the control of a small, mysterious, and bloody organization called the "Stranglers," whose summary methods came in for harsh criticism before a less secretive committee took over.

Third, vigilante organizations could quickly deteriorate into havens for some of the very people they were organized to control. In late 1852, a group of "rowdies" dubbed the "Moguls" caroused through the streets of one California mining camp night after night, "howling, shouting, breaking into houses, taking wearied miners out of their beds and throwing them into the river, and in short, 'murdering sleep,' in the most remorseless manner." The "poor, exhausted miners" complained but did nothing. "You will wonder that the Committee of Vigilance does not interfere," wrote one resident. "It is said that some of that very Committee are the ringleaders among the 'Moguls.' "

In a similar situation, in Montana, after the vigilantes acted decisively to halt the Plummer gang's activities, a vigilance organization continued, the membership clinging to power yet changing so that one old-timer complained, "There is not a thief come to the country but what 'rings' himself into the present committee."

Few vigilance groups persisted; most flared in the crisis-response pattern of the miners' court, and again and again, they demonstrated their perception that "mob violence and drastic action [were] necessary" because "in those days it seemed an utter futility to await the legal process and uncertainty of the law." Thus, one California argonaut commented on a mob action in taking an accused murderer from the sheriff and hanging him, "This was a rapid and harsh administration of justice, but it was the unanimous decree of more than one thousand persons."

Front-line law enforcement officers often bowed to public opinion simply by looking the other way. In Bodie's three documented incidents of vigilantism, "noninterference on the part of lawmen" was a common feature. How direct was the vigilante's challenge to regular government can be gauged by their hanging of a Montana murderer who had received a pardon from the acting governor. The pardon was still in his pocket, and a note was fastened to his coat: "If our Acting Governor does this again, we'll hang him too."

Only the weight of public opinion could buttress such deeds. A minister in Virginia City in 1864 marveled at the vigilantes' "dreadful government," so secretive, brutal, and controlling, but cited a "most perfect acquiescence" and his own gratefulness for "the existence of such a Committee."

Even though vigilante justice relieved most of the citizenry of both fear of crime and civic duties, it thrust violent retribution into everyday lives in a way that courts did not. A gold town resident might stroll down a street or venture into a nearby gulch and find blue-faced men hanging from trees, windlasses, roof beams, corral bars. Walking home from school in Virginia City in January 1864, young Mollie Sheehan came upon a vigilante hanging in an unfinished building.

The air was charged with excitement. I looked. The horror of what that look photographed on my memory still sends a shiver through me. The bodies of five men with ropes around their necks hung limp from a roof beam. I trembled so that I could scarcely run toward home.

Mollie had recognized two of the men: Jack Gallagher, who had journeyed to Montana with her family, and "Club-foot

George," who had been kind to her. "I did not know," she wrote, "that he and Jack were 'Bad men.'"

Mollie soon saw another man hanged, fled from a "Street ruckus," and finally saw the infamous Jack Slade hanged when his captors used a corral close to the school. "A man in a black hat standing beside Slade made an abrupt, vigorous movement," Sheehan wrote, "I turned and sought the refuge of home."

Residents of other gold regions would attest to a quite different reality. William Goulder, who toured gold camps in Oregon and California and settled in the early gold camps of Idaho, asserted after nearly fifty years in Idaho, "I have been neither a prisoner nor a recluse, and in all this time I have never yet seen a man killed, nor have I seen the corpse of a murdered man."

Recent studies based on newspapers, extant court and public records, contemporary accounts, and gold region memoirs support Goulder's contention by concluding that most residents found the gold frontier peaceful and orderly most of the time. Some modern studies even judge the agencies of the law generally effective from their initial presence in boom areas, while others show some pioneer memoirs that portray a violent environment to be drawn more from shared myth than from reality. Still, in the early rush days—and sometimes long after—residents of gold regions were compelled by circumstances to participate actively and extralegally in attempts to create a secure and democratic environment. Often, they met the challenge. Even the Chileans in gold rush California, frequently the target of Anglo-American racism, "were very much impressed by the democratic ideals, the constitutional principles, and the political sophistication" of the dominant group.

Thus, many stampeders evinced the pride of the Colorado gold camp lawyer who asserted that "we have just reason to congratulate ourselves that we . . . furnished the best practical illustration of the average American citizen for self-government." The stampeders' pride was reinforced by the fact that their mining law in particular "found ample recognition from territorial and state legislatures, from Congress, and from the Supreme Court of the United States," while miners' court decisions on other issues were made legal and binding by territorial legislatures.

But whatever the protections for life and property the gold-seekers provided, some segments of the population through no choice of their own continued to experience greater danger and uncertainty and often found themselves excluded from the democratic process and from democratic safeguards in general. Foreigners such as the California Chileans and certain American minorities faced not only the challenges shared by all stampeders but special challenges imposed by their Euro-American counterparts.

Chapter Ten

Natives and Strangers

The streets are thronged with robust miners, hailing from all climes and countries.

—*Newspaper correspondent, Bear Gulch, Montana, 1868*

No Chinamen need apply, for they don't allow them here.

—*Leadville miner, 1879*

Each stampede of any size produced a vivid medley of nationalities and ethnic cultures. At the same time, Americans of northwestern European extraction predominated in the gold regions in sheer numbers and power, just as they did in the broader American culture.

Sometimes, men of different nations and ethnic origins formed alliances in working a claim or starting a business, but groups with common origins usually self-segregated for mutual support. Anglo-American argonauts did this as readily as their minority counterparts, clustering with others from their own home town or region. This inclination to "[seek] the contact and friendship of others of similar origin" manifested itself even in the names of saloons and boarding houses—Joplin Headquarters, Texas House, Boston House.

Some did not have the luxury of choice in banding together, as they were both openly and subtly segregated by the larger population. Anglo-Americans might grumble about competition but readily accepted and absorbed into their gold rush fraternity most northern and western Europeans. At the same time, they increasingly treated American minorities and foreigners from less-favored countries with disdain and tried to exploit them or eliminate them as competitors for goldfield resources and opportunities. Most maligned were the Chinese and the native Americans, with African Americans and Latinos also struggling against the limited and diminishing opportunities allowed by the dominant culture. Yet minorities played influential roles in the finding and mining of gold, making their presence felt, and frontier gold rush culture was not without an appreciation of diversity and a sense of democratic fair play.

Again, California established the precedent. As soon as word of Marshall's strike started spreading, Mexicans and Chileans poured in by the thousands. They were soon joined by large contingents of German, English, Irish, and French gold-seekers—four thousand of the last by April 1851—as well as "Kanakas" (Hawaiians), Australians, Italians, Portuguese, and Chinese, "a boiling mass of people of all nations."

The Chinese migration started slowly. Only fifty-four resided in the territory in February 1849, but 1851 marked the beginning of major, long-term Chinese immigration through San Francisco. In 1852, as many as twenty thousand undertook the three-month, seven-thousand-mile voyage. Meanwhile, other foreign immigrants continued to arrive in smaller but still significant numbers. One prospecting Swede recorded meeting "some 260 fellow Swedes."

Such migrations resulted in the foreign-born comprising half or more of some camp populations. Of course, among that number were those who had already chosen to seek their fortunes in America before the gold rush. Thousands of Irish natives had fled to America during the devastating potato famine in their country in the mid-1840s. German immigration, too, had been heavy, and many Cornishmen and Cornish families had escaped "disastrous depression" in Cornwall by migrating first to Illinois and Wisconsin, where the men worked in the lead mines.

Subsequent stampedes boasted an equally cosmopolitan array of national origins. Among the Fraser River argonauts were "Poles, Italians, Scots, Mexicans, Welshmen, Californians, Yankees, Metis, French-Canadians, and Englishmen." In Virginia City, the foreign-born population included Irish, Germans, Russians, Poles, Swedes, Norwegians, Danes, French, French-Canadians, Scots, Welsh, and the reviled yet determined Chinese. In the Klondike, the number of argonauts from Australia earned mention, as did Scandinavians, and a community of "several thousand" on Dominion Creek was christened "Paris" because so many of its citizens had French roots. Like their predecessors, many of these foreign stampeders were not only responding to the lure of gold but were escaping political and economic instability in their homelands.

Although measuring a gold rush population was like trying to hold water in a sieve, the estimates of some California camps as half or more foreign-born held true for many later gold rush sites. In the decades following the first big rush, emigrants from the countries of northern and western Europe came to predominate numerically among the ethnic groups, in part reflecting the heavy migration of the Irish, Germans, and English to the United States, in part reflecting the general acceptance and ease of movement of those groups within the dominant culture. A Rocky Mountain census showed nonnatives primarily from England, Germany, and Ireland. In Bodie's foreign population, the Irish led in numbers. Even after American immigration patterns had shifted, with southern Europeans predominating after 1890, gold rush populations reflected the northwestern European immigration.

Many foreigners were no doubt discouraged by unfamiliarity with cold climates, with the distances of gold sites from international sea routes, and with the cost of travel, but growing American nativism of the late nineteenth century also played a part in dictating who would appear in the gold regions and how much of a chance they had to glean the gold and to find social acceptance. It is no accident that a Montana mining camp newspaper of 1868 proclaimed the camp's diversity by citing only the mingling of Europeans and Americans and of Americans from different regions. It did mention southern Europeans, but in some gold towns both they and the Chinese were unofficially "taboo."

This is not to say that gold centers lost their multiethnic flavor—far from it. With few United States restrictions on immigration until 1882, when the Chinese migration was severely curtailed by law, foreign miners were likely to be present in numbers whether a mining town was developing or fading. In those that became quartz mining centers, both southern and northern European immigrants with hard-rock mining experience—especially Cornishmen—made up a large part of the industrial labor force: "Wherever there is a hole in the earth, you will find a Cornishman at the bottom of it." As stampeders began to move away from what they considered poor and exhausted placer diggings, less-favored foreigners, especially the Chinese, moved in to rework the earth.

For two racial minorities, the rushes had come to them, whether they wished to be involved or not. The Spanish and Mexican residents of California in 1848, called "Californios," and the native Americans and native Canadians living across the gold rush frontier would find their lives irrevocably altered. The story of natives and strangers on the gold frontier rightly begins with them.

By the time of James Marshall's discovery in the Coloma Valley, more than a century of limited Spanish and Mexican settlement in California had yielded a culture of vast ranches, relaxed living, and pastoral abundance. All of that would have eventually changed after the ceding of California to the United States in 1848, but the mid-century gold rush swept away this pastoral existence within a few years.

Stampeders brought a burgeoning population, challenges to the huge land grants held by the Californios, a hurry-up attitude in direct contrast to the slow-paced nature of Californio life, and a conviction that Californios, lacking Yankee enterprise and Anglo blood, were relatively primitive people upon whom the charmed light of manifest destiny did not fall. Though the lives of the ranchers resembled in ease and abundance the lives many stampeders hoped to lead when they accumulated their millions, argonauts found them far too lackadaisical, their "quiet homes and drowsy hamlets" a direct contrast to "the promptness, struggle, and rush at the mines."

Initially, the Californios profited by providing beef to feed the miners and by selling them horses and mules or by digging for gold themselves. Some used the Indian peonage system employed in their ranch operations, paying for native mining labor with food, clothing, gold dust, "geegaws," or some combination of the four. But Californios quickly found themselves pushed aside by Anglo enterprise. Even in employing natives, the real masters were the Anglo-American and European adventurers who had established themselves as California rancheros. One American who married a Miwok Indian leader's daughter had an estimated six hundred Indians digging for him in 1848 and 1849 in present-day Calaveras County. Another American ranchero realized eighty thousand dollars from the gold placering of sixty-five Indians working on the Trinity River. On the Feather River, a third was said to collect one hundred dollars' worth of gold daily from each of his Indian diggers.

A military official visiting the gold region in summer 1848 made a typically brutal nineteenth-century distinction in observing "a great many people and Indians" on one stream, and he figured that half of the estimated four thousand miners were Indian. Most of those were working for the rancheros, and some hired out to stampeders coming in with money as well as grand dreams.

An American's claim that the natives in his work force were "well treated and well fed," supplied with copious amounts of fresh beef and the opportunity to laze in the sun for most of the day, sounds a bit too much like the southern plantation owner's picture of the fine living standards provided his "darkies." Forty-niners noted that "the Indians on the ranchos . . . are considered as stock and are sold with it as cattle, and the purchaser has the right to work them on the rancho, or take them into the mines." They were often paid a pittance, if at all, and some were violently restrained from attempts to leave employers.

A number nonetheless took up their woven willow baskets and worked independent of the newcomers, whole families washing the streams. Fathers scooped up "the invisible mud and sand of the river bed," mothers carried it to the riverbank and performed "skillfull gyratory manipulations," and even the children often showed more skill than the white miners. The Indians worked unencumbered by rocker or long tom, occasionally employing pick and spade but usually digging in exposed bedrock with hunting knives and washing in pans.

Yet "there were Americans who thought nothing of driving the natives from any productive area." And even when Indians were gathering the dust unchallenged, "whites seemed almost to compete with each other to see who could make (or at least relate) the most lopsided deals" with them. Men claimed to have traded glass beads weight-for-weight for Indian-dug gold dust, and storekeepers adopted inflated "Indian prices" for goods. A merchant from Illinois was observed instructing a native to pay for a box of raisins by placing it on a balance scale and pouring out an equivalent weight of gold dust. The man did. Merchants justified such chicanery by arguing that Indians had "no more bissness with money than a mule or a wolf," having "no religion, and tharfore no consciences."

Exclusionary attitudes and even violence toward Indian diggers started with the influx of settlers from Oregon in 1848 and 1849. The Oregonians naturally resented the advantage of the rancheros with large Indian work forces at their command, but they also arrived with a pioneer heritage of Indian hating. Most Oregonians had emigrated from the western border states, with a particularly large contingent from Missouri. Quite used to seeing Indians as treacherous enemies, they mounted a "war of extermination" against the California tribes, one in which no native was safe. One observer noted that the Oregonians "had rather shoot an Indian than a deer any time."

As some natives resisted, an all-too-familiar pattern in Indian-Anglo relations surfaced in California: escalating violence taking the form of retaliation against any member of the other race. In addition, as the gold frontier spread, a marginal class of toughs blatantly exploited and attacked Indians. Near Denver in 1860, two "Bummers," or urban loafers, raided a Cheyenne and Arapaho camp, terrorizing the women and children while the men were away. In Idaho in 1861, a "class of degraded and lawless white men" who were "neither miners nor prospectors" supplied Indians with liquor, cheated them at gambling, and abducted and raped Indian women at will. In Montana in 1863, two "rowdies" in an argument with Indians shot to death two adults and two children. As other whites tried to calm the understandably nervous remaining natives, the murderers returned with a friend and fired again into the lodge, killing an Anglo acting as interpreter.

Anglo responses to these crimes say a lot about the relative powerlessness of Indians. In the first instance, Denver citizens sent former mountain man Jim Beckwourth to the camp to appease the residents and promise "full restitution for all stolen property and for any damages claimed," but the action seems to have been made less out of a sense of justice than out of a desire to keep the Indians pacified. In the Idaho case, statutes against selling liquor to Indians and other crimes existed, but there was no one to enforce them, and the prospectors regretfully but firmly kept their distance. In the third case, a posse apprehended the murderers, but in a decision disliked but accepted by the miners, the culprits were only banished "on the ground that it was not murder to kill Indians, and the white men [*sic*] were killed accidently."

If Indians retaliated against white chicanery and violence, they stimulated old fears of "savages" in even the most judicious of whites. If they did nothing, they remained a target for every unscrupulous con man, rapist, and murderer. Voices were raised in their defense—"He who has shot an Indian has shot a man," argued one San Francisco newspaper—and sometimes they received justice, as had the Yaqui Indians caught burning the corpses in California. But they could never be assured of consistently enjoying the most basic of legal rights.

Nor could they act effectively to turn a situation to their advantage or to assert their own claims. In Alaska, Eskimos and other natives "were barred from working gold claims . . . even on their own tribal lands." Other initiatives were discouraged as well, although the Yuma with their fording service on the Colorado and the Tlingit packers on the Chilkoot Pass enjoyed some success, and natives sold food and clothing to stampeders or to middlemen, soon " 'catching on' to trading schemes" and "asking exorbitant prices for everything." Despite these inroads, as independent entrepreneurs, natives entered only marginally and temporarily into a gold region's economic life, and gold-seekers often ignored their attempts to impose tolls or to collect for services.

In addition, the newcomers persisted in ignoring or misreading any Indian resistance to white appropriation of resources. When Paiute near booming Bodie tried to intimidate woodchoppers into sparing trees, a Bodie newspaper attributed their action to "chronic jealousy." Instead, the Paiute "depended for their very survival" on the trees' fruit, the pine nut, as a dietary staple.

Indians on the gold frontier were also excluded from the legal process, unable to vote or to testify against Anglos in court. Disputes between Indians were virtually ignored; all of Deadwood took notice the first time a native was hauled into Anglo court for killing another Indian.

The natives' lack of rights was ironic in that many natives had initially proven instrumental in gold discoveries and helped argonauts survive. In fact, one of James Marshall's Indian workers might have discovered the first gold in Sutter's millrace. Yokut Indians in 1848 made the discoveries that would lead to development of California's southern mines, and natives continued to locate and share information on gold sites, as well as provide other services. In Idaho, the Nez Percés "remained friendly and helpful" long after Anglo prospectors had made a travesty of the tribe's land rights, the natives repeatedly providing shelter, directions, and food to the invaders. An Alaska stampeder found Eskimos "always ready to help when they see us in trouble."

Many tribes grew hostile as the era progressed. Longtime western hunter Mountain Bill Rhodes had in the late 1840s camped among the Sioux, Blackfeet, and Crow, enjoying "the best they had" and retiring at night "with just as little fear as I would in St. Louis or any big city—and a d—— sight less." By 1868, however, he concluded "these same Indians, if I got near them, would scalp me." Rhodes was dead the next year—killed by Indians. And matters would soon grow worse with the invasion of the Black Hills; the Sioux raided every trail around Deadwood during its first year.

Even when Indian-white relations did not turn violent, natives lost in the end, for they lost the patterns and significances of their ways of life, with every aspect of their cultures eradicated or changed forever.

Disease and alcohol ravaged tribes. As a consequence of the viruses carried by "Nomads," one stampeder found natives along the Alaska coastline "dying like poisoned rats" and predicted there would soon be "hardly any left." Nearly fifty years earlier, an argonaut had saved the residents of a California Indian village from a similar fate by vaccinating them against smallpox, only to watch many descend into alcoholism. He concluded gloomily that "it seemed as though it would have been far better had they all perished with smallpox."

Other gold-seekers observed with some dismay the general corruption of native values and increasing dependence on white culture. An Alaska prospector complained that the "confiding, generous, helpful, simple-hearted" Eskimo learned treachery from the whites and concluded that gold-hunters had "already done more harm in a few days than the missionaries can make up for in years." A journalist in Dawson watched a group of adult natives cluster around a toy display in the Alaska Commercial Company store and ruminated, "Scarcely six years ago these same fierce Ayan Indians [probably of the Tutchone tribe] had met me near Pelly River attired in mooseskin garments. Now they were sadly degenerated, wearing store clothes and an Americanized aspect."

The comment has validity in that natives were increasingly compelled to adopt an unnatural, and often bastardized, form of Anglo culture. At the same time, the observation partakes of the nineteenth-century romanticism surrounding the "noble savage." The two most successful gold-hunting natives were Tagish Charley and Skookum Jim Mason, who shared with George Carmack in the discovery of Klondike gold and with indisputable claims actually grew rich from it. Photo portraits of these two in their suits and stiff collars, watch chains looped across their chests, dispel the notion of the pathetic pseudo-Anglo; the men face the camera with assurance and stylized grace.

But perhaps this was an illusion built on plenty of money. Either way, Charley and Mason were even more of an anomaly than others in the exclusive ranks of lucky prospectors. Throughout the gold rush era, those natives who did want to prospect were systematically excluded from working claims of any value through social pressure, outright coercion, and laws denying them the right to file a claim. (Even Charley and Mason were compelled to yield the privilege of the extra "discovery" claim to their companion Carmack.) And if natives wished to have nothing to do with this business of digging gold from the ground, the influx of stampeders still turned their lives upside down, hastening the decline of the tribes and their eventual dependence on the federal government.

At first, it seemed as if the presence of gold would have the opposite effect on another American minority, the African Americans. Black sailors were among the crew members deserting New England whalers in San Francisco Bay early in the California rush. Southern stampeders brought slaves with them—sometimes with a promise of manumission—and free blacks joined the stampede, spurred by black abolitionist leaders who saw in the California rush an opportunity to capture a piece of the American dream and gain respect in American society.* By 1852, "2,000 black men and women lived in California," most of them male and most of them miners. Some of the slaves were hired out as cooks or launderers, thus supporting both their masters and themselves.

Schemes involving the use of large numbers of slaves in mining were devised but generally fell through, in part because of the resentment stampeders directed against anyone who monopolized a placer mining area through his ability to command a work force. The American emphasis on democracy and individualism was quickly translated in the California diggings into an active hostility toward anyone using what was perceived as an unfair advantage. Furthermore, stampeders disliked the association of digging with the institution of slavery. This attitude was not necessarily born of abolitionist fervor. Rather, "free white diggers" would not "degrade their calling by associating it with slave-labor." Besides, for a slave owner, the logistics of moving, providing for, and controlling a large group of slaves were for-

---

* Most traveled in racially mixed companies, although one all-black mining company sailed from New York in November 1849.

midable. Thus, slaves usually arrived "with their masters in groups of three at the most."

Many immigrants anticipated that California would validate slave labor by entering the Union as a slave state. Instead, it entered as a free state, but its laws favored slavery, allowing owners to "retain slave property if they were 'in transit' through a free state" and holding in its 1852 Fugitive Slave Law that any slave brought into the state before its admission to the Union remained a slave. Further, widespread disregard for laws allowed continued slaveholding in many locations, whatever the slave's date of arrival. At the same time, lack of enforcement made it conceivable for a slave simply to strike out independently without bringing the law down upon him.

Masters still had brute force and legal power on their side. In one instance, antislavery miners convinced a slave "to tell his master that he was a free man in California and ask for a grubstake so that he might go on his own as a miner." The master responded by proclaiming that he would whip the slave for this insurrection and daring the other miners to try to stop him. No one did.

Some slaves were able to work for their freedom but remained in thrall to owners who retroactively added room and board to the price of liberty. And even if the slave won freedom, he or she might well find it difficult to retain. A prominent case was that of Carter Perkins, Robert Perkins, and Sandy Jones, who were hired out by their young Mississippi owner to John Hill with the understanding that they could receive their freedom by paying their mining earnings under Hill to Perkins. Accordingly, Hill eventually released the three as free men, and they began mining on their own, hauling freight, and possibly blacksmithing. But when the California fugitive slave bill passed, the Mississippian filed to have the men returned. They were arrested as fugitives, their $850 worth of gold dust, mules, wagons, and other property confiscated, and hauled in one of their own wagons to Sacramento, where a proslavery justice of the peace ruled against them without allowing them to testify. A long legal battle in California ensued, but the three were eventually shipped back to Mississippi, although they apparently never arrived and may have found freedom in Panama.

Another celebrated California case involved former slave Stephen Hill (no connection to John). An agent of Hill's former master appeared in the state in 1854 claiming that he was to take Hill back as a slave, even though Hill insisted his master had given him his freedom in California. Hill's "freedom papers" were missing from his cabin—stolen, his sympathizers surmised. The sympathizers harvested Hill's crops, hid his cattle from the agent, and distracted the latter in Stockton "with drink and conviviality" while Hill made a successful escape.

During the three years the California fugitive slave law remained in effect, former slaves endured terrific uncertainty, their only insurance the slim white freedom papers. Yet despite the machinations of former owners and a discriminatory legal system, hundreds of slaves purchased their freedom through labor in gold rush California and managed to hold on to it. Meanwhile, free blacks continued to make their way to California. Some had compelling personal reasons for making the journey. "I saw a colored man going to the land of gold prompted by the hope of redeeming his wife and seven children," wrote one Ohio forty-niner. "Success to him. His name is James Taylor."

Most free blacks simply wanted economic opportunity, a chance for which they had even more reason to yearn than their white counterparts. And to a certain extent, they did find a freer climate, if only a "fragile and spotty democracy" in regard to race relations. An English observer found in the diggings "more tolerance of negro blood than is usual in the states," attributing this not to a more democratic spirit but to "the exigencies of the unsettled state of society . . . and to the important fact that a nigger's dollars were as good as any other."

Whatever the reason for the easing of discrimination, blacks shared with their white peers a belief in "pluck, tenacity, and perseverance" as a means to success. And African Americans enjoyed some visible early mining successes. They predominated in the prospecting party discovering the strike that gave birth to Downieville, and they had a reputation for luck and skill. "There are but three classes of individuals that can even work these mines," wrote one southerner, "the Negro, Indian and Irish."

In 1848, a black ship deserter showed up in Monterey with $4,000 worth of gold, and another African American reportedly accumulated $100,000 in gold digging in Tuolumne County before blowing it all in typical prospector fashion in a San Francisco gambling spree. That same year, a white man arriving in San Francisco called preemptorily to a black man to carry his luggage. In response, the black produced a pouch of gold, saying scornfully, "Do you think I'll lug trunks when I can get that much in one day?"

Even long after the rush had begun, there were reports of black men doing phenomenally well. No wonder that a Missourian could still write with unbounded optimism in 1852, "This is the best place for black folks on the globe. All a man has to do is to work, and he will make money."

Unfortunately, it wasn't that simple. Like Indians, black miners enjoyed little or no safety from claim-jumping bands of whites. Sometimes Anglo miners would come to the rescue, but generally " 'decent' white men" would "deplore" the jumping of blacks' claims without providing any restraining action. For protection, blacks could work in company with sympathetic whites, but even that strategy might prove useless. A band of white southerners took by show of force claims being worked by black and white New Englanders on the Tuolumne River in 1849.

In addition, miners' law might be twisted against the African Americans. In Amador County, one cabal of white miners "tried to call a miners' meeting to organize the expulsion of black miners nearby who had well-paying claims."

Again, the government and courts, when established, remained singularly indifferent or even hostile to the minority group. California's constitutional convention almost included a provision prohibiting free blacks from settling in the state and slaveholding Anglos from freeing their charges there. Those who had come as free blacks to California had almost as much reason to worry about proving their status as did former slaves. And like Indians, blacks in 1850 were barred from testifying against whites. Thus, when a slave woman was beaten by a southern saloonman who was not her owner in Nevada City in 1850, she could not testify against her attacker; nor could the family of a black barber who saw him shot and killed by an intruding white "gentleman" in their family quarters adjoining the barber shop.

In both cases, whites stepped in to address the injustice, a miner confronting the saloon owner, a mob demanding the "gentleman" from the sheriff. But both perpetrators got off, the first legally (even after shooting and wounding the defending miner), the second by having friends spirit him from the local jail.

Blacks in gold rush California responded to threats and miscarriages of justice by forming all-black communities in the camps and towns, by establishing formal and informal mutual-aid societies, and by gravitating toward service jobs and small businesses in which they were more likely to be accepted.

Early in the rush, San Francisco, Sacramento, Marysville, and Stockton all had clearly identifiable black communities, and mining locations had names reflecting the presence of one—and usually more—African Americans: Negro Bar, Negro Hill, Negro Slide, Negro Tent, Negro Flat. Such names would signal the presence of this minority across the West.

From California's black communities stemmed a variety of self-help efforts. In early 1850, thirty-seven black men in San Francisco published their announcement of a mutual-aid society, "not just for themselves, but also for 'new comers.'" With the Perkins-Jones case in 1852, Sacramento African Americans instituted a defense fund, began speaking out against poor laws prejudicially applied, and began building a political coalition with San Francisco blacks, thus inaugurating "organized black activism in California." Fraternal organizations, too, helped maintain pride and solidarity; San Francisco had a black Masonic lodge in 1854.

Meanwhile, blacks carved niches for themselves as boardinghouse operators, restaurateurs, stewards, launderers, and particularly as cooks, the last "in tremendous demand in California by 1849," receiving pay of $125 a month and being perceived as "the most independent of men." Women cooks shared in this popularity; a boardinghouse operator named "Aunt Maria," a former slave in Sonora, was revered as "an excellent cook . . . always well paid for her services, especially at weddings and banquets."

These choices were limiting, of course, and one wonders upon reading of a black miner who took in the white miners' washing at night whether he wanted to or felt compelled to meet cultural expectations by "doing for the white folks." But the service industry, such as it was, tended to be much more valued and profitable than the same activity in the East. And many blacks catapulted into the ranks of the small or mid-level business owner in this way. In 1854, one black group proudly reported that San Francisco African Americans "owned two joint-stock companies with a combined capital of $16,000," as well as a number of stores, shops, and "public houses," even an eight-hundred-volume library and a brass band.

Despite these achievements, a proposed bill in California in 1857 would have excluded black immigrants from the state and regulated those already there, restrictions "suggestive of some of [those] placed upon Jews in Nazi Germany." Blacks expected the bill to become law. After all, Oregon had succeeded in prohibiting free blacks from holding real estate, bringing suit, and entering into a contract within its borders.

The significance of these exclusionary tactics in terms of the gold rushes is that although California's bill failed to pass, it nonetheless spurred blacks to join the Fraser River rush, with an estimated eight hundred African Americans in Vancouver in 1858. British Columbia officials encouraged many to stay in Victoria and take service jobs vacated or created by the rush, but some would push on all the way to the Cariboo as miners in the early 1860s.

As long as there seemed to be plenty of gold to go around, many other argonauts wished African Americans well. At Missouri Flat in the Colorado diggings in 1860, one Anglo prospector condescendingly noted "a trio of Afric[a]'s sunny children digging for 'golden sand' " and politely acknowledging that they were doing all right—"making grub, sah." The three Georgians "had had some ' 'sperience' in mining, but 'day did not like de country, and 'spected for to go back sum time in de fall.' " The Anglo prospector concluded his observations by "hoping that they might make a nice thing before 'dey leff.' "

Beyond such patronizing good wishes, black prospectors were still likely to have to fight to keep a good claim; in French Gulch in the Colorado diggings, a group of blacks formed a self-protective working company, but whites still drove them from their profitable claims, forcing them onto "Nigger Hill," which fortunately turned out to be a sweet-yielding location anyway.

African Americans never knew when such perseverance would be used against them. An Anglo party traveling the Yukon came across two Anglos who claimed that a black man had stolen their boat and provisions, but the party later learned the two had given up "and turned their outfit over" to the more determined stampeder.

Throughout the era, scattered blacks managed to make good on the promise of western gold. Former North Carolina slave John Frazier hit a rich lode on Brown Mountain in Colorado, dubbed it "Black Prince," and found himself riding in the carriages of rich capitalists come to court him. Former slave Clara Brown settled in Gregory Gulch, where she used her earnings cooking and washing to invest successfully in Colorado mining property, eventually developing a statewide reputation as "a successful mine owner and philanthropist." In the Klondike, a black couple, Charles and Lucille Hunter, after following the terrible Stikine Trail and enduring childbirth on the trail, had their struggles rewarded with success on Yukon-area claims.

But despite such stories, black participation in the stampedes appears to have waned. And just as blacks could not expect to enjoy any respect for their rights as Americans, Latinos could not expect any respect for the mining expertise many of them possessed—this despite their substantial contributions to American gold mining. Chileans and Mexicans had often worked in or observed the silver and gold mines of their countries. The Chileans, or "Chilenos," brought with them long, arced knives "well adapted to picking deposits from the cracks in rocks" and other tools. They knew how to dry-wash for gold and how to build stone-wheeled "Chile mills" for "grind[ing] resistant materials." Mexicans introduced the arrastra for crushing quartz and provided key terms in the mining vocabulary.

But Anglos treated the language barrier as an unnecessary and exasperating failing on the part of the Latinos, addressing them with a form of broken English "murder[ed]" with a few garbled Spanish words, then opining "that it is nothing but their 'ugly dispositions' which makes the Spaniards pretend not to understand them." In addition, Anglos lumped all Latinos together for denigration, whatever their national origins or class, complaining "They ain't kinder like *our* folks." Chileans in particular felt singled out for persecution because they refused to be intimidated.

As early as summer 1849, Anglos were actively seeking to force Latin American miners from the California diggings and passing resolutions for their expulsion. But Latinos refused to give up the gold quest, banding together with French argonauts in reaction to one of the first acts of the new California legislature in 1850, the Foreign Miner's Tax. Imposing a twenty-dollar-a-month license fee, it was aimed at Latinos but required of all foreign miners.

The precariousness of the Latinos' position is evidenced by their being required to obtain permits from and to turn weapons over to Anglo committees, and by a call that went out publicly around Sonora in spring 1850 for "volunteers to form a company for robbing returning Mexican miners of their gold dust." Fortunately this plan met with general public disapproval and only one volunteer, but animosities continued. When an Anglo prospector newly arrived at Mokelumne Hill in 1852 walked into a store and responded to a Spanish greeting in English, he was attacked with a knife, only afterward learning that, as a result of claim disputes, Latino and Anglo sections of town had become enemy territory.

Latinos repeatedly found Anglo hegemony too difficult to overcome. At American Flat in 1852, a group of Chileans and Mexicans in good faith purchased a claim from some charlatans while the Anglo owners were away. After the original claimants returned and found communication impossible, they collected other Anglo prospectors in the area and marched on the camp in such force that, although the Latinos initially "took up their guns and . . . formed a line of defense," they were compelled to surrender.

When the Latinos were able to communicate through an interpreter, they were encouraged to look among the men present for the ones who had duped them. They did so with no success, but described them well enough for the other miners to recognize them as two transient gamblers. Apparently, some effort was made to find the perpetrators, but a miners' edict requiring the newcomers' immediate expulsion stood; the Chileans and Mexicans "lost their money and were driven like criminals from the community." One prospector pronounced himself "decidedly ashamed" of his participation in the incident, stressing "the perfect innocence" of the Latino group to his acquaintances. Yet the injustice remained.

One sensational punitive incident rooted in racism is that of "Juanita," or "Josepha," a Mexican woman in her mid-twenties who was lynched in summer 1851 in Downieville. Latino women had steadily trickled into California, both as prostitutes and as members of the Latino stampeders' families, for Mexican men in particular were accustomed to having their wives with them in the mining areas. The Mexican, Peruvian, and Chilean women had gravitated toward the predominantly Latino Sonora and the southern mines in general, but Josepha was living in the strongly Anglo-American far northern mining area with a Mexican man named Jose, who may or may not have been her husband. Either way, she did not have a loose reputation.

The incident started when a miner named Fred Cannon apparently broke down the door of the cabin Josepha shared with Jose—either in a drunken Fourth of July frolic or in an attempt to make sexual advances to her. Cannon then refused to pay for the broken door and argued with both Jose and Josepha outside the home. When he called her a whore, Josepha challenged him to come inside and repeat such aspersions. As he entered, by her own admission, she stabbed him.

The popular Cannon's wound was fatal. Despite a near-veneration of white middle-class women in the California camps, Josepha's lowly class and Mexican ethnicity negated any sympathies based on gender. A miners' court hastily found her guilty and sentenced her to hanging. Josepha went composedly to her death, adjusting the rope herself, and announcing, "I would do the same again if I was so provoked."

Miners' actions such as Josepha's hanging—which excited critical comment even among Anglos in California—were meant to root out not only crime but any hint of minority insurrection. Sometimes, however, they had the opposite effect, as in the case of "a very gentlemanly young Spaniard" sentenced by a vigilance committee to be whipped. He requested death rather than "the vilest convict's punishment" of flogging, then, rebuffed, vowed to "murder every American that he should chance to meet alone," leaving at least one observer with the conviction that he would do just that.

Such incidents encouraged social banditry, illegal acts carried out in retaliation for obvious injustice. The legend of Joaquin Murieta, the fabled "Robin Hood of the West," grew out of the California rush. According to some accounts, as a youth in 1850 Murieta had been forced from a good claim on the Stanislaus River and relocated to the northern mines, only to be ejected there as well. There are also stories that Americans raped his mistress before his eyes and hanged his half brother. Whatever the truth of the matter, Murieta symbolized Latino frustration.

With the fading of the California rush, South American stampeders receded from the American gold frontier. Mexicans never participated elsewhere as fully as they had in the California rush, but a steady number of Mexicans—and Mexican Americans, usually lumped indiscriminately with them—continued to join the stampedes. Recent studies have indicated that once legal systems developed in the gold areas, Mexicans and other minorities could expect to enjoy about the same rights and protections as their Anglo counterparts. Such a conclusion, while valid in terms of greater personal safety and equitable sentences, ignores the fact that Mexicans and other minorities had simply "learned their place" in the West, and it wasn't all that different from the place they were to occupy throughout America.

If a Mexican or Mexican American wanted to dig for gold after California's heyday, his trails generally circled through New Mexico and Arizona and possibly into southwestern Colorado, through a region where the placers yielded little, and only by patient effort—that is, a working area not coveted by impatient Anglo prospectors. More culturally disposed to communal effort than their Anglo counterparts, Mexicans worked together informally with simple improvised tools and "gained control" of many southwestern placer areas in the 1860s. Mexicans were said to "dominate" Arizona's placer districts in 1871, and in 1881 were still highly visible in southwestern placer mining.

However, in the truly rich placer and quartz areas, Americans felt "Mexicans, as foreigners, had no right to American mineral wealth." (And even if a Mexican American could prove citizenship, that only thrust him or her into another disfavored

minority status.) Further, although Anglo-Americans didn't necessarily wish to work for someone else in quartz mining operations, outside the Southwest they didn't care to see Mexicans
taking the jobs. A whites-first policy applied, whether the white
was an Anglo-American needing a grubstake or a "Cousin Jack"
fresh from Cornwall.

All of these restrictions applied to the ubiquitous Chinese argonaut as well—in fact, applied with special intensity to this "most
exotic, most transient, least free, most brutalized, and most clannish nationality."

After Chinese began pouring into California in the early
1850s, most of them under a Chinese-controlled credit-ticket
system "which paid their passage in turn for a stated term of
labor," they quickly fanned out, "penetrat[ing] to practically every mining district" in California by 1854. Chinatowns sprang up
in even the moderate-size mining camps. By 1857, an Oregon
newspaper was commenting with irritated alarm, "The Chinamen are about to take the country. There are from one thousand
to twelve hundred in this county [Josephine] engaged in mining.
They are buying out the American miners, paying big prices for
their claims."

Large numbers of Chinese went early to the northern Idaho
diggings, surging with the prospecting flow into central Idaho. By
1870, they "held the greater portions of the creek and gulch
claims of the Boise Basin." They moved into the Montana diggings, "outstaying" other placer miners, and even ventured into
New Mexico and Arizona, where they met with Mexican competition. They tackled the Klondike, too. "Your true 'John Chinaman' is an adaptable Celestial," wrote one Yukon contemporary.
"He often made good in the North."

The Chinese were represented in smaller numbers in Colorado, deterred by limited placering locations and the early hegemony of the "overwhelmingly white" citizenry. Their penetration
of the Black Hills, too, remained relatively limited, but even so,
Deadwood's Chinatown was rumored to have "the largest Chinese
population ever achieved by any town of the size of Deadwood,
outside of China."

The Chinatowns throbbed with mysterious life and color, the cabins smelling of incense and displaying "mysterious Chinese inscriptions" tacked over their doorways, the "shabby little shops" resplendent with exotic woods, silks, and ivory. There were "joss" houses (Chinese temples), and the opium dens that served the functions of a saloon for the Chinese.

By 1870, in some western areas, the Chinese "accounted for one-half to almost two-thirds of all miners." These men "were often skilled miners, trained in the California tradition of the old placers" and able to "[make] money where anyone else would starve." Working in relatively large companies, sometimes fifteen or twenty together, provided both a measure of mutual safety and an adequate work force for locations requiring tedious gold removal procedures. Like the Mexicans and Mexican Americans, the Chinese had a strong communal approach to work and relied on simple tools, piling their dirt into rickety wheelbarrows and employing the rocker "long after their white counterparts had abandoned it." By keeping their operations simple, they could move on quickly when forced to by poor yields or other miners' hostility.

Anglo-Americans and even other foreign and minority groups saw the Chinese as the "yellow peril"—impossibly alien, cunningly crooked, morally deviant, and even "ungodly"—this in a frontier society that often prided itself on its ungodliness. Everything about the Chinese seemed outlandish, either comical or sinister, and westerners refused to distinguish one Oriental from another—every Chinaman was "John," every Chinese woman "Mary." The women almost invariably arrived as indentured prostitutes and found themselves pawns of both the larger western culture and their own countrymen. Unlike the men, they often had little hope of ending their indenture, instead being kept in virtual slavery by the practice of adding time to the indenture with working days lost to menstrual periods.

The men might enjoy some power within the insular Chinese community itself, but could seldom if ever aspire to equality within the larger culture. Any attempt to "fit in" was "resented even as much as . . . Oriental ways"; a Chinaman offering to "buy the house a drink" in standard western fashion met with cold disdain, and one celebrating in the wild and woolly fashion of his Anglo counterparts was likely to feel the full weight of whatever law existed.

Two factors that contributed to the disdain and suspicion displayed by other gold region residents were the Chinese reputation for clannishness and the sense that they posed an economic threat. In condemning the Chinese on the first point, Anglo-Americans and other westerners failed both to recognize their own tendency to congregate with others from their home of origin and to see the part their prejudices played in reinforcing the insular nature of the Chinese, with informal or formal town government mandates often dictating that they live only in the local "Chinatown."

The Chinese seemed to pose an economic threat in three ways: in their efficient gleaning of the gold from American soil, in their availability as a cheap labor force, and in their saving and spending patterns. Other inhabitants of the West tried to meet the first two perceived threats by informally and formally barring Chinese from the mines, from any job Anglos might want, and even from those they didn't. Foreign miners' tax bills were introduced in western legislatures with the Chinese as their primary targets, just as in California the tax had been aimed at the Latinos. The sponsor of one such bill in Idaho Territory frankly stated, "The contemplated law was intended entirely for the benefit of the natives of the Flowery Kingdom."

As the placer opportunities diminished, such exclusionary tactics would become even more pronounced. The 1872 Montana territorial assembly actually voted in a retroactive provision that no person from China could hold any mining claims in the territory, or any real estate at all. And a Montana newspaper editorialized, "We don't mind hearing of a Chinaman being killed now and then, but it has been coming too thick of late. . . . Don't kill them unless they deserve it, but when they do—why kill 'em lots."

As for hiring out to others, Oriental immigrants faced a roadblock here as well since many westerners were convinced "the Chinese were going to take away jobs from American miners." A widow operating a California gold camp boarding house found her boarders so adamantly opposed to her hiring of a Chinese cook that only her threat to move convinced them to accept him.

If a cook's job could stimulate such feeling, the idea of employing Chinese labor in placer or quartz mining proved volcanic. Debate over Chinese mining labor was so fierce that it sometimes led to bloodshed between opposing whites and outright hiring bans. About the best employment many Chinese could hope for was temporary nonmining labor, best represented by the Chinese construction crews on the transcontinental railroad in the late 1860s but also in smaller, equally arduous undertakings. For example, although Chinese were unofficially banned from California's Meadow Lake during its boom, a group of forty were employed to haul in the five-hundred-pound casting of the hand press for the town's first newspaper. They "lugged it on long poles to a point somewhere near the town, then exhaustedly collapsed."

Even if the Orientals proved no competition, they were still resented for their perceived lack of commitment to the economic development of the region; one Bodie newspaper proclaimed them "a drain upon the interests of the country, as their accumulations are removed from our shores." If the Chinese spent, westerners believed, they did so within the tight-knit Chinatowns, with much of the money filtering back to the homeland.

Actually, it could be argued that the Chinese had a greater commitment to the economic development of the gold regions in that they continued working placers, and thus maintained the lifeblood of camps and towns long after other stampeders had departed. And while they did not spend money with the liberality of the Mexican in California, they did put more gold dust into the economy than other westerners were willing to concede.

Still, anti-Chinese riots and smaller violent confrontations erupted at various times in "dozens of towns," among them San Francisco and Denver. In the mining town of Rico in the San Juan Mountains of Colorado in 1882, forty to sixty masked intruders entered two Chinese dwellings at midnight and dragged the occupants from their beds, kidnaping and beating them, throwing some of the victims into an icy creek, and ransacking the two homes.

The law proved a shaky refuge. A Chinese immigrant, like blacks and Indians, could not testify against a white man in a frontier court. Further, the courts often treated disputes between Chinese with blatant unconcern, as evidenced by a Montana assault-and-battery trial in which a number of Orientals testified. The judge "with complacency, singled out the ugliest pig-tail, imposed a one dollar fine and costs and dismissed the outfit"; a local policeman then "drove the band downstairs by order of the court."

Yet some gold-seekers and frontier courts demonstrated concern that Chinese receive justice. In the Idaho diggings, a miners' court found a Chinese prospector guilty of rigging a stick of stove wood with powder in order to injure white miners. They gave him twenty-five lashes and felt their action justified, but word spread that "the white miners at Pierce City had brutally attempted to murder an unoffending Chinaman." The court members found that when the bulk of the mining population, those who had departed in the fall, returned to resume mining in the spring, the returnees "could hardly look at those . . . who had been actors in the rumored outrage."

Formal courts, too, often refused to champion white over Oriental automatically. In gold rush Deadwood, where the federal judge served as "unofficial counsel for the Chinese" to ensure fair trials, an Anglo gambler wounded by a Chinese native in a fight over a woman took his opponent to court; the gambler "had the face of an angel but a record in the badlands that acquitted the Chinaman." In gold rush Bodie, where the best attorneys routinely represented Chinese as well as white clients, a white man beat an elderly Oriental, and the justice court jury accepted the word of the victim's Chinese friends over that of the Anglo's friends and found the man guilty of battery. Further, a study of law and the Chinese in Montana has revealed that the courts of the gold rush era took seriously their protection of Chinese rights as well as those of other westerners. However, as the gold boom waned and social strictures slipped into place, they bowed to public opinion and "became anti-Chinese, too," reflecting the trend in the larger culture that intensified racism and restrictions against this ethnic group in the 1880s and 1890s.

Ultimately, anti-Chinese racism rubbing against a democratic frontier sense of fair play produced a strangely warped version of tolerance and sympathy for Chinese on the mining frontier.

They were welcomed in some camps, but always within a strict set of parameters. When the Chinese men were pulled from their beds and beaten in the incident at Rico in 1882, the town council dispatched a wagon to collect the victims, and the local paper editorialized with spirit that the violence was "one of the most shameful affairs that ever disgraced any so-called civilized country," called for the "severest punishment known to the law," and even insisted ringingly that if the attackers "were half as good citizens as these Chinamen have been, Rico would be a better town to live in." Yet the writer obviously considered the Chinese an inferior service class, and he pontificated about the right way to "clear a Western town of its Chinese" when "necessary"—"Give them a reasonable time to settle up their business and leave." In boomtown Bodie, laws were evenly administered and whites responded with approval to a Chinese who fired at a dog sicked on him by an Anglo bully, yet whites also voted 1,144 to 2 against allowing further Chinese immigration, and they flew flags and celebrated around a Main Street bonfire when the Chinese Exclusion Act was passed.

The Chinese naturally responded to such attitudes by becoming even more clannish and narrowly focused, proving masters at "attend[ing] strictly to business—their *own* business." They formed mutual aid societies, and they imposed their own structure onto their frontier society with "tongs," mafialike attempts at community power and self-government that often resulted in Chinese fighting Chinese.

For most other minorities, some degree of assimilation seemed necessary, desirable, and attainable, at least while in the goldfields. Some Europeans changed their names and tried to blend into the Yankee majority; others clung to home country associations and ways, and still others—probably most—made no secret of their origins while entering fully into the dominant culture. Gold town fire companies and guard units are a good measure of this latter group's success; both usually carried some prestige, composed as they were of the more stable, community-minded element. European immigrants often appeared upon their rolls; in Bodie's National Guard company, "nearly every member . . . carried an Irish name."

The French, like the Chinese, were accused of clannishness; in gold rush San Francisco they were perceived as excluding themselves from Anglo society by exhibiting "chauvinism . . . at least equal to American." And the San Francisco vigilance committees of 1851 and 1856, although ethnically mixed, possessed "a strong anti-Irish bent."

But stampeders of European extraction usually found some of their cultural traits valued by the Americans. The French were lauded for their industry and enterprise, the European Jews for their industry, thrift, and business skills, the Germans for their industry, "soberness," and "honesty."

Of course, some of the very same characteristics were demonstrated by American minorities, by Latinos and the Chinese, but Americans usually refused to see or acknowledge the parallels. A sober and industrious Chinese prospector was simply trying to undercut American enterprise. Any values not shared by the American culture were equally suspect; the Mexicans' more communal approach to mining was seen as alien and unfair by individualistic Americans.

Another way in which European stampeders enjoyed greater autonomy was in the strength of their associations. Knowing the dominant culture's modes of operation and being generally accepted by it, they established not only mutual aid societies but everything from fraternal organizations to political interest groups, from hospitals to mainstream business networks. They were also able to provide their members better protection. When a man shot down an unoffending German in Denver in 1860 and bragged of killing a "Dutchman," local German-American immigrants passed a set of resolutions, raised money for the man's capture, and sent armed members of their Turner's Lodge to supplement the forces guarding him, taking the same kind of concerted formal and informal actions their Anglo-American counterparts would take.

Those with European origins also made their own contributions to the economic development of the gold frontier. In Helena, Montana, Jewish merchants built informal cash-flow networks with their counterparts in other urban areas, gaining "access to cash and credit beyond Helena's entrepreneurial community," thereby serving as "a stabilizing element in a vulnerable marketplace." The Cornish, who contributed greatly to quartz mining, even pioneered a contract mining system in which prospectors "assessed the possibilities of a lode, contracted with the miners to get the ore out, and earned a proportion of the value of the metal." But the influx of a large number usually signaled a "settling down" phase—in fact, the "sober, family-oriented 'Cousin Jacks'" represented the mining bedrock of the post-stampede industrial town.

While a boom was still on, foreign miners of all nationalities might at any time encounter the attitude that they could "till the soil and make roads, or do any other work that may suit them; but the gold mines were preserved by nature for Americans only." Yet apparent abundance and a fluid society weakened the vociferous Americans-only attitude in favor of European minorities. In the laissez-faire atmosphere of the booming gold camps, Norwegians and Belgians, Nova Scotians and Scotsmen were usually welcomed to try their luck along with everyone else and to accumulate what precious dust they could. Irish immigrant Thomas Francis Walsh sparked hope in many a gold-seeker's breast with his reworking of a "played-out silver mine" to produce more than $7 million in gold in the 1890s.

Because a common language and cultural heritage played a big part in such acceptance, it is not surprising that two men of Scotch origin were among the biggest Klondike winners: "King of the Klondike" Big Alex McDonald, whose coffers for a time seemed bottomless, and former Scottish coal miner William Scouse, who took "the first bucket of rich pay dirt from the richest creek in the world" and maintained "one of the most famous claims in the Yukon," Number Fifteen on Eldorado. Yet Birch Creek, Alaska, gold discoverers Sergi Cherosky and Pitka Pavaloff realized little from their find; as a relative of Cherosky explained, "Not knowing anything about mining and how to stake claims, they lost all their claims to the white men."

One might say that the most successful foreigners were Americans themselves, those who became bonanza kings along with McDonald and Scouse in the Canadian Yukon. Only one other ethnic group became highly visible winners, and they did so—to Americans' consternation—as a group.

As mentioned earlier, the Anvil Creek claims of Scandinavians in gold rush Alaska—many of them reindeer herders from the ill-starred Dawson relief expedition—had some Americans looking for ways to divest them of their holdings. Not only did the Americans argue with the Scandinavians' method of staking claims, but they unsuccessfully contended that the Lapps in the group "were simply European Indians" to whom the restriction against natives working claims applied, and they "lumped Norwegians and Finns with the Lapps to allow them to confiscate everything that the former herders had staked."

Such challenges encouraged a continued ethnic unity that kept a variety of intriguing ethnic traditions and styles flourishing. Even the despised Chinese could provoke a sense of pleased wonder in other argonauts. One California stampeder responded with obvious fascination to the sight of Chinese prospectors traveling near Bidwell's Bar in 1852 "with their mining tools such as cradles, picks, pans and shovels, all tied to the middle of a pole and a man at each end of the pole," all with "broad brimmed hats, short breeches, wooden shoes, pipe in the mouth, and long hair plaited and lying down on their backs," providing "quite a show for a green Californian."

The Chinese and other ethnic groups also brought color and foreign flair to the physical surroundings in the gold camps and towns. The exotic Chinatowns with their odd jumble of shops and opium dens and temples, the modest Mexican enclaves with their bursts of bold color and open-air markets, even the sturdy, quaint cottages reminiscent of homes the Cornish miners had left behind in Cornwall, signaled a generally pleasing aesthetic and cultural variety. The strains of a German music concert or the vitality of a French theatrical production entertained all.

Another manifestation of diversity, the elaborate ethnic celebrations, drew stampeders of every race to see or join in the fun. At a celebration of Chilean independence on the Feather River in 1851, the Chileans and the Anglos with whom they often came into friction found a temporary harmony in the liquor bottle, leading Louise Clappe to comment wryly, "Though the Chilenos reeled with a better grace, the Americans did it more *naturally!*"

Across the gold frontier, the Irish staged parades to mark St. Patrick's Day, the French to commemorate the French Revolution. The Germans celebrated Mai Fest, the Mexicans Hidalgo Day in September and Cinco de Mayo in May. The Chinese welcomed the Chinese New Year in February, and at a rousing Fourth of July festivity in British-held Dawson, the crowd alternated between singing "The Star Spangled Banner" and "God Save the Queen."

Besides those public events, private celebrations maintained the fragile yet prized ties to the homeland. Cornish immigrants gathered at Christmas for their traditional holiday meal of roast goose and roast pork, and Swede Carl Allvar Kullgren bent much effort toward a traditional "Swedish Christmas Eve" celebration for nineteen of his countrymen on his second Christmas in California.

Foreign argonauts like Kullgren acutely felt their distance from their native land. The hardworking Kullgren's cry after an arduous and unsuccessful mining experience carries a special poignancy: "Oh, I must have been a madman for ever having left you, my homeland!" But the difference between Kullgren's distance from home and that of a New Englander in California or a midwesterner in the Klondike is moot when one is so completely removed from valued associations and facing repeated frustration and disappointment. Most of the argonauts, whatever their origins, felt like strangers in a strange land, writing letters to and eagerly awaiting letters from home. And the folks back home waited and wondered and wrote and pleaded for the argonauts to do the one thing that many would not, could not do: come back.

Chapter Eleven

Home Ties

I am a stranger in a strange land, with the bonds of friendship, the endearments of the home of youth and the fond ties of kindred all exerting their influences upon me, and like the pole to the needle, they attract all my thoughts and preferences back to the land of my home and family.

—*California argonaut William Swain to wife Sabrina in New York State, February 17, 1850*

Any image of stampeders as antisocial loners eager to escape home associations quickly crumbles in the face of the majority's obvious longing for word from those left behind. "I would cheerfully pay 100 doll[ar]s for a Letter from Home," wrote one Californian, while another urged his parents to "write me at least 6 pages full no matter if it is gossip all is interesting to me."

Stampeders treated the arrival of letters as they did a lucky strike and hoarded them as they did their gold dust. A California miner recalled scenes of "hoorahing, jumping, yelling and screaming" when the mail arrived, while a member of an Alaska prospecting company wrote his mother, "We all read our letters over again and again," men pausing periodically in their daily work to "pull out a letter [and] read it." A Yukon prospector at the mining outpost of Rampart, finally receiving letters from his home in Maine more than a year after leaving it, seemed "enraptured," "transfigured," even unsteady on his feet.

Although—as will become apparent—letters from home could cut and wound, they were simply "treasures richer than any the mines can yield." A Klondike argonaut declared a friend's missives "the most expensive letters I ever received but . . . worth every cent and more." A Californian confessed that "I live upon your letters, with a small sprinkling of pork and bread," while another vowed that "a letter from Home . . . will sustain my courage more than anything else on Earth."

Newspapers, too, helped the argonauts connect with events in the distant world. Men perused the Civil War news in the early 1860s and gathered to listen to and cheer the Spanish-American War dispatches conveyed by a coveted paper in the Yukon. But *any* news from the home regions was prized; one San Franciscan in the Yukon paid a dollar for a San Francisco paper "and read every line, advertisements and all," thus rendering himself "better informed of the doings of the world, for that day at least, than anyone in Frisco." With such intense examination by a whole string of eager readers, the papers often frayed into fragments.

Still, letters were what argonauts craved most. They gave reassurance of a family's welfare and their continued loving support and the longed-for details of family and community life: how a young child was developing, what was happening to a neighbor's business, who in town was getting married.

Meanwhile, the families back in Vermont and Virginia, in Minnesota and Mississippi—even in Sweden and South America—also bore the tremendous strain of waiting for word of the argonaut's physical, moral, and economic well-being and of coping at home in his prolonged absence. Both the adventurers and their kin found themselves continually frustrated in their attempts at communication because of poor mail service and the tensions that grew out of their differing experiences and differing roles.

Each new rush precipitated a demand for fast, efficient mail service from home to diggings and back, yet mail lines were tenuous, rudimentary, overloaded, and expensive. Any town or city serving as jumping-off point for the diggings became a mail center, but no one could be prepared for the avalanche of letters that resulted, and the federal postal service took some time to catch up with the massive population shifts the rushes caused. Private "express" companies jumped in to fill the void, carrying mail by boat, horseback, wagon, stagecoach line, even sled, and such urban centers as Denver and Virginia City had to rely upon their shaky and exorbitant services for many months.

United States post offices were established in San Francisco and Sacramento fairly quickly, but by October 1849, the San Francisco office was buried in "more than 45,000 letters 'besides uncounted bushels of newspapers' " and the clerks locked themselves in against "the shouting, pounding, threatening crowd." In one month alone in 1850, forty-five thousand letters flowed into San Francisco on ships and forty thousand were sent out.

Argonauts grumbled that their letters failed to reach home, but they had even more trouble receiving one. Mail often ground to a stop along the way to the gold sites. Sometimes the postal service, federal or private, bore the blame for this delay. Yukon stampeders protested mightily as their mail sat for months hundreds of miles upriver, the "mountains of mail sacks piled up under a tent at Tagish waiting for the proper officers to forward them" while "the men of Dawson almost went mad with impatience and worry."

A Virginia City newspaper advised "nervous individuals who rush frantically to the door every time a wheelbarrow passes, expecting the mail coach" that four incoming stages were bogged down in mud so thick that even foot travel was impossible. In the early days of Deadwood, the wagon trains carrying mail were repeatedly stalled by fears of Indian attack.

As a result of such conditions and delays, mail arrived in the gold centers smudged, torn, mangled, mixed this way and that— and in overwhelming piles. One wagon train might bring "tons of delayed letters" to Deadwood, and when a pony express firm replaced the wagon trains, an expressman "often carried three thousand to forty-five hundred letters" at a time. In the summer of 1900, postal employees at Nome found themselves deluged with an estimated four pounds of mail per resident.

As the clerks worked steadily, a gold town phenomenon developed: the mail line. People camped all night to obtain or keep a spot in such a line, then waited—sometimes all day—to reach the postal window and see if they had anything from home. In gold rush San Francisco, the lines sometimes spread "six abreast and several blocks long." In Sacramento in 1851, even though the mail center stayed open from 8:00 A.M. to 8:00 P.M., with "a delivery window for nearly every letter of the alphabet," at each window lines stretched "more than around the block." Those near the front of the line when the window closed simply waited out the night in place.

In Denver, "long strings" of people were gathered before the delivery windows of the private express "at all hours of the day." In Deadwood, where the mail arrived in the wagon trains from Sidney, Fort Laramie, and Fort Pierre, thousands would gather, and after an initial sorting a man would announce the names of those to whom the letters were addressed. In Dawson, when the Canadian post office opened in fall 1897, virtually every claim "within 20 miles" was represented in the line of "unkempt and shaggy men" hoping to find their names on one or more of the thirteen thousand letters already jamming the post office. In Nome, where twenty-four federal clerks worked valiantly sorting the mail "around the clock . . . in three eight hour shifts," mail lines "wound around the block."

Those lines took on a life of their own, with food vendors working the crowd and the destitute or lackadaisical making money by taking a place in the column, then selling it. Anyone approaching the postal window for news from home also had to save some money for the postal delivery charges. Even twenty-five cents was a strain on many a stampeder's dwindling purse— especially when multiplied by the number of letters he hoped were waiting.

For those in the urban areas, home delivery was a distant dream. Few gold towns managed this service in the early years, a surprising exception being Nome. There postmaster John Clum quickly employed two former Portland mail carriers who had had the foresight to bring their uniforms with them. They began making rounds "to the amazement of everyone who saw them," the sheer ordinariness of their movements in this topsy-turvy environment "stop[ping] traffic almost as quickly as a fight would have done."

Prospectors in the diggings actually had a better chance of obtaining home delivery, but for an express company price. Some chose to avoid the high charges by sending one of their number into the nearest urban center to get supplies and mail. However, travel was often difficult, they lost the man's work for the time he was away, and, considering the long and often futile waits at the post office, his absence could be extended and fruitless. In Deadwood, a rule was strictly observed: One could "inquire for mail in no more than one name" upon reaching the postal window, thus making it necessary to return to the end of the line if asking for a friend. The "unsophisticated tenderfoot" who offered to pick up his companions' mail "never made the promise more than once."

Californians in the diggings paid as much as two dollars per letter and fifty cents per paper just to get their mail delivered from San Francisco; Montana and Wyoming prospectors, a dollar each or more for letter delivery. The Montana-bound letters traveled "by way of the Isthmus of Panama to San Francisco," to Oregon Territory, and then inland hundreds of miles by express, often completing the journey to the diggings "by any reliable person coming this way."

The expressmen gained respect for the hard job they performed. Their routes led over unsettled territory, the difficulties heightened by climate and vagaries of weather. Clubfooted Idaho expressman I. B. Cowen endeared himself to the prospectors of Pierce City one winter by repeatedly tramping the eighty snow-covered miles from Lewiston carrying "between sixty and seventy-four pounds of express matter on his back, in addition to his blankets and provisions." The prospectors would sometimes impatiently "break the trail through the heavy masses of newly-fallen snow some ten miles out," meeting and accompanying Cowen to camp.

On the Kobuk River in Alaska, stampeders of 1898 gave "the Flying Dutchman" Carl Knobelsdorff almost as eager a welcome, well aware of the winter dangers to which he exposed himself as he skated over the rough river ice. Once when he had to leave part of the mail sixty miles downriver because of his load and the depth of the snow, the prospectors dispatched two of their number to retrieve it, one later deeming this "the hardest journey I ever hope to make" and judging Knobelsdorff as having "more grit" than anyone on the river.

In the urban areas, people had expected instantaneous government mail service and resented private company prices. In Denver, the cost of claiming letters from the express company was so high that "addressees would often take possession of the letter, then avoid the charge by returning it and claiming it was for someone else." Clerks tried to foil such claims by reading the beginning of a letter to the would-be recipient to assure ownership, but the caller would ask to hear more—and more, tying up the clerk for long periods and getting his news for free. Finally, "a notice was posted stating that mail would be distributed only to those who identified parties from whom they expected mail."

Starting letters on their way could prove even more expensive, with $2.50 the charge from Nevada City, California, to San Francisco. In Alaska, stampeders on the Kobuk River paid $1.00 each to send letters to Cape Blossom on the coast, where they could be picked up by revenue cutter. In 1897 in Dawson, which despite its urban status more nearly paralleled the remote diggings, it cost $1.00 an ounce to mail letters out. "This does not guarantee delivery," wrote one resident, "but only reasonable exertion to forward the letter to its destination, the sender taking all risks." Some Dawson residents paid as much as $100 per letter in an attempt to ensure delivery.

The hardest thing for many was the wait for that first letter from home. Typical was the lament of California argonaut Charles Hinman from the Feather River to his wife in Groveland, Illinois. He had "sent two and three times a week to Sacramento and San Francisco for Letters and papers," paying as much as five dollars for the service, only to be "disappointed every time." He could only conclude that the mails had been held up, "for I can hardly believe there is not one in Groveland that is willing to spend an hour in writing to me."

If others from one's hometown were not receiving mail either, a fellow could at least easily believe mail service at fault. If one stampeder from a given area began receiving letters, others could get news of home from him, but his good fortune also increased their impatience and sense of personal isolation. Sometimes, maddeningly, a man might learn from another's letter that his own family was writing regularly, but still no letters appeared.

Like the Rampart prospector, many argonauts went as long as a year on the trail and in the diggings before the first letters filtered through. After that, the mails usually remained erratic for two or three years, never stabilizing for the duration of the rush. One letter from home might take only a month or two, another remain in transit for six months or fail to arrive at all. Some mail reaching prospectors in Circle City, Alaska, was two years old.

Although the advent of the telegraph during this era seemed fortuitous, telegraph lines couldn't be strung fast enough to keep up with the adventurers. When they were, the service proved expensive—a dollar a word in Virginia City—and unreliable, with the fragile lines often down. Some gold-seekers turned to the fortune-tellers and spiritualists who appeared in the mining camps, using their services not only in quest of a strike but in quest of news from home. In gold rush Dawson, a woman who used spiritualism to predict correctly that a stampeder's wife had safely delivered a baby boy hosted a series of "spook meetings," with the men "drifting in, asking their questions and seeming to get the homesick pains out of their hearts."

The most reliable communication came in the form of travelers arriving from the old home regions. Even unfamiliar arrivals were welcomed for their experience of the world beyond the diggings; in Deadwood, the stagecoach brought "passengers who as recently as last week had seen a locomotive and a train of cars."

As new adventurers poured in, others wended their way homeward carrying messages and letters from and information on those still in the goldfields. A Stockholm newspaper in 1854 listed the addresses of eighty-four Swedes in California, courtesy of a recently returned argonaut.

As the Swede's list hints, the lack of a specific destination in the new gold region and a propensity to move with the strikes added greatly to the problem of mail delivery. Men wrote home from the trails asking families to write them at two or even three far-flung gold locations, then bounced among entirely different sites. One California gold miner acknowledged the difficulties in a letter to his wife and mother: "I find this voyaging life disagreeable, as it is here today and there tomorrow and prevents me from hearing from home regularly or writing regularly and gives me a disagreeable feeling about the folks at home." Almost fifty years later, an Alaska prospector also alluded to the problems of the "voyaging life": "Each letter you get from me has almost a different plan, but ideas change as I hear new things."

Mail puzzles were further complicated by the shifting western geography as huge unorganized lands became territories, which were then given statehood status or divided into smaller territories or states. James Fergus in Virginia City in 1861 chided his wife, Pamelia, for sometimes addressing his letters to Washington or Nebraska territories, announcing, "We are now in Idaho." But Idaho Territory had just been carved from Washington Territory and bordered Nebraska Territory; further, Virginia City would become part of Montana Territory the next year. James earned Pamelia's unapologetic, pointed response: "The reason we directed so many ways was in hopes you would get some. Maybe James Fergus did not know where he was."

Argonaut letters contained a number of common subjects and themes. Of course, the gold-seekers had to say something about their economic fortunes, and most did so by reporting themselves close to achieving mining or entrepreneurial success. Some predicted boldly; forty-niner Lucius Fairchild informed his parents in 1850, "I am just as certain of a fortune this summer as though I had it in my pocket now." (He was wrong.)

Others wrote more guardedly, one Californian after a successful day of mining fearing that this success would "not be equaled or exceeded soon again." Most found a position between Fairchild's vaulting optimism and such foreboding, reassuring themselves and their families that hope had not died. "While I believe we shall make money, I don't look for any great amount," admitted one Klondike prospector. Another, leasing three promising creekside claims, noted diffidently, "It is all chance and there may not be much in any of them." Henry Rogers attempted to console his mother with his lack of quick success in Alaska by urging, "Do not feel disappointed at my not striking anything yet, as things are looking better than I expected and everything seems to work around to our advantage."

If most were not doing nearly as well as they had hoped, they hastened to assure their families that they were doing as well as anyone else from the home area and to challenge any news to the contrary. When Jasper Hill read in a family letter that two other Mount Pleasant men had each cleared two thousand dollars over the summer of 1849, he wrote, "Well, I would no more believe that than I would believe that I am in Mt. Pleasant." Faced with the news that another Mount Pleasant company had indisputable riches to show, he managed a "Glad to hear it," accompanied by a wistful "wish we had been with them."

California argonaut Alonzo Hill repeatedly lessened his own failures by pointing to those of his Boston company members. He had "more [money] than all the Bay State Boys have got," he assured his father in March 1850. The next month he reported that he had been gathering news of other company members, and "I have done as well as many of them tho' I have done nothing." Five months later, he seemed to take some solace in reporting most company members "struggling to get to the States this fall" but hampered by lack of money.

Challenged in his pronouncements by the news from home that a traveling companion had sent home three and a half pounds of gold when he had only sent fifteen ounces, Hill pointedly remarked, "I do not hear of any [of the original company] returning with their expectations realised" and tartly noted, "I know to a cent how much Gold 'George the Third' [Duell] sent Home, it was 15 ounces—and not 3½ pounds."

Henry Rogers in Alaska was more subtle, but the message was the same. "Do not be too free in comparing notes with the girls in the other boys famil[ies] as to our success," he wrote his sister. "Let them tell their own story to their folks."

Men also reminded the folks back home that some of the success stories were built on activities their families would not condone. "When a man is reported to have returned from this section to the States with $6,000 or $10,000 honestly acquired since the first day of November last," wrote one Californian after a winter in the diggings, "you may just post such a man as the ninth wonder of the world."

As the vast majority of argonauts from California to Nome struggled with lack of economic success, they found themselves engaged in a subtle competition with relatives who had stayed behind. When a brother wrote of obtaining a good job at home, when word arrived of a cousin's promotion, argonauts' congratulations carried a wistful, pained note. Even the moneymaking activities of sisters and wives could make the gold-seekers' choices look desperate or simply unnecessary. Henry Rogers in Alaska wistfully remarked on his sister's success as representative of a textbook firm, "If I could have done as well at home it would have been nice." And some wives proved particularly adept at running absent husbands' businesses, bringing in as much as their husbands were making in the West—or more.

Again, argonauts might admit to others in the diggings and gold towns "that had they remained at home, directed their attention to their farms and saved the expense of their outfit, that they should have been much better off." But it was harder to admit that to the family and friends who had sent them off with high expectations and had often funded their expedition.

The proof lay in how much money a man could send or bring home, and most mailed or carried precious little. An occasional $100 or $200 remittance dribbling its way eastward stood in stark contrast to the quick thousands the argonaut—and often his family—had expected. By 1852, forty-niner Charles Churchill seemed to give up on realizing enough to help his family, noting to his brother concerning their widowed mother and sister, "I hope that if I ever get out of this country that I may be able to do something for them myself."

Far from being the family's economic hero, many remained dependent upon it. William Swain in California, concerned about his wife and daughter, wrote his brother George, "I hope that you will see that they are provided for" and promised to remunerate him "and ever feel grateful for your kindness." Alonzo Hill told his father, "I intend if possible to send you funds sufficient to cover all the expense I have been to you, and more if possible, but I am not making much money now."

Even if they did not have to call upon family members for funds, argonauts had to marshal every cent to pay the high cost of living and make something of their quest. His hopeless statement to the contrary, Charles Churchill kept trying to succeed in California. Engaged in a mercantile firm, he wrote his brother Mendal in 1854 wishing him success in his own enterprise, a furnace company, and wishing "that it was in my power to render you some assistance." But he could not even send the "small remittance" he had hoped to provide for their sister's millinery shop venture.

In particular, men who attempted the leap to quartz mining found themselves in a bottomless pit of expenditures. Here, they reported truthfully, every cent's worth of gold taken out of the ground went into assay tests and mining equipment and transportation costs, leading a Colorado entrepreneur to complain "the longer I stay here the poorer I get."

Sometimes stampeders simply avoided telling their families that they were saving their funds or using them on another chancy goldfield enterprise. They were breaking a miners' commandment to maintain their families "in food and raiment during thy sojourn here," yet far removed from any privations their loved ones experienced, they believed every day that the next morning might bring the financial bonanza that would validate their choice to go gold-hunting and secure the family's fortune.

As they struggled to maintain their hopes, pride made men testy. When Alonzo Hill's brother Luther wrote mentioning that he was about to collect a debt of a thousand dollars and also urging Alonzo to send money home to their father, Alonzo exploded, notifying his father, "I have wrote him by this mail that if he will get that money [the debt] and loan it to you I will send as much. If he is not willing to do so why should I. . . . write me when Luther lets you have $1,000." Colorado stampeder Robert Russell, upon learning that a well-off Minnesota neighbor had expressed concern for his family's welfare, railed angrily, "Now she need not give hir salf any bouthar about my wife or Childrens rags i belive that the[y] never cost hir eany thing."

Meanwhile, families continued to hear exaggerated tales of other successes and to read glowing fictions in the press. "I understand by some letters from Iowa that there has been a Mountain of Gold found in this country, also a Lake of Gold," one California miner wrote dryly. "I should not wonder if the next news we get from the states would be that California has turned into a solid mass of gold." Alonzo Hill wrote his father, "You say you see by the papers the prospects are good at the mines, allow me to say that much paper information is false," and noted bitterly, "I suppose you will always be of the opinion that to obtain a fortune in California was sure, and easy."

When argonauts did enjoy successful mining or a taste of good gold camp wages, that experience also widened the gulf in understanding between home and gold region. A California adventurer from Ohio responded to his wife's suggestion that he and his partners return home to work for the railroad with bold scorn and a disregard for cost-of-living comparisons: "Well, at what wages? Perhaps *one dollar*. Ha, talk to Californians about one dollar per day!"

In addition to voicing their frustration with family expectations regarding work and wealth, argonauts repeatedly expressed their disappointments with other company members. One forty-niner "was thought to be such an awful stout & good hand" but instead "proved and showed himself a mere nothing," sitting and talking about inconsequential matters while others were working. An Alaska prospector complained of a companion "all the time kicking about this, that and the other" and proving "a regular baby when things go a little hard."

Lucius Fairchild in California wanted those at home to write and convince a friend to return, as "he is doing no good here he will not work nor will he do any thing else. . . . the longer he stays here the less money he will have until he will not have enough to get home with."

Gold-seekers wrote to friends and families about who among their mutual acquaintances had headed home, moved on, or just arrived, who was "making grub" and who was not. They also conveyed messages. In California, Jasper Hill wrote his storekeeper father that a Mount Pleasant acquaintance he encountered "wished me to say to you that if his wife needed anything out of the store, that he would settle it when he came home about one year from this time," a promise Hill considered sound. In Colorado, James Fergus wrote to wife Pamelia in Little Falls, Minnesota, "Mr. [Daniel] Bosworth . . . says tell his wife for God *sake* to send him a letter" and later instructed, "Tell Mrs. Bosworth that [Daniel] is well and will write as soon as he receives a letter."

Stampeders usually responded reluctantly and cautiously to families' and friends' entreaties for advice to share with those contemplating joining a rush. Men who felt foolish for having succumbed to the gold lure were tempted to encourage others, giving truth to the maxim that "misery loves company." As one Californian put it, "I really hope that no one will be deterred from coming on account of what anybody else may say. The more fools the better, the fewer to laugh when we get back home."

Others talked of the difficulties, not only in the hunting of gold but in living conditions. "If some of our friends could see how we have to work and live here and see how we look sometimes they would not be so fast for coming out," wrote James Fergus. A Klondike gold-seeker echoed this theme: "I have so little under me that I think of my own bed at home every time I get into mine here. Don't let father talk any unsophisticated youths into coming into this country, for it's a terror." Others flatly advised all "friends and kin" to "stay where they are." A Dawson stampeder wrote, "If you see anyone who is intending to come here—just tell them don't," as "no one can figure out the difficulties from Michigan." A Californian tersely instructed his friends to "stay at home," his enemies to come; while a "Nomad" concluded: "I wouldn't advise anyone to come here if they have any way of making a living at home."

At the same time, argonauts attempted to reassure their families, declaring they did not regret their own choices. A forty-niner admitted that he would not again travel overland to California "under the same circumstances" even if he were assured of a "princely fortune." But echoing Lucius Fairchild's "It is the journey to learn human nature," he judged the trek and his five-month experience in the diggings had "learnt me to have confidence in myself . . . disciplined my impetuous disposition and . . . learnt me to think and act for myself and to look upon men and things in a true light." A Klondike prospector concluded with less assurance that "[in] this, as in many other things, one has to 'go a journey to learn,' I suppose."

Gold-seekers presented camp life as healthy and exhilarating. They wrote of fine mountain rambles, of energizing climate, of their prodigious appetites and blooming health. They described the scenery and the pleasant aspects of gold-hunting, its freedom and excitement, and the tremendous vitality of the camps and towns.

Men also reassured home folk that they had not abandoned moral principles. They acknowledged the tremendous amount of gambling and drinking taking place, usually remaining discreetly (and perhaps guiltily) silent on the presence of prostitution. And they wrote of taking mother's and father's, sibling's and sweetheart's and wife's moral precepts to heart, of avoiding the many shafts of sin. "I shall read the passages in the Bible you have pointed out with great pleasure," a California miner pledged to his brother. Another wrote to his sister, "I am sorry that Mother feels uneasy about my drinking. She has no cause, tell her, to fear of my ever being attached to it." No doubt his mother wanted to hear the same words a Rocky Mountain argonaut offered *his* mother: "In this country where every body drinks, I don't taste a drop." A fourth argonaut was probably more realistic in reassuring his father that he was trying to "leave this country as pure and innocent as I came in," but adding, "that ain't anything to brag on, is it?"

Gold rush participants also took pains to note that they were living fairly comfortably. That usually meant getting enough to eat, and many reported weight gains accompanying the appetites stimulated by their strenuous outdoor lives. Those who had no money or who were camped far from urban resources stressed—sometimes unconvincingly—their ability to survive. Although Henry Rogers maintained a chipper air in his Alaskan letters to his mother and sister, he and a partner became nearly destitute, living in a tent outside Juneau, and his wistful reference to an Indian's luck in killing a black bear surely gave his mother pause: "I wish [the successful hunter] could have been one of us, for the skin would have brought $15 in the market, and the meat would have been useful."

Another concern that argonauts felt compelled to address was that of safety, particularly from Indian violence. "You must not be frightened at the hard stories you hear. They all come from men who are made of milk and water," wrote one forty-niner. A Colorado stampeder "got a glimpse at a Chicago Paper in Which I See it Says the Miners and people in this country are anticipating a good deal of danger from the Indians," but insisted "there is nothing of it." Others played down the Indian threat by describing encounters with natives in a light, entertaining vein or by carefully selecting their audiences; one Californian omitted a close call with Indians from letters to his mother but shared the story with his brother.

Yet another point of reassurance involved friendships and partnerships. Even if some companions from home had proven ill suited to the adventure or contentious, men hastened to assure their families about the sterling qualities of the remaining company members—or they sang the praises of their newfound prospecting partners. The usually upbeat tenor of gold-seekers' comments often was genuine in that it reflected high hopes. Sometimes, of course, it was forced, contrived to convince the writer as well as his family that everything was working out. The adventurers couldn't help but draw a contrast between their own lives and those of the loved ones to whom they wrote. One forty-niner deemed his family "lucky" in "enjoying yourselves at home among your friends & acquaintances while us California boys are roaming about, from one place to another sometimes comfortably situated & at others out in the rains & storms."

Sometimes argonauts simply vented their pique at their penurious, relatively comfortless lives. After a bout of typhoid fever, Alonzo Hill grumbled to his father that he had in the California diggings denied himself what healthy food was available "on account of the exorbitant prices" and in the city eschewed "Balls, Bullfights, Theatres, Circus &c" because "money comes too hard to me." James Fergus in Colorado responded to his wife's complaint that she was lonely by inveighing to her about his circumstances: "If you was shut up in these mountains among the snow, half clad and half fed, working day and night except when asleep or eating, no acquaintances, few newspapers, and the prospects gloomy ahead, then there would be some chance of feeling lonesome and having the blues."

As the quest continued, men experienced a growing sense of isolation. A Californian complained mildly at not receiving letters from "any of the young folks" but soon had enough from young and old to "really thank you all for the kindness you have shown in writing to me as I think I have been more fortunate in getting letters than any other one from your town." Still, he admitted, he "would like to have them oftener."

By contrast, Alonzo Hill was repeatedly disappointed. He initially adopted a bantering tone, writing his sister, "Now Some I Know receive their letters regular every mail and now what's the [*sic*] why Dont I[?]" But soon he was pointedly (and poignantly) signing himself "Alonzo Still" and complaining, "It is cruel after one has struggled so long and hard to get to the [post] office, then received for an answer 'nothing Sir.' " By the time he had been two years in California, he responded to his father's plea to write with open fury: "You say it is a duty I owe *My* Family & Friends to write often. I suppose I have a Great many *relations* and some *friends* who would be glad to hear from me, but What! What!! is the *Family*." He still loved his sister, who had neglected to respond to his plea, but had "lost all desire to write to her."

Stampeders who heard regularly from home became all too aware of how things were changing in their absence. A Californian wrote his family after examining a hometown paper, his use of "your" highlighting his sense of alienation, "I see that your town has quite a number of persons there, strangers to me. I suppose things look quite different to what they did when I was there."

Single men worried about romantic chances lost. "Society will be changed when I return," wrote one. "I am afraid few familiar faces will greet me. The girls will all be married and the young ones that fill their places will grow out of my Knowledge."

Married men worried about missing their children's development, and comments from home did nothing to alleviate that concern. Nine months after William Swain had departed for California, wife Sabrina wrote of their only child, "Little Eliza has grown nearly half in this time, and she has long ago forgotten that she has or ever had a father."

Argonauts' worries were compounded by the fact that child mortality rates were still high in mid-century, older relatives might go at any time, and epidemics could winnow the population of the most settled, apparently secure areas. One California argonaut, alarmed by rumors of "some thirty to forty deaths" near his Iowa home, fretted that it would take two months for him to hear by mail whether he had lost family or friend.

From afar, fathers tried to advise and extend a protective arm, urging children to "be kind and pleasant to each other and help Mama all you can." Most of their advice went to or through mother. James Fergus fretted to Pamelia that he had forgotten to warn their son "not to go gunning when it snowed nor to go so far away from town that he could not find his way home."

Fathers inquired repeatedly after the children's welfare, with the comments in letters from home obviously proving haunting. "You mentioned in your July letter that [son] Willie had crooked legs—but have not alluded to it in any of your subsequent letters that I have received," Frank Kirkaldie wrote to wife Elizabeth from Montana. "Have they—or are they likely to become straight?"

Another worry was family economic needs. Charles Hinman, accumulating gold dust in California but with no way to send it, wrote his wife, "I fear you will need some before I get Home" and urged her to ask for funds from a neighbor with sons in Hinman's company, as the sons would have charge of Hinman's earnings if he should die before returning.

Family men also bothered about children's moral development, urging overworked wives to impart proper guidance and serve as models of rectitude. But distance and hard work precluded much direction of any kind. In Colorado, James Fergus gave up advising Pamelia on "pressing matters," as "it is so long before I get your letter and you get mine." After locating in Bannack in the Montana rush, he concluded, "I ought as a duty to send much good advice to the children but hard labor every day leaves but little time for thought or reflection." Instead, they were to rely on their mother, "their own judgment and good books."

Whatever their family situation, gold-seekers exhibited signs of longing and nostalgia. "You don't know how homesickly it is," mourned one resident of gold rush San Francisco, while another wrote of the song "The Old Folks at Home" that "it is a very sweet thing, but it is sung and whistled about our streets so much, it has become actually sickening." One Klondike company half a century later banned the singing of "Home Sweet Home" in their cabin, as the feelings it aroused were too painful.

The homesickness and sentimentality were demonstrated in part by the attitude toward children, even scarcer than women in the early stages of any rush. Bret Harte's "The Luck of Roaring Camp," the tale of a rough California mining camp's charmed response to the birth of a baby, has its roots in argonauts' near-veneration of young children. When Martha Purdy gave birth to a son in gold rush Dawson, her all-male community, both traveling companions and neighbors, "took full charge," keeping the cabin fire fed and supplying her with "fresh-baked bread, cakes, chocolates, ptarmigan, moosemeat, every wild delicacy of the country." Meanwhile, "miners, prospectors, strange uncouth men" came tromping in, eager to talk of "their own babies so far away" and to hold hers, "to see his toes, to feel his tiny fingers curl in their rough hands, to see for themselves that his back was straight and strong." Some came just to watch his daily bath; many left gold dust and gold nuggets as gifts.

The longing for children also manifested itself in the popularity of child performers. In California, miners showered the child actors known as "Fairy Stars" with gold dust, and in one night alone, the most famous of this group, the pixielike Lotta Crabtree, raked in more money than her gold-seeking father "had made in the mines in four years." In the Yukon, child star Margie Newman "sometimes stood heel-deep in nuggets after she rendered a sentimental song" and stimulated a miner to compose a heartfelt blessing.

God Bless you, Little Margie, for you make us better men.
God Bless you, Little Margie, for you take us home again.

Even when a stampeder chose long-term prospecting over family life, even when one's children became only strangely maturing faces in occasional photos and characters in a distant drama, gold-seekers felt the tug of home. A Sweetwater rush lament went, "They have not seen their daddy/For many and many a year/'Cause I couldn't make the riffle [earn enough in gold dust]/And had to stay out here."

Repeatedly, those who persevered wrote of their mission in terms of duty, to the family and to themselves. "Jane I left you and them boys for no other reason than this," wrote a California argonaut, "to come here to procure a little property by the sweat of my brow so that we could have a place of our own." Jasper Hill admitted, "My thoughts are oft-times that I could be with you and partake of some of your good meals, pies, cakes, &c., and afterwards have a merry chat. But I feel it a duty devolving upon myself to take advantage of my situation and try to make a little of a raise so that I can have a pretty fair start in the world hereafter."

Past failures still haunted some. "You say you want me at home in your business," wrote Alonzo Hill to his father. "I was there once, but did not have a chance to do business much because I did not know enough, and I am not very much improved by California. . . . I am certain I should never be worth a d——d for any thing in the Atlantic states."

Fear of rejection was strong. A California adventurer, unsuccessfully prospecting the Sierra, dreamed "that I was abandoned by my family, my friends and the whole world because I had not found a gold mine." An Alaska company wanted to go home but anticipated a "cold welcome among our town's folk, who will probably ridicule us as 'fake gold-hunters,' 'prodigal sons,' and all that."

Even when gold-seekers were assured of a sincere welcome, they backed away from returning empty-handed. Jasper Hill, encouraged by his parents to return to Mount Pleasant by obtaining funds from two hometown men succeeding in a California store, responded somewhat huffily, "I thank you for the privilege but do not think I will ever be under the necessity of calling on any of my friends for money to return with." Lucius Fairchild responded to his parents' urgings to return by reasoning, "Just think to have it said that Lush [Lucius] come home with out making any thing and those *Mutton headed B- Boys made a pile.* You would be ashamed of me and I am sure I should be ashamed of my self so I am bound to have the pile if possible."

Another argonaut reasoned similarly, but with great bitterness, in telling his wife in Ohio how much he missed her.

But I am willing to stand it all to make enough to get us a home and so I can be independent of some of the darned sonabitches that felt themselves above me because I was poor. Cuss them, I say. I understand they prophesy that I will never come back. Darn their stinking hides. If God spares my life, I will show them to be false prophets.

As one man put it in an equally bitter vein, he stayed and worked "that I might not be a dog for other people any longer." One Californian mused, "I do not know what I shall do when I go back home if I cannot buy a good farm, for I don't think I will ever again be willing to work for fifty cents a day as I used to in Ohio." Three years after leaving home, an Alaska prospector informed his family that he was now "a bull-necked, horny-handed, whiskered backwoodsman" unable to function "under the eye and hand of a boss."

Sooner or later, however, most revealed that they had run out of funds, begun to regret their rough and uncertain existence, or simply concluded that "if I ever was going to make anything I should have done it before this." About eleven months after leaving home, a Yukon gold-seeker figured up his losses and announced to his family that by the time he paid his way home he would have "less than I could have made at home playing marbles or shooting craps." He judged his Klondike experience "a bitter one, but instructive." More obliquely but tellingly, a Kotzebue Sound adventurer wrote of his formerly merry group, "Poetry doesn't come to any of us anymore. The poetry is wearing off the L[ong] B[each] & A[laska] M[ining] & T[rading] Co."

When men reached this point, most started the long journey home to families who had been suffering their own frustrations, heavy workloads, and heartaches.

Like the argonauts themselves, families waited for what seemed an eternity for communication to be established. When William Swain went to California from New York State, his family received letters he had written on the Overland Trail as far west as Fort Laramie, but then spent seven months looking for further word. Brother George expressed his frustration at one point by writing, "May you be happy. We are not. We have not heard one word from you. And we are thus left to live on anxiety and conjecture alone. What the dickens is the reason you don't write, Boy?" None of William's company had been heard from, leading George to note with a certain lack of logic, "If you are all dead, it would be no more than civil to send us word." When the void was finally breached and communication established, George was still urging William to "take a good deal of pains to be punctual in writing, for you have no idea of the anxiety we feel if we do not receive news from you as we expect."

News of epidemics in the gold regions frightened families just as similar tidings frightened the travelers concerned about the folks at home. A New York City woman, learning of cholera in California and not having heard from her husband in some time, begged him to "come directly home *if you love me. . . . I cannot* endure this suspense."

Even without epidemics, stampeders' living conditions gave families cause for worry, as did the idea that their loved ones were living among strangers in an immoral environment. The folks at home had felt some comfort in sending their loved ones off with friends from the same area, but when those groups split up—often rancorously—families felt the fallout. Relatives shared not only community ties but bonds forged in the gold-seekers' absence and reinforced through their concern and vicarious sharing in the gold enterprises. After Henry Rogers's company disintegrated, he wrote to his mother and sister in an apparent response to their queries, "Don't give that crowd a second thought, either collectively or individually . . . for they are not worth it; but do as I am trying to do, i.e. forget the whole thing." Rogers failed to realize that his relatives might not be able to isolate themselves in this way and that they had been regarding the company as his surrogate family, obliged to care for him if he became ill or impoverished and to provide the social pressure necessary to keep young men on the straight and narrow path.

Even thousands of miles away from their families, men knew their chances of getting away with unacceptable conduct were poor if other people from their home were in the area. A forty-niner rejected a lucrative opportunity to deal monte in California because he judged the word would get back to his wife "as fast as steam could carry it." In some cases, fellow stampeders did cover for the errant sheep or try to spare the family heartache; a California miner revealed to his wife in Wisconsin that another man from their community was living with a prostitute and keeping a San Francisco brothel, but he instructed her to tell the fellow's wife that he was in the grocery business.

Rumors that apparently had no basis in fact reached the folks at home. California prospector-turned-freighter John Callbreath appeared to be responding to one when he wrote his family, "You seem to have a strange idea of my standing here according to what Bob Morison told me." He assured them that if they "could only know my real circumstances and standing amongst business men and my reputation in good society it would dispel the veil of mystery that seems to exist between us." There was little "veil of mystery" in the reports that floated back to Pamelia Fergus that husband James had been "a spreeing in the mountains" of Colorado. Such conduct would have been out of character for James, but "distance and changed circumstances had been known to affect the habits of more than one wandering man."

Some of the mental tortures that wives endured are captured in Sabrina Swain's letters to William in California. She was ever aware "that while I am writing these lines to you, your body may be moldering back to its mother's dust from whence it came," and repeatedly lay awake nights envisioning William "in the deepest trouble and distress." At one point she wrote, "O! that I could tell you the feelings and anxiety that I have had for you since you left home—none but God knows or ever can."

Some wives and children were left under the protective eye of other family members. Frank Kirkaldie moved his wife Elizabeth and four children to her mother's house in Illinois when he set off for the Montana goldfields. The arrangement apparently did not suit Elizabeth. Frank wrote with somewhat formal and essentially useless sympathy, "I am very sorry if any new cause of unpleasantness has arisen in your intercourse with your mother or anybody else which renders your stay in Joliet more irksome to you than formerly."

William Swain's wife Sabrina and daughter remained on the family farm with William's mother and brother George, an arrangement of which Sabrina apparently had no reason to complain. But she sent a sly appeal to her husband in early December 1849, only days after he had finally reached El Dorado: "As to a man for the winter to do chores, I do not know what we shall do yet. We do not know of anyone that would be faithful. I think if we could get William Swain we should be very well pleased." Despite the coyness of this appeal, the need was real and immediate, as "Mother and I cannot do both the out-chores and what there is to be done in the house."

A recent, ground-breaking study of the families left behind in one small Minnesota town during the Colorado and Montana rushes, *The Gold Rush Widows of Little Falls*, by Linda Peavy and Ursula Smith, reveals the scope of women's responsibilities in their husbands' absence. A large number of married men who went gold-hunting left their wives "in the very situations their husbands sought to escape"— with faltering businesses and hard-to-manage farm operations. James Fergus's departure for Colorado left Pamelia in full charge of their four children's "education, discipline, health, and moral development," of business matters including "a house and lots, properties held in trust for others, and a failed business about which she knew virtually nothing," and of the complete range of farm chores usually divided along gender lines.

The study of the Little Falls "widows" shows that they took on these myriad responsibilities reluctantly and gained little power from them. Many did grow more decisive, confident, and independent in handling matters in their husbands' absence, but the economic struggle was fierce. "To be so cramped for a little money, as you have been, I know is extremely embarrassing, annoying and unpleasant to say no worse of it," Frank Kirkaldie wrote his wife from Montana. "And I regret very much that you should be reduced to such straits."

Families waited in vain for money, even when it had been sent by whatever means at the argonaut's disposal—government mail, express carrier, or personal acquaintance. A nearly destitute Little Falls wife wrote James Fergus wanting news of her husband; Fergus responded that the man was indeed working in Colorado and had written her and sent money twice. Fergus himself sent treasury notes and gold dust home, but Pamelia waited with growing despair, needing the funds to pay property taxes, then received only part. At one point, a letter arrived with the envelope cut open and the ninety dollars Fergus had placed inside missing.

Most wives received little or nothing from their husbands in their absence. Despite Robert Russell's angry rejoinder that a neighbor "need not give hir salf any bouthar about my wife or Childrens rags," Pamelia reported Agnes Russell and the children looking "very destitute." She told James, "Their neighbours say they do not know what she will do this winter although they have raised enough wheat for their bread only, they have seven children and she looking for the eighth."

Just as the gold rush communities were called upon to make some provision for destitute stampeders, communities at home were compelled to come to the aid of stampeders' families. In doing so, they did have reason to speculate on whether the absent men would ever resume their moral obligations at home.

Talk could be raised by the length of a stampeder's stay in the gold regions, already a point of contention between anxious wives and the husbands determined to remain until they had something to show for their venture. Pamelia Fergus responded to James's plan to extend his stay in the Colorado diggings in hopes of making a fortune with the comment, "Why you *might* make a fortune in a few years almost any where." She quickly repented, concluding, "You done what you thought for the best and it is right to me." But soon she was entreating him to come back "and make a home for us in Illinois or Iowa," using the arguments that "our time is short here at best" and that she wanted only "enough to be comfortable." James finally heeded her pleas, then indirectly "shift[ed] some of the blame for his failure" in the diggings by writing, "I could do something . . . if I was to stay here this winter, but I promised you I would come home and I will."

A few years later, Elizabeth Kirkaldie, trapped in her mother's home in Joliet, wrote in an apparent response to Frank's enthusiastic assessments of Montana that "the water, the fish, the timber, and the gold are all very fine—get as much of the latter as you can and hurry back and get us a home any where but in Joliet." Frank either diplomatically or obtusely answered, "Perhaps I am mistaken, but it seems as though I detect in the tone of your letter ... an impatient or discontented state of mind" and attempted to explain why his search was taking so long.

Another problem that plagued argonaut-family relations was the deceptively higher wages in the gold regions. When the normally supportive Mendal Churchill in Ohio learned that brother Charles was earning a hundred dollars a month in California, he could not resist pointing out that it was twice his own salary. "I should suppose on that salary you could save 7 or $800 p[e]r annum," he wrote, but admitted, "I do not know anything about the expense of living in California."

In general, families tried to be supportive. If some had sent their adventurers off directing them to bring back a fortune, most were still ready to accept them without one. Few expressed this willingness with as much simple, good-hearted eloquence as did George Swain to William.

> If you fail there, you are not to blame. You have tried your best to do well, and if you can't do it there, you are better off than many who have gone there with their all and left nothing behind to fall back on. You have something, and friends who will meet you just as cordially unsuccessful as successful—and more so, for we are sure you have suffered.

George offered to send money for William's passage home, pledging himself "bound to share with you in misfortune as well as good fortune."

Taking such an offer—or asking for money—was simply one of the most painful things an argonaut could face. When Henry Rogers wrote his mother and sister from Alaska requesting "about $125 to get me home," he admitted that "to write this letter has been harder for me to do than anything else that I have been through since I came to Alaska."

Some families, however, would have been happy to receive such a letter. For of course the rushes did contain men who would fail to maintain their family ties or fulfill their moral obligations, Timothy Smith of Little Falls being a case in point. His letters to his wife from Colorado grew infrequent—and usually entreated her to send him money. He borrowed funds from James Fergus to go home, but used them for other purposes, popping up here and there—anywhere but Little Falls, apparently. He occasionally made noises about having Mrs. Smith and the children join him, but, virtually abandoned, they were evicted from their home and forced to sell off possessions bit by bit.

A similar situation occurred with a Little Falls man who had opened a tavern in Bannack. "Mrs. [Margaretha] Ault wanted us to tell you next time we wrote to be so kind as to write to her and tell her where Mr. [John] Ault is and what he is doing," Pamelia wrote James. "She has not heard from him for three months [although] she writes regular once every two weeks."

Tavernkeeper Ault invited his wife to join him in Montana but never sent her money to do so. Her midwifery in Little Falls could not sustain her, and she moved to Ontario, complaining that she had received no support from her husband since his departure. When Ault died in Bozeman, Montana, he left only two hundred dollars' worth of property, and James Fergus provided a eulogy for those who succumbed to the hedonistic allure of the gold camps: "Poor John" had made money and could have lived comfortably with the wife whom he "thought a great deal of," but "with him it was out of sight out of mind"; he had "lived fast, been always in debt, spent his money in eating, drinking and with women."

Of course, for most families, the news of an argonaut's death far from home plunged them into shock and sorrow, heightened for some by revelations of the manner or cause of death. A Montana argonaut died in a fight over a prostitute, leaving a wife and two small boys in Minnesota. When Charles Churchill died after a few years in California, his brother Mendal learned with dismayed incredulity that drinking had hastened Charles's demise.

Families were also frustrated by the lack of particulars, by a very fuzzy sense of their loved one's life and passing. Mendal Churchill asked his informant, a stranger to him, for details of his brother's death, for a daguerreotype, and for special attention to the California grave site, but he garnered little information and had trouble learning whether his wishes were being carried out.

Seldom did any appreciable amount of money reach the family of the deceased. When a Swede named John Malmberg was killed in California, he had hidden three hundred dollars' worth of gold dust so well that searches by his friends did not uncover it for his next-of-kin, his mother in Karlshamn, Sweden. The gold-seeker's situation was judged "worse than war, for there is no pension."

Many families lacked the consolation of knowing their adventurer was dead. All through the late nineteenth century, stampeders left home and were never heard from again, or wrote a letter or two and then seemed to plummet from the face of the earth. One Oregon couple whose eldest son had gone to the California goldfields heard that he was coming back and had taken sick and died, that he had been robbed and murdered, that he had been seen still alive and roaming (and thus had abandoned his family). The mystery was never solved.

Gold town newspapers carried queries from families and friends—"Has anyone seen my son—or my husband—or my brother." Typical was the comment of a woman looking for a relative through a Skagway newspaper: "We have gotten many conflicting reports, some that he was drowned, some murdered, and others that he died of starvation and insanity, but 'the still small voice' has said to me all the time he lives in earth life."

One California wanderer who did return home many years later found himself frequently queried as to the whereabouts of a relative—"my mother's brother, who went to California in '49 and we never heard from afterwards." The inquirer "would tell his name and describe his looks, although the party giving the description was not born when the uncle left, but they had heard him described so many times by their mother or an aged grandmother."

The uncle may well have been the "Last Man of the Dead Camp," a type described by Prentice Mulford in 1871 as having "ceased years ago to write home," refusing to return poor and instead "cling[ing] to the deserted bar ... so long as the oft-turned, oft-dug-over dirt will yield a dollar or two per day." When this man died, he entered the ultimate isolation of an unknown grave in an abandoned camp while "in a far-away Eastern home mother, wife, daughter, sister, still long, and wait, and weep, vainly hoping" for his return.

But not all women sat weeping "in a far-away Eastern home." Not only did a small but significant number of wives and daughters accompany or join their stampeders, but independent entrepreneurial women participated in each major rush as entertainers, prostitutes, laundresses, journalists, even freighters—and yes, occasionally as prospectors and miners. They brought their own vitality and dreams to the gold regions, and experienced in their own ways the excitement and disillusionment of gold rush life.

Chapter Twelve

Gender and Gold

What I wanted was not shelter and safety, but liberty and opportunity.

—*Martha Munger Purdy Black, Klondike stampeder*

The scarcity of women in the new gold regions was noted by all and sundry. "You have no idea how few women we have here," a young man wrote from San Francisco in 1850. A Nevada City, California, youth "got nearer to a female" than he had in six months and professed he almost fainted at the encounter. One Klondike stampeder on Eldorado Creek termed women "scarcer than gold."

In many camps native women initially provided the only female companionship. Men readily sought those women as dancing partners, as bed partners, as experienced survivalists and suppliers of food and adaptive clothing. Yet they regarded the women patronizingly. Those events in which the miners of a region gathered to tread the boards (or dirt floors) with a handful of Indian women were termed "squaw dances," the " 'Society' girls" present characterized by names revealing of the men's attitudes—Lucy Baggage, Scarface Ellen, Thompson's Maggie.

In many camps—and even some towns—the presence of a real female immigrant at first drew crowds, with prospectors pressing gold dust and nuggets upon her in homage.

Entrepreneur Eliza Farnham must have doubly regretted the failure of her scheme to bring a vessel filled with marriageable

337

women to California in 1849 when she arrived in San Francisco, for she found that with the appearance of a woman on the street, "doorways filled instantly, and little islands in the streets were thronged with men who seemed to gather in a moment, and who remained immovable till the spectacle passed from their incredulous gaze." In Georgia Gulch, Colorado, the news that a saloon had acquired a waitress brought a crowd of miners to gawk at her "like 'boys who go to see a monkey.' " The first few women in Nevada City proved "a great attraction, and had they put themselves up on exhibition, they would have drawn great houses."

One might assume that the women causing such a stir were prostitutes quick to seize an opportunity, but a few "respectable" women, those who represented the traditional values of home and family, arrived early, too, providing more of a surprise— sometimes even to their menfolk. Johnson Barbee in Cripple Creek wrote his wife that the camp was "a godforsaken hole" not fit for her or their daughter, but wife Kitty immediately joined him anyway rather than continue living with an uncongenial relative.

Women such as Barbee came in small but significant numbers as the goldfields formed—wives and daughters accompanying or joining husbands and fathers, entrepreneurial widows, even some single women who maintained respectability by being identified with female support roles and middle-class status.

As in the larger culture, women were categorized as "good" or "loose" based on their sexual behavior and—to a certain extent—class standing. Thus, women in general throughout this era remained more constrained by social convention and social divisions than the male rush population. Nonetheless, women of different classes and occupations found their presence and skills valued and enjoyed some of their new environment's freedoms, buying these freedoms with plenty of uncertainty, hard work, and rough living.

Typically, a mining camp of hundreds, whether in California, the Yukon, or at a gold strike in between, would first boast only three or four women residents—perhaps a woman running a store or saloon with her husband, a young woman still living with her parents, a widow operating a boarding house. The number grew slowly but stayed low through the boom period. An

1850 California census, perhaps overlooking prostitutes, revealed about 8 percent of the population to be female. In Bodie during its boom, about one in ten residents was female. In the Rocky Mountain mining camps, the number climbed to one in five only after the rushes had peaked.

Among these small female rush populations were women who participated in the finding and mining of gold. Elizabeth Bays "Jennie" Wimmer and her husband were both employed in James Marshall's millrace crew, Jennie as cook and laundress, when Marshall made his famed discovery. To Jennie, it was her discovery: "I said, 'This is gold, and I will throw it into my lye kettle . . . and if it is gold, it will be gold when it comes out.' " The next morning, she found in the bottom of the soap kettle "a double handful of potash" containing "my gold as bright as could be."

Women might have had the right to a proprietary air in other gold strikes as well. An Indian woman's discovery of gold nuggets on an island in the Fraser River is said to have ignited that 1858 rush. And George Carmack possibly appropriated the glory of Klondike discovery not only from his native brothers-in-law but from wife Kate, who reportedly made the stupendous Bonanza discovery as she cleaned camp pans in the stream.

Sometimes women proved instrumental in the gold search in indirect ways. Salome Lippy reportedly convinced husband Tom to move to claim 16 on Eldorado Creek in the Klondike because the timber at the location looked better for cabin building; number 16 would produce more than $1.5 million in gold for the Lippys.

If women such as Wimmer, Carmack, and Lippy were unlikely discoverers or agents of discovery, self-avowed women prospectors seemed even more unlikely in a culture that encouraged women to devote themselves exclusively to the welfare of husband and family and promoted an ideal of refined, passive "ladyhood." Louise Clappe sounded the proper ladylike note from California in November 1851 after trying her hand at placer mining. "I have become a *mineress*," she announced, but "I can truly say . . . that 'I am sorry I learned the trade,' for I wet my feet, tore my dress, spoilt a pair of new gloves, nearly froze my fingers, got an awful headache, took cold and lost a valuable breastpin." Other women approached the business of gold-hunting with more gusto. The rush record is lightly studded with tantalizing references to women miners. As noted earlier, Mexican and Indian women worked as members of mining families. Europeans and Anglo-Americans were known to do so as well. A French couple at Sonora in 1851 was observed digging gold together, the woman dressed "exactly like her husband— red shirt and pants and hat." In another California location, a married couple and the woman's sister worked a claim, the man scooping the earth into a pail, the wife transporting the pail to the rocker, and the sister operating the rocker.

Women repeatedly showed themselves capable of gold digging in the literal sense. A Marysville, California, woman advertised for a husband in 1849 by noting that among a host of other skills she could "rock the cradle (gold-rocker, I thank you, Sir!)." A dozen or so women were among the estimated four hundred prospectors at Clear Creek, Colorado, in 1859. In Montana, women "worked side by side with their menfolk in the claim, skirts hiked up, shoveling tailings all the long day." From the Yukon and Nome rushes, photographs show women clad in Victorian long skirts pushing wheelbarrows and digging sand, even pausing at ease in trousers and work boots.

Most of those women could defend themselves against charges of unwomanly conduct by claiming the familial role of "helpmate"; digging and rocking were simply one step outward from the more wifely task of cleaning and weighing the gold a husband brought home daily.

But a limited number of females mounted independent prospecting efforts. One Mexican woman even brought peons from her native country to work a California claim. In fall 1848, a Monterey official visited the diggings and "found several women gathering gold." A Mexican woman in the Tuolumne County dry diggings was reported to have gleaned more than two thousand dollars' worth. But a San Francisco paper dismissed as implausible a report that "two intelligent and beautiful young ladies" from the South were busily mining, their "old grey headed negro" servant serving as guard and cook.

A few women did become long-term prospectors, taking on some of the eccentricities common to dedicated male gold-seekers. In Deadwood, the reclusive Henrico Livingstone jealously guarded from her "shabby little cabin" the claim she worked. In 1898, an independent woman in her mid-forties was reported to have fifty-eight claims on Alaska and Yukon creeks and was well known to prospectors from western mining camps. She hired and supervised men to work her claims, wore "tight knee pants and long stockings," and spoke and acted in "just as tough and rough" a manner as her colleagues, although "swift and graceful" in her movements.

Other women made gold-hunting a sideline, including a Californian who "daily, after finishing her housework, took her 'crevicing-spoon' out among the rocks and searched for gold." Still others found that they didn't even have to go out looking. A boardinghouse keeper in Downieville was sweeping her dirt kitchen floor when she discovered a lump of gold, then a number of particles. She and a friend dug five hundred dollars in gold from the floor in one day.

Some women, like many men, were willing to stampede a long way for gold. About thirty prostitutes made the journey from Dawson to Nome to stake claims in July 1899, "hik[ing] inland in their long dresses" but arriving too late to grasp any good locations.

It is unclear whether the prostitutes planned to work the claims themselves, to hire workers and oversee them, or simply to speculate in hopes of selling at a profit. Certainly, most of the women in the diggings and gold towns who became involved in mining did so as claim owners and speculators. Virginia City storekeeper Mary Agnes Hamilton's husband reported to relatives that Mary Agnes had "a Severe Attack of quartz on the Brain." Putting the storekeeping profits into mining claims, she had acquired "Four Hundred feet of quartz" but wanted "ten thousand."

Claire Frisbee, wife of a Colorado Springs upholsterer, was termed Cripple Creek's "first lady prospector," apparently on the basis of her mining speculation rather than claim work. Frisbee knew how to turn a profit by selling only partial interests and holding the rest.

Women with means, like men with means, grubstaked miners, assuring themselves as much as half of the proceeds—if there were any to be had. Others took claims or partial interests in claims in payment for services rendered. Everyone had to be wary in this practice. Cripple Creek brothel madam Blanche Barton was given so many eighth interests in claims by one customer that "she figured she must own most of the camp," but they turned out to be 27 eighth interests to the same claim.

Some women not only owned claims but oversaw long-term development of their mining interests. In gold rush Dawson, influential trader and mining investor John J. Healy's wife was recognized as "a capable business woman"; she handled a number of claims independently, and her attractive nineteen-year-old daughter Alfreda was beginning to direct on-site work on her own claim.

Another Dawson woman who oversaw her own claims was Martha Munger Purdy, whose husband had opted to head for the Sandwich Islands while she continued on to the Klondike and gave birth to her third child. Purdy staked on Excelsior Creek, then was persuaded by her parents to leave her gamble in charge of her brother George and join them and her older children in Kansas. She promised her parents that if the claims "did not yield at least $10,000 before the next year," she would give up all thoughts of return. Soon Purdy received the welcome news that the claims had paid off. She hurried back to Dawson, where she joined two men in a claim-working partnership and located at a nearby mining camp, doing both the unconventional—overseeing her own investments in the workplace—and the traditional—cooking for the crew of sixteen that she and the other partners hired.

Purdy smashed stereotypes in a number of ways; not only was she a woman in a man's world, but she was a mother and a woman from an upper-class family. Gender, motherhood, and social standing seemed to dictate that Purdy would make her presence felt only as a rather retiring, "civilizing" influence. That is what the gold population both hoped for and dreaded: that women would bring order, stability, refinement, comfort, and morality. The *Alta California* newspaper editorialized on this point in October 1849 by likening women to the cement necessary in building, noting, that "the society here has no such a cement; its elements float to and fro upon the excited, turbulent, hurried life of . . . gold hunters."

In fact, the late nineteenth century saw American women adopting the belief that their special skills were needed to extend the concept of a well-run home to the world beyond it. Middle-class American women became cultural missionaries. The wife of a newspaper editor in boomtown Helena decided that "there is a wide field for usefulness here, and, entering upon the work earnestly and prayerfully one need never be lonely or disheart-ened." And when Hannah Gould mounted her Women's Clondyke Expedition reminiscent of Eliza Farnham's California effort, Gould vowed that her women would provide Dawson with the cultural institutions so sorely lacking—"library, church, recreation hall, restaurant, and hospital."

Not all women went to the gold regions with such altruistic community plans. Some simply wanted to fulfill their culturally dictated role by finding a husband—preferably a rich one. This was no idle dream in light of the scarcity of women. Even Latino prostitutes, marked by both their devalued minority status and their occupation, married successful miners in early-day Califor-nia frequently enough for observers to take note.

En route to the Klondike, one gold-seeker sailed with a blue-blood mother and daughter of "impoverished fortunes" from Virginia and queried the comely daughter as to their motivations for such a perilous journey. She explained their circumstances were such that she had told her mother they could "take in washing if necessary." But she had laughed and concluded, "I don't expect to." The girl soon married a successful Klondike miner.

Many already married women simply wanted to keep their families together or to reunite them—and, like their husbands, to improve the family's economic fortunes. One woman, camp-ing through the northern California winter of 1849–50 with her husband and children in the wagon that had brought them from the States, commented that "comfortless as it was," it was "better by far than luxury with one member of [the family] missing." Pamelia Fergus could not help wistfully observing upon a visit to her mother in well-settled, pastoral Geneseo, Illinois, "If we had our house here in this country I should want to stay I have lived a mong the Indians and the frountears long enough and like improvements and good society." But she resolutely set off to join husband James in Montana, which became her long-term home.

Not all wives were so accommodating. A reporter found Wyoming prospectors asserting that "they would not on any account, permit their better half to undertake such a life as they have to lead," but, he concluded, "May it not be that they would not, because they did not happen to get the kind of better half that would consent to it!"

Having consented, some reached a point where they dug in their heels. Abigail Arcan joined the overland stampede to California with her husband and got as far as Santa Cruz, where she announced, "You can go to the mines if you want to. I have seen all the God-forsaken country I am going to see for I am going to stay right here as long as I live." She did just that.

Often, women missed the companionship of other women. When young Kate McDaniel's family reached Grass Valley, California, a female boardinghouse keeper rushed out and eagerly assisted Kate's mother from the wagon, hugging her "as though she had been a long lost sister" and declaring, "O my dear, you seem just like an angel come to me in my loneliness."

Women also mourned the loss of other comforting, familiar associations. Estelline Bennett recalled of her mother's life in Deadwood, "She would have preferred to live where she had a large elm-shaded yard for her children to play in and long stretches of tree-shaded streets and country roads along which she could drive with her horse and buggy, her children piled in around her."

By contrast, the gold regions seemed impossibly isolated, raw, and harsh. Malinda Jenkins was a practical, adventurous working-class woman, but she still judged Dawson "no place for a woman," maintaining that "hell and Dawson was one and the same." Laura Berton, a schoolteacher who arrived after the rush had deflated and married one of the remaining stampeders, concluded that "the men all loved the country, but I do not think many of the women really did." The whole environment was too alien, too threatening, and Berton aptly summed up the social conditioning many women had to overcome.

> I had been brought up by parents steeped in the Victorian Tradition and early in life they had perhaps unconsciously inculcated in me the deadly fear of two bogies: first, a strange Man who might do dreadful things to me, and second, The Woods, where dreadful things might happen. Now here I was surrounded on all sides by vast quantities of both.

Yet even wives who felt such tensions most keenly responded to gold frontier challenges with spirit and relish. As Louise Clappe wrote on settling in at Indian Bar, "This strange, odd life fascinates me." Behind such acceptance and enthusiasm was both a traditional adherence to the marriage vows—Laura Berton titled her memoir *I Married the Klondike*—and a mounting sense of independence from some of the old, constricting modes of living.

This sense of freedom and anticipation was felt even more strongly by the women who arrived independently. Martha Munger Purdy, chafing during her self-exile in Kansas, dreamed with longing of the natural wonders of the "vast new rugged country." As her mining experience shows, however, Purdy was no mere seeker after sublime vistas. The women who joined the stampedes, whether as wives, as husband hunters, or as entrepreneurs, did so with many of the same hopes that men had. Even so, the very presence and activity of middle-class women often did signal a shift to "civilized" ways, including an emphasis on material comfort.

Observers on the gold frontier repeatedly noted the markedly better nature of the dwellings inhabited by women. At the stage stops on the route to Deadwood, "wherever there was a woman, the accommodations were a little more elaborate and comfortable." In Frances Gallagher's "modest cabin" at Miner's Delight, "the presence of a woman (with a capital W.) was touchingly manifest in all the little household arrangements," including a curtained window, buffalo skin carpets, and a side table bedecked with books and photo albums. An argonaut visiting a bonanza claim on Eldorado Creek found "a two-storey house—not a cabin, but a good two-storey log house," and concluded correctly that a woman was in residence.

Sarah Royce in a California mining village cloth-and-frame home provided the illusion of different rooms by grouping her furniture into kitchen, dining room, and parlor areas and curtaining a bedroom space. She lavished special attention on the "parlor," converting rough boxes into plush ottomans. A half century later, Martha Purdy—over the objections of the male members of her party—insisted on carrying linen tablecloths, silver, and a bolt of rose-patterned cretonne to Dawson. In their cabin there, she fashioned curtains, packing-box covers, and cushions from the cretonne, "laid the table with coloured oilcloth" and scalloped the edges, and generally made the space "as pretty and homey as possible." On Eldorado Creek nearby, Edith Berry not only presided over the two-story house but offered amazed visitors fresh milk from the Jersey cow she kept in a sawdust-padded stable.

Women also—indirectly or directly—caused a return to some of the strictures of "civilized" life. When a family with four children, one a fetching sixteen-year-old girl, reached a Montana camp, the miners shaved and put on fresh shirts, "all trying to appear like civilized men." Soon there were regular shaves for the whole camp, haircuts for a few, and a general shift from buckskin shirts to blue flannel ones, from moccasins to shoes. Then one female immigrant "decreed [the miners] should wear starched shirts." A camp resident dutifully fashioned a "bosom board" on which to starch shirtfronts, and the woman busied herself daily "explaining the process of ironing shirts to the 'native Americans.'"

The prospectors of Gold Creek apparently took this woman's "innovations" with good cheer, but if such social pressure seems like nothing more than cultural indoctrination, "good" women also brought a genuine spirit of caring and social activism to the gold frontier.

"Bad" women did, too—the stereotype of the whore with the heart of gold is built upon various prostitutes' willingness to help others in need. Dawson prostitutes in 1898 turned into heroic nurses as sickness ravaged the town. But generally speaking, prostitutes were too firmly caught up in the disordered, shifting frontier environment, too clearly regarded as outside the boundaries of "true womanhood," and too closely identified with their often extralegal profession to become agents of civilized order and welfare assistance.

"Good" women at first seemed just as unlikely as a power for social welfare and order, simply because they appeared so out of place. Many women on reaching the gold frontier acutely felt the force of a society based on everything contrary to their moral and civic impulses; those middle- and upper-class women who could do so often entered an isolation both "self-imposed" and subtly encouraged by that society. As a male Dawsonite observed, "real" social functions were few, theaters and public dance halls "alike impossible for a decent woman," so that "whatever social thrills or successes she knew must come within her husband's or brother's small circle of friends."

But many working-class and some middle-class women could not afford or abide such seclusion; not only were they "anxious to make money by honest industry," but they took seriously their cultural role of improving society. Thus, women plunged into fund-raising activities for benevolent societies, beginning churches and schools and identifying and meeting individual welfare needs.

As intimidating as the social environment could be for a woman, the men who created and contributed to this climate regarded crusading "ladies" with a bemused, deferential respect from California to the Klondike, even the gamblers meekly falling into line to pay for the women's community efforts. Stampeders testified again and again to "good" women's role as the social and moral conscience of the gold regions. One practical, decidedly unsentimental California prospector described the first women in Nevada City as "angels of mercy" to "many a poor suffering soul."

Women themselves regarded their activities more matter-of-factly. In Dawson, the no-nonsense Malinda Jenkins and her grown daughter joined a phalanx of women that confronted the prospectors on their claims and even in the saloons seeking money for Father Judge's hospital. These women also responded to the news that a sick Dawson wife needed help by visiting her cabin. Finding that her husband had not been adequately tending the scurvy-ravaged woman, they enlisted the help of the North West Mounted Police in forcing the belligerent spouse to move away temporarily, then tended to her themselves until she could get on her feet again.

The Dawson women's effort apparently represented an alliance between working-class women such as Jenkins and the "ladies" of Dawson's limited society. But however much class lines might blur, "good" women fought fiercely the boom-camp tendency to accept prostitutes as desirable symbols of womanhood. Prostitutes were a threat and an affront to the ideal from which middle-class women, lacking economic and political power, drew what social power they could. Although respectable women sometimes entertained the idea that prostitutes were more sinned against than sinning, that did not change attitudes of exclusion. They were particularly apparent in the San Francisco incident in which U.S. Marshal William H. Richardson, accompanied by his wife, asked gambler Charles Cora to remove Cora's mistress Belle from a public event. After Cora was hanged for fatally wounding Richardson, a local newspaper ran a notice from "Many Women of San Francisco" announcing that they bore Belle no ill will, but "every virtuous woman asks that her influence and example be removed from us."

An occasional "loose" woman challenged this forced segregation. In Deadwood, one crashed the first society dance, pulling a gun and firing wildly to protest the double standard that allowed her paramour to attend.

As gold rush societies developed, only the most socially secure, both female and male, could publicly acknowledge prostitutes as people. In Deadwood, one venerable widow in her fifties (the age itself unusual in a gold town) was confident enough of her status to befriend the women of the red-light district, paying them social calls and occasionally helping one make a fresh start. Another member of Deadwood's social elite, federal judge Granville Bennett, graciously greeted a notorious prostitute on the street one day. His male companion asked incredulously, "Do you speak to that woman?" Bennett's reply: "Yes, I can afford to."

Any woman—prostitute or not—who did not fit the standard

definition of womanhood was viewed by the class-conscious with disapproving or patronizing eyes. A woman traveling on the Panama route to California in 1849 noted of another female passenger that she was "going independent and alone, to speculate in California—of course, no very agreeable person." Louise Clappe clearly felt superior to the handful of women sharing Rich Bar on the Feather River with her, especially to the "Indiana Girl," a noisy backwoods belle shod in miners' boots. Sarah Royce found only one other woman in residence on her arrival in Weaverville, California—"a plain person . . . from one of the western states . . . acquainted only with country life"—and reported with bemused class-consciousness the woman's enhanced frontier social standing. A male traveler judged the female population of Miner's Delight to be composed of one cultured lady, two relative nonentities, and "Dalilah," or "Candy," who "was not a respectable person when she first struck the camp" but was "studying virtue" from a no doubt chaste miner named Jack Holbrook and demanding that she be called Mrs. Holbrook.

Times did change rigid class judgments somewhat—an independent woman going to the Klondike was not looked on with the automatic disfavor afforded the Panama traveler. The turn of the century, after all, was bringing an increased acceptance of women venturing into the public sphere. But standards of propriety were far from lost.

One unwritten gender rule that remained in force was the exclusion of "nice" women from saloons, gambling halls, and certain dance halls and theaters. In only two known instances in Deadwood's early existence did respectable women enter saloons—in one, an absentminded spinster school administrator entered by mistake; in the other, an irate wife found her husband losing at the gambling table and scooped the jackpot into her apron. In both cases, the men in the establishments simply froze.

Women sometimes chafed at their restrictions. Edna Bush found Sheep Camp on the Chilkoot Trail a fascinating place and looked with intense curiosity on a dance hall operated by two women who could "dance, sing, swear, play roulette, shake dice and play poker." But Bush didn't get to see those feats for herself—the dance hall remained off limits for her.

Even when they wanted to maintain proprieties, women found it hard to do so. For example, standards of dress gave way to frontier realities. Women were achingly conscious of what long journeys had done to their dress and grooming standards. One recent arrival on the road to Sacramento "met a man wearing an honest-to-goodness white shirt" and "shrunk from being observed in her threadbare shabbiness." Fifty years later, poor Emily Craig, having endured two years in the Canadian wilderness on the road to Dawson and having seen her clothes burn in a tent fire en route, was embarrassed to enter Dawson in her buckskin dress purchased from natives. At a "roadhouse," or overnight bed and board, in Alaska, one prospector observed a diffident young woman dressed in men's clothes, her hair tucked into a hat, and was told that she felt humiliated at being among a group of men in this attire. (In fact, the young woman might have worn men's clothes as a means of deflecting male attention—the prospector did not know she was female until she was pointed out to him.)

Even the roughest, most independent men displayed rigid ideas about how women should dress and behave. A group of tough, oath-spouting bullwhackers agreed to allow a persuasive young woman to ride along with them to Deadwood, then became "shocked and horrified" when she joined them not riding sidesaddle but "*a la clothespin.*" They were embarrassed to enter Deadwood with her. Almost a quarter century later, when Malinda Jenkins's scow approached Dawson's riverfront, her waiting husband's first words were, "Why didn't you put on a dress?" She responded, "I come the whole way in pants." But he persisted, chiding, "You could a-put a dress on to land in." Her response: "Aw, shut up, will you?"

The idea of a respectable woman in a camp full of men seemed "absolutely indelicate" to some. A Klondike stampeder wrote in a tone more worried than outraged, "The men with their wives with them have my sympathy, unless they have located or know where they are going. Fancy a woman being left alone in a camp with a dozen or 20 men while her husband goes up a river . . . for two or three weeks."

Sleeping arrangements were often unconventional, to say the least. The most pious of women found themselves lodged above bars and sharing their accommodations with a variety of strange men. When Emma Feero and her children joined John Feero in Skagway, they stayed in the main "hotel," in the sleeping loft over the public dining area, the six of them sharing three single mattresses in a curtained sleeping space accessible only by a ladder and with a drunk customer on either side. "We had a ball, us kids. We thought it was just loads of fun," daughter Edith would remember. "But not Mother. Mother, she didn't sleep."

On the trails, too, necessity created some odd situations. Dawson entrepreneur Belinda Mulrooney, on a trip between Skagway and Dawson, discovered a woman from Seattle collapsed in the snow, having attempted to reach Dawson on her own to meet her husband. Mulrooney, unable to get the woman to budge, loaded her onto her sled and took her to her own camp, where she found two old-timers had sought shelter for the night in her tent. Mulrooney tried to roust them, but they urged her to place the half-frozen woman between them, promising, "She'll be as safe as if she was in God's pocket!" When the party reached Dawson, the woman located her Norwegian husband and cried, "Honey, Honey, come and take a look. Those are the men I slept with on the trail!"

In a typical scenario in the diggings, a Klondike gold-seeker found a young woman from Oregon living with her miner husband and two hired men in a one-room cabin. The visitor couldn't hide his curiosity—wasn't it "awkward to sleep in the same room with two other men besides your husband[?]" The young woman seemed unperturbed—she had a curtain—but admitted that the previous winter had been hard, with five hired men sharing their quarters: "They were very good, and I didn't mind it after a while. Still, I don't think I would like it again unless we just had to."

Married women sometimes found their position made more difficult by husbands who lost their moral bearings away from civilized codes of conduct or who became demoralized and disappointed in their gold hunt; those men could turn their frustrations on the wives who had stuck with them. One Grass Valley sometime miner, informed by his wife that they had no food, flew into a violent rage and drove the family from their house. In another California camp, a dissipated doctor in delirium shot and killed the wife who had followed him from Kentucky in hopes of renewing their marriage and reforming him. And in a similar incident, a California miner shot and killed his Australian wife. Whether he killed her out of jealousy over a rival lover or over her relative earning power as a washerwoman is unclear, but both he and the doctor were hung by lynch mobs. The miners "said justice must be done if there is no law, and that no man could kill a woman and live in California."

Stampeders showed themselves willing to intervene in cases of abuse. In Summit District, Montana, a miners' court found a man guilty of mistreating his family and banished him from the area. When he returned with lawyers, the miners obligingly re-tried him and again reached their original verdict. With the man gone for good, his judges periodically appeared to help the wife with household chores.

"All good women and little girls" were a "sacred trust" in Deadwood, while in Dawson a " 'square' or 'straight' woman . . . was as safe as, or safer than, in Washington, D.C." The harshest sentence for drunk and disorderly conduct or disturbing the peace in Bodie during its boom—thirty days in jail—was meted out to "a drunk whose only crime was 'using the vilest kind of language before ladies and children.' "

The safety of children proved particularly vexing. As the childless Louise Clappe put it, her home of Rich Bar was "an awful place for children; and nervous mothers would 'die daily,' if they could see little Mary running fearlessly to the very edge of, and looking down into [abandoned prospecting] holes—many of them sixty feet in depth." As for the moral climate, because gold rush urban areas were jammed into gulches and stream intersections with no regard for zoning, children in even the most elite families lived in tantalizingly close proximity to the click of the roulette wheel, to the shouts and bawdy oaths ema-nating from the saloons, to the heady music pouring from the dance halls, to the sight of fleshy, provocatively dressed and painted women lounging on balconies and in front of "cribs."

And children loved it all—the deep, mysterious pits and im-posing gravel heaps, all the noise and mess of mining, the color, spectacle, and informality of community life, the joys of discov-ering berry and flower, glade and brook, on the hillsides. Their memoirs brim with remembered excitement and delight, lending support to a study of childhood on the frontier, Elliott West's *Growing Up with the Country,* that demonstrates that young people generally responded positively to their new western environments, imbibing a spirit of freedom, fearlessness, and liberality and often adapting far better than did their parents. Emma Feero, who had refused to consider a second night in Skagway's rude hotel, would spend the rest of her life in Skagway but remain apprehensive, always checking for intruders. " 'Course, where she was raised back in the state of Maine, [there was] no drinking, no wild life. I always figured that's what made her that way," explained daughter Edith, who was not "that way" at all. And Estelline Bennett would write of the mother who longed in Deadwood for carriages and country lanes: "She was one of those who . . . 'never can understand the way we have loved you, young, young land.' "

Some women did love the "young, young land" for the very real employment opportunities it afforded them, one concluding that "it is the only country that I ever was in where a woman received anything like a just compensation for work." In gold rush Virginia City, jobs for women were "plentiful," the *Montana Post* promising that women "who can use their hands smartly, are sure of immediate employment at high wages." One man wrote home from Bannack, "Women's labor pays as well [as] if not better than men's."

Most found the domestic duties they had always performed for free could bring undreamed-of compensation. Washing, for example. Stampeders fantasized not only of comely maidens but of "a good faithful washerwoman with her suds, and her arms akimbo." A woman made fifteen to twenty dollars a week at this activity on one California bar, leading another to write, "She has all she wants to do, so you can see that women stand as good chance as men." In 1849 San Francisco, washerwomen received five to eight dollars for a dozen items "just dipped in to the water and rung out at that," leading a resident to conclude "female labour is above every thing else."

Washing became such a lucrative occupation that various groups wound up jockeying for the trade. In gold rush San

Francisco, the original launderers, Mexican and Indian women who used a pond called Washerwomen's Bay, faced competition both from newly arrived Anglo working women and from male stampeders looking for ways to make money. As Chinese men took over laundering here and across the gold frontier, they were perceived as robbing women of jobs. The laundresses of Helena, finding themselves undercut by the Chinese men's "washhouses," placed a notice in the local newspaper "warning the Chinese washermen to leave town or be visited by a committee of laundresses who would enforce its wishes." The laundrymen responded by reminding Montanans that they were only requesting "that the good people of Montana may let us earn an honest living by the sweat of our brow." The whole brouhaha pointed up the struggle of minority groups in general, including the overwhelmingly Anglo and European working-class laundresses, to grab and hold on to a piece of the economic pie.

One way for women to capture a piece of the pie was to make pies—and bake bread and cook beans and otherwise feed the ravenous male population. Upon her 1849 arrival in California, Luzena Wilson was approached by a miner and offered five dollars for the biscuits she was cooking for her family. Interpreting her amazed stare as reluctance, he then "doubled his offer." He was determined to have "bread made by a woman" and pressed "a shiny ten-dollar goldpiece in her hand."

Men with the prodigious appetites born of outdoor living and hard work proved a bonanza for female cooks. In California, an argonaut in 1850 found women "doing full as well as men," earning "as high as $30 per day" cooking meals for the miners. The African-American woman spotted that same year "tramping along through the heat and dust of the desert" on the Overland Trail carrying a "cast-iron bake oven on her head" had the right idea.

Pies *were* the ticket for many cooks, perhaps because, as one Alaska stampeder observed, "there is nothing like pie to bring a fellow to his home senses." One woman who transported a stove cross-country set up a "pie house" in Nevada City, California, producing dried apple pastries that were eagerly snatched up at a dollar each. The stampeders in the diggings would play euchre each night "to determine who should pay for the pies when they went to the 'city' " on Sundays, when her business would be "literally thronged with miners waiting for her pies to come out of the oven." Even charging five dollars a pie, some women could not keep up with demand.

One woman "pulled her own sled weighing 250 pounds" into Circle City, Alaska, and established a bake shop and laundry service. Another, "a motherly, determined old lady who with great courage, ventured into the Yukon country to make her fortune," cooked for one of the Klondike kings at a salary of one hundred dollars a month and board—eight dollars' worth of food a day. Another Dawson woman baked beans and filled old butter tubs with them, then sold them to eager miners "starved for good, well cooked food."

Other traditional skills and roles were rewarded as well. There were stories of gold camp seamstresses making a dollar for each button sewed onto a garment, and in Virginia City a "servant-girl" could make fifty dollars a month—"as much as a teamster." But perhaps the most accepted and popular form of employment open to respectable women was that of boarding-house operator. It was a "natural progression" for women cooking for miners to start taking them in as boarders, and in 1850, "nine out of every thousand persons gainfully employed in California ran boardinghouses or hotels."

The boardinghouse—and even the hotel—drew upon the culture's understanding of woman as homemaker, as provider of food and domestic comforts. It was, in one sense, the easiest line of work for women to enter, in that they might simply open up rooms in their home and add mouths at their tables. But establishing and running a boardinghouse required a tremendous amount of work. Some women had to start from scratch. Luzena Wilson began her El Dorado "hotel" in Nevada City, California, by buying two boards "from a precious pile belonging to a man who was building the second wooden house in town," cutting and driving stakes into the ground, placing the boards atop them, and preparing a meal that attracted twenty miners.

Even those who already had a decent-sized dwelling to house their enterprise faced the constant demands of feeding and otherwise seeing to the needs of family and boarders or guests— without the aid of well-stocked markets, without enough utensils and linens and other household items so taken for granted in the East. They also had to play nurse to sick boarders—an added burden indeed when, as in a Black Hills camp in 1879, "every boarding house contain[ed] several sick men." One boarding-house keeper concluded, "If I had not the constitution of six horses I should [have] been dead long ago. . . . I am sick and tired of work."

No wonder that women, like men, were drawn not only to mining speculation but to other speculative ventures as an easier form of moneymaking. A California-bound woman over her husband's objections bought fruits and vegetables in one Pacific port and realized a nice profit from them in San Francisco. In the Yukon, Malinda Jenkins launched a booming business selling doctored alcohol as whiskey and hired a bartender from Seattle to help.

If speculation—especially liquor speculation—seemed to run counter to the nineteenth-century image of womanhood, some women challenged the stereotype even further by hiring on for others in traditionally masculine employment, working as tollgate operators, even as blacksmiths, the latter leading a Montana editor to moan, "When feminines turn Vulcanists and broncoshoers the worser half will get their deserts."

Women also set up their own businesses, from soda manufacturing to saloon operation, from farming and livestock enterprises to real estate sales. During Dawson's rush, "an Illinois society lady" garbed in furs, a dark wool dress, and "dainty boots" bought and sold real estate from a tent in which the temperature hovered at 20 degrees below zero. Martha Munger Purdy not only bought into various mining ventures but managed a sawmill for her father, despite the fact that the foreman didn't want to work for a "skirt." Belinda Mulrooney in one year in the Yukon "made enough money on restaurants, roadhouses, and mining" to erect the sumptuous Fair View Hotel in Dawson.

Perhaps the most startling independent trade pursued by women was that of bullwhacker, or freighter. A young Spanish woman in the Weaverville area in California owned a mule pack-train and bought and loaded goods herself. A woman known as Madam Canutson freighted to Deadwood with her own ten-yoke oxteam, her freighting skill garnering the admiration of other bullwhackers.

Most of these entrepreneurial women were wives and mothers. A Weaverville merchant ready to propose to the young Spanish freighter was told "she had a husband somewhere in the mines" and a five-year-old child. Madam Canutson, whose husband was homesteading ranch land for the family, carried her baby with her on a six-week freighting job between Pierre and Deadwood, referring with pride to the "awful toney bed" she had created for him in the oven of a stove being hauled in for a hotel.

Wives who embarked on nontraditional ventures tended to behave very independently. Some did so by controlling the family finances. A San Francisco woman "set her husband up in business" producing soda. An observer noted, "They are making money fast. She doesn't trust him with the profits 'nary time' but keeps the stuff in her own hands." Others pointedly maintained their economic autonomy. Malinda Jenkins at first ignored her saloon-operator husband as a client as she built her whiskey supply business. When he asked why she wouldn't sell to him, she responded, "Pay me like everybody else and you can have all I make." She reported with satisfaction, "He sure had to come through for every keg delivered at his place."

Most working wives, however, remained traditional enough to see their income as family funds, to serve as economic helpmates in addition to their other duties. Yet recent studies indicate that frontier women in general—and gold rush women in particular—"were often the breadwinners and wage earners who provided crucial capital for male accomplishment."

This fact was recognized, if not articulated in these terms. In 1853, a San Franciscan wrote, "The better work here for 18 Dollars a month until they bring their wives and keep Boarders." In gold rush Denver, a family was able to hire workers to build a cabin on the basis of the mother's promise to keep boarders. When the father died of consumption shortly thereafter, the mother and daughter worked "almost day and night, sewing, and mending" while the two sons went out gold-hunting. The daughter later recalled that her brothers "were unsuccessful in their search for gold and in order that they might be able to continue their work, every cent which we could accumulate, over and above enough to buy us flour, was given them."

The essential nature of women's earnings was such that it caused a frontier departure from the high valuation accorded upper-class women, ladies who, in the words of Jack Crabb in Thomas Berger's *Little Big Man,* "was useless for all practical purposes." Nowhere is this more compellingly demonstrated than in Louise Clappe's encounter with a male resident of Rich Bar, who effused over the earning skills of another Rich Bar wife. "Magnificent woman, that, sir," he informed Louise's husband. "A wife of the right sort, *she* is." As a laundress, the woman had earned her husband "nine hundred dollars in nine weeks, clear of all expenses." The admirer declared that, with such an uncommon woman, "a man might marry, and make money by the operation." He looked at Louise, she reported, "as if to say that . . . I was a mere cumberer of the ground; inasmuch as I toiled not, neither did I wash."

Nonetheless, Clappe enjoyed a favored status as a representative of civilization and femininity. By contrast, the "bad" women, the dance hall girls and prostitutes of the gold frontier, led a more uneven and precarious existence. Feted and reviled, rewarded and victimized, they shared more fully than did other women in its golden excesses and in its squalor.

The initial ready welcome of prostitutes is reflected in a San Francisco newspaper item of May 1850 titled "ENLARGEMENT OF SOCIETY." The paper was "pleased to notice" the arrival of fifty or sixty of "the fairer sex," including French beauties, and reported the bay "dotted by flotillas of young men, on the announcement of this extraordinary importation." But only four months later, the same paper editorialized, "We must confess our regret at the perfect freedom and unseemly manner in which the abandoned females . . . are permitted to display themselves in our public saloons and streets."

Although most elicited some respect simply for braving the rigors of the trail or voyage and for "offer[ing] the lotus of femininity" in the masculine environment, "loose" women were divided by status. At the top were the dance hall queens—sometimes "loose" only in the sense that being a performer was still looked upon with suspicion by moralistic society. The queens made the most of their morally ambiguous position; Cad Wilson, "the most successful and celebrated of all the Dawson dance-hall queens," was regularly introduced by manager Eddie Dolan with the intelligence that her mother had written telling her "to be sure and be a good girl and pick nice clean friends." Then, motioning Wilson onto the stage, he would evoke a roar of appreciation with "I leave it to you, fellers, if she don't pick 'em clean."

Also at the top, at least fleetingly, were the most favored prostitutes, often the madams of the classiest bordellos. Representative of this group was red-haired Pearl De Vere, the discreetly powerful and elegant madam of Cripple Creek's luxurious Old Homestead. When she died of an overdose of morphine on the night of a grand Homestead party, the town gave her a splendid funeral.

Next in the pecking order were the theater and dance hall girls, the former performing onstage, then joining miners for drinks in curtained boxes, the latter dancing with all comers for a fee, then parting them from more of their gold dust with drink purchases. Those entertainers were usually considered a cut above their "frail sisters." Some observers judged them simply working girls; others intimated that they had been or continued to be prostitutes. One Dawsonite concluded that "some had a history and some not before they came, but all had a history after arrival."

They ran the gamut in terms of looks and accomplishments. There were beautiful and able entertainers adept at performing vaudeville stunts and being "entertaining companions—the kind 'a fella would like to buy a drink for.' " There were also girls "stolid and Dutchy in expression, beery and festive in habit, with shiny foreheads and a disposition to sweat at the sound of a fiddle, as fickle as fortune and homelier than original sin."

Many of these performers resented "the stigma generally attached to their profession." They worked hard and enjoyed "the freedom to come and go as they pleased and to pick and choose among the men who lavished attention upon them." Gifts and marriage offers often poured in, and a popular dancing partner in Virginia City earned "more in a week than an equally well-favored girl in the East could earn in two years." Two dance hall girls in Dawson earned fifty thousand dollars from a besotted miner by alternating dances with him until his money ran out.

Still, in a world in which "nice" women did not enter the dance halls and theaters in which these women worked, dance hall women could not escape the sexual stigma, especially since some did work as prostitutes or became prostitutes. Further, sometimes—especially as a boom began to fade—known prostitutes began working the dance halls to drum up business.

In every rush of any size, the prostitutes appeared in an ethnic, national, and regional diversity almost rivaling that of the gold-hunters themselves. Barely had the California boom begun than prostitutes from New Orleans, from Chile, and from Mexico were heading for San Francisco and the mining regions. Prostitutes came from the Sandwich Islands, from France, from Australia and China. As the mining frontier grew, "abandoned" women from all regions of the United States began to appear.

Like the prospectors, and like the gamblers and other mining camp hangers-on, the prostitutes soon established a pattern of moving from rush to rush. In Idaho, one Boise resident wrote, "As soon as the miners began to flock to this country these women began to come out of Portland, the Dalles, Walla Walla, and other places, sometimes a dozen in a drove, astride an Indian pony," and wearing men's clothes. In Helena, Montana, they "appear[ed] to have migrated from other mining camps or bonanza towns." In Cripple Creek, the female parlor house operators were "old hands from Leadville, Aspen, and Colorado City."

Not only did prostitutes follow the rushes, but many ebbed and flowed on a seasonal basis with the miners themselves. Just as many miners during the winter left the remote, colder camps for warmer climes with better economic opportunities, so did the prostitutes, who became known as "summer women."

The characteristics of the women who sold sex for a living in the gold regions ranged from "the elegance, the social intelligence and the real and regal-courtesan-like charm" of the top prostitutes to "the tawdry, cheap, vulgar and disgusting, sheer dirtiness and sluttish habits of crudely painted creatures." A more useful way of categorizing these women, however, is by the degree of control they were able to exercise over their own lives. At the bottom on this scale were the women literally held in sexual slavery, then the women who operated independently but led marginal existences, and finally, those women who accumulated enough capital to exercise some autonomy and power in the gold regions.

Indentured prostitution was a feature of gold rush life from the beginning; women paid with their bodies for transportation to the goldfields. Mexican prostitutes coming into California did so under six-month contracts of servitude, while a shipload of French prostitutes, unable to pay their passage, were reportedly auctioned off to the highest bidders in San Francisco. Almost half a century later, "white slave girls," primarily Belgian, appeared in Dawson under the watchful management of pimps known as *macques*. Many of the prostitutes owed passage money to these men.

But the most abused group in this regard had to be the Chinese prostitutes. Respectable Chinese women seldom immigrated to the gold regions at all; for example, even after the huge immigrations of the 1850s and 1860s, "Chinese men outnumbered Chinese women in the United States by more than twenty to one," and Estelline Bennett could remember only one socially acceptable Chinese woman, a Chinese merchant's wife, who made her home in Deadwood. With a demand both from Chinese men and from other frontiersmen, Chinese girls were kidnaped, sold, or taken in feudal wars in their homeland, forced into prostitution, usually controlled by the tongs or by individual Chinese brothel-keepers, and consigned to lives of misery across the gold frontier. Chinese men found them "as valuable as real property in securing a mortgage."

In theory, the girls were to work out their passage money and then be liberated, but few came even close to freedom. Most independent prostitutes fared little better. Although living conditions were hard for all on the frontier, for prostitutes, they were often abysmal. In Dawson, the "common prostitutes" lived first in open tents lining the roadway, their quarters resembling "a small-town annual fair for breeding animals." They then inhabited bleak "frame shacks" in an alley behind the dance halls and were finally moved by the authorities "to a section of swamp-land well back from the business section."

The prettier girls or women could improve on these living conditions by entering one of the better brothels, but prostitutes in general, alternately sought after and scorned, were unable to create any kind of stable environment. They endured abusive relationships with both customers and lovers and were controlled and harassed by city authorities when town governments took hold. They fought back however they could, further widening the gulf between themselves and "proper" women but getting in their licks. When one young female operator of a "questionable boardinghouse" was arrested for selling liquor without a license in Dawson, she responded to the question of "Guilty or not guilty?" with asperity: "Me plead guilty? Why should I plead guilty when half the men in this town . . ." As she began pointing at men in the courtroom, the judge called an adjournment and later reinforced the hegemony of the male power structure by compelling the woman to plead guilty or face a "blue ticket" ejecting her from the territory.

Perhaps the greatest challenge prostitutes faced was the loss of whatever physical attractiveness they possessed. Banking primarily on their beauty, they saw it flee as disease, hard living, alcohol, and drug abuse took their toll. In Deadwood, Estelline Bennett watched the cream of the red-light district, "the lovely light ladies—pretty, beautifully gowned, and demure man-nered," descend within two or three years into shadows, "pallid and shabby, with dingy old 'fascinators' over their heads," steal-ing down alleyways in search of money for alcohol or opium. Tuberculosis periodically attacked western prostitutes, its spread fed by "too close housing, too little light and air and exercise, too strenuous and gay a night life." But often it was not disease but despair that killed these women; across the gold frontier, prostitutes took an overdose of laudanum, shot them-selves, or drank themselves to death in the manner of failed prospectors.

Despite the many negatives that drove women to suicide, prostitution consistently beckoned as a means to riches. In an environment in which a woman-starved prospector would pay an ounce of gold dust, or sixteen dollars, simply to have a female sit with him at a card table or bar, the sky seemed the limit. One prostitute who made the rounds of the California gold camps on horseback in men's clothes "said she had made $50,000 and regretted that she had not double the capacity for increasing her gains." Common streetwalkers from France increased their fees a hundredfold in California, enjoying the stampeders' predilection for the "ease, taste, and sprightly elegance" a French origin seemed to provide. When Charles Cora shot William Richardson in the dispute over whether Cora could bring his mistress Belle to a public event, the alacrity with which Belle "advanced fifteen thousand dollars in gold to a well-known criminal lawyer" for her lover's defense said something about the economic means of at least some women of the demimonde.

A recent study of women in Helena's early years provides further evidence that some prostitutes managed to gain a large measure of economic and even social freedom—at least until the pioneer period gave way to the imposing of eastern structures and standards. Some prostitutes and madams arrived with money, purchased business properties, and generally integrated themselves into Helena's "legitimate economy," making almost half of the "265 property transactions undertaken by women [in Helena] between 1865 and 1870" and claiming "all twenty mortgages made to women during the period." Their business acumen was such that more than half of them were financially sound, "in some measure controll[ing] the terms of their exploitation and profit[ing] accordingly." For example, saloon dancer and "occasional prostitute" Dolores Jarra bought Helena properties and purchased a homestead from her lover, an Anglo clerk. Departing the town, she married and returned as a member of the "economic elite," sending her daughters by the Anglo clerk to a private academy and seeing one make a socially advantageous marriage.

Jarra had transcended the traps of prostitution and proven a worthy businesswoman, all without even being able to write her name. But hers was still a transition few women could make given the social and economic realities of the time. It remained easier for men to remake themselves on the gold frontier than for women to do so.

* * *

Still, women of every class and occupation—with the understandable exception of the indentured prostitutes—found this frontier liberating. Californian Mary Jane Megquier asserted, "The very air I breathe seems so very free that I have not the least desire to return [to Maine]."

The sense of liberation came in part from the greater legal freedoms accorded women. Because females were so scarce, frontier governments wanted to offer inducements to their immigration. That desire was in large part the reason that courts and congresses of territories and states born in gold hopes were more likely than were their eastern counterparts to affirm married women's property rights, to establish permissive divorce laws, and to favor suffrage.

Contributing to and growing from such legislative progressivism was a relaxing of the codes of conduct between "good" women and men and a sense that even married women retained many romantic options. Informal social visits and outings involving single men and married or unmarried women were common, as evidenced in a young Virginia City man's journal entry: "loafed around with Mrs. B. all day." The shift in social convention could be a little troubling; one married woman in Helena advised her absent husband that she had gone riding to a mining camp with a bachelor doctor and wrote uncertainly, "Was there any harm in doing so? This town is so different from my old home that I need one to advise me often. I have always depended so much on mother's judgment in matters of this kind." There were still some who professed shock at such freedoms, as did the Virginia City arrival who in 1863 noted with consternation "two women driving along, apparently unattended by any male escort."

Women learned to rely upon their own judgment, including the decision to divorce and cohabit with another man or remarry. After ranch hand John Grannis of Denver "spatted" his thirty-six-year-old wife on the mouth, he reported that she "concluded to make her Bed by her Self & leave me for ever & take her chances in life with Mr. Griswold on the free love Principal." As if that were not enough for the hapless Grannis, when he joined the gold rush to Virginia City, he found "Mrs. Grannis" and Mr. Griswold in the immigration as well. His wife divorced him in Virginia City and, apparently tiring of the free love principle, married Griswold.

Grannis's wife was representative of many gold frontier women who determined they "had to put up with mistreatment or neglect no more than [they] had to be dependent upon a man for an income." A Protestant minister in Virginia City described the result, not so strange to the eyes of Americans at the end of the twentieth century, but startling for its day.

> There were a good many weddings at that time in Montana, not that there were many persons to marry, but because the few married often. The newspapers gave "divorces" a standing heading, the same as "marriages" and I have counted the same number of records under each heading. In some instances, a name would appear simultaneously under both headings.

Or, as a Helena matron observed, "It is a common comment that a man in the mountains cannot keep his wife."

The independent actions of individual women did not usually lead to any concerted efforts for the vote and other women's rights—women on the gold frontier were usually too busy, too isolated, and too caught up in a world "geared to men's social and economic activities." In fact, men and women alike on the gold frontier voiced a very traditional aversion to the "radical" feminism introduced in the Seneca Falls women's rights convention the same year Marshall discovered gold in California. But what people say and what they do are often two different matters entirely. When it came to women's rights, the population of the gold regions "admit[ted] in detail what it denied in general."

Estelline Bennett probably caught the frontier situation and mood best in her comment that most western men were for women's suffrage, as the men were "doing the sort of thing that brought them into no competition with women who were so scarce anyway that men thought they should have whatever they wanted and more too, for that matter."

In response, women became more confident of the value of their skills, more likely to reject the restricting tenets of the "Cult of True Womanhood" that assigned women a limited, subordinate role. Martha Munger Purdy, for example, resolutely forsook "shelter and safety" for "liberty and opportunity." Looking back, Purdy would explain in terms that the culture tended to associate with masculinity, "I liked the life, the vigorous challenge of it—the work and play of it. I had faith in myself—that this tide in the affairs of my life would lead me on to fortune."

Purdy was right, although in a much more traditional sense than she realized at the time she entered the Klondike pregnant and without her husband. Her mining profits slipped away in unsuccessful grubstaking of prospectors and extraction costs, but after her divorce, she married George Black, who became commissioner of Yukon Territory and then speaker of the Canadian House of Commons. Martha Munger Purdy Black became the doyenne of the territory, enjoying a long and full life as a statesman's wife.

Yet that life was initially made possible by the challenging of gender stereotypes, by independent action and hard work. Purdy, like so many other gold rush women, contributed to a gender-free frontier legacy of tremendous effort, adventure, and freedom. Some of those who contributed, male and female, were lucky, and some were not. We cannot leave them without taking a last look at how they assessed their stampede experience—and how we can do so today.

Chapter Thirteen

Perspectives

Booms are fun; the briefly rich try so gallantly to extend the small reach of human extravagance that they tend to lead the eye away from the lapsarian nature of it all.

—*Larry McMurtry*

I fell in love with the life, and I want to live and die in a mining camp.

—*California veteran W. G. Searles, in Goldfield, Nevada, 1907*

The gold frontier had been stingy in fulfilling the seekers' dream of golden riches but almost too accommodating in providing them with their second desire, adventure in a new country. The results could not be measured on a set of gold scales but they did weigh in the lives of the stampeders themselves and in the life of the American West and the nation.

Few grew rich from gold digging. A survey of the ranks of the "bonanza kings," the late-nineteenth-century barons of western mining, lends credence to the idea of frontier opportunity for the self-made man, but only a handful of those men had more than passing ties to gold stampedes and gold prospecting. Using the kind of shrewd business sense valued in the East, most built their empires on careful investment in diversified long-term mining interests. For them, a rich placer claim was only one of various potential tools in establishing and consolidating economic, political, and social power.

Still, one genuine treasure claim could provide a provisional admission ticket to the world of great wealth made visible in ostentatious living, wide-ranging investment, and grand philanthropy. Starting with the Nevada silver kings in the 1870s and continuing through the heyday of the Klondike kings, the biggest lottery winners etched themselves into the public consciousness as western models of success. They built elaborate homes in San Francisco, in Denver and Seattle, furnished their mansions lavishly, made grand tours of Europe, bought art and racehorses, rode about in elegant carriages, married their daughters off to the American or European aristocracy. They launched transportation companies, bought up western real estate, and extended and varied their mining investments. The former prospectors among them not only grubstaked old friends and lavished gold nuggets on hotel bellboys but also endowed hospitals and pumped money into charitable organizations.

But men catapulted into the mining aristocracy on the basis of a rich claim—primarily the Klondike kings of 1897 and 1898—seldom knew how to maintain their golden fortunes. The live-for-the-moment mentality that often pervaded the mining camps pervaded the successful prospector's subsequent dealings as well, so that these lottery winners flared like meteors across the western skies. "Expense! Expense! That word is not understood in the north," railed Klondike millionaire prospector Pat Galvin at a nephew who dared to question his freewheeling ways. His philosophy: "If you have money, spend it; that's what it's for, and that's the way we do business." Galvin quickly lurched into bankruptcy. Whether a fellow dug out a hundred or a hundred thousand or a million in gold, most demonstrated the truth of Dawson dance hall queen Diamond-Tooth Gertie's words: "The poor ginks have just gotta spend it, they're that scared they'll die before they have it all out of the ground."

Claim-holders who struck it rich usually retained a certain rough naïveté, a simple understanding of business matters, a sense of delight in playing the grand tycoon. They invested part of their money, but tended to do so haphazardly. Some focused on the local enterprises that spelled status in the raw gold-rush community—ownership of dance halls, of steamboats, of area real estate and mining properties. Buoyed by their sense of luck and power, they treated the manipulation of wealth as a short-term game. At the same time, many continued to succumb to the lure of the gambling table and particularly to that of the liquor bottle. Typical was the rise and fall of prospector Dick Lowe. While serving as a chainman on a government surveying crew on the Yukon creeks, he acquired a fraction of a claim that fell outside survey lines, a fraction so stupendous that it was "for its size the richest single piece of ground ever discovered," yielding half a million dollars in gold. Yet Lowe drank and spent and within a few years was "peddling water by the bucket in Fairbanks, Alaska."

There were exceptions to this pattern of careless management and unrestrained living. Some long-term prospectors showed business acumen and steady, if unspectacular, success. Some maintained the true Midas touch. The Berry family—Clarence and Ethel, followed in matrimony and affluence by Clarence's brother Henry and Ethel's sister Edna—demonstrated that small-scale western entrepreneurs could parlay a rich Klondike claim into lifelong wealth through wise management and investments. Prospector Winfield Scott Stratton in Colorado became and remained one of the western mining kings, demonstrating spectacularly that a stereotypical prospector—independent, eccentric, rambling, reclusive—could also be a canny, farsighted, community-minded mining magnate. At least this one could. Stratton's business prowess and philanthropic activity were legendary. His largesse extended to providing bicycles for Colorado Springs laundry girls and hiring men "to feed the horned larks" during a Colorado blizzard. More important, he tried against impossible odds to maintain a western democratic environment in a rapidly industrializing mining center. Against his will, he was drawn into the bloody labor wars in Cripple Creek in the 1890s, but he remained an exemplar of western gold rush culture—solitary and self-sufficient, yet eager that everyone get a fair shake and willing to spread his good fortune around. "There ain't one of them [Cripple Creek] millionaires that ever done a thing for me except Win Stratton," Cripple Creek gold discoverer Bob Womack would remark.

The discoverers who spurred stampedes seemed to share a curse not unlike that looming over those who disturb a mummy's tomb: Finders of rich deposits were said to come to "untimely or violent ends." But most seem to have simply faded from the scene early, their pittance from sale of the claim in hand, and continued their rambling lives.

Some became bitter: California's James Marshall inveighing against the "grabbers," Womack complaining that "they have lawed me out of every cent I ever had," the Klondike's Robert Henderson carrying a grudge against the Rabbit Creek discoverers who had followed his tip to glory without sharing the good news with him.

And for every Womack or Henderson, there were hundreds of thousands of gold-seekers who also could justify bitterness at their own economic failure. Some managed to resist the culture's equating of great financial wealth with success and stopped with modest gains. Through steady toil in the diggings, one California stampeder cleared a thousand dollars and departed for home declaring "that was all he wanted": funds to pay off his farm and enable him to support his wife and child. But others, less easily satisfied or less fortunate, sooner or later found "poverty and discouragement . . . the only fruits they reaped from their hard and thankless labor."

In economic terms, the entire mining West even with its bonanza silver and copper reserves "scarcely broke even." True, Marshall's California discovery led to the doubling of global gold stores in just ten years and provided "a highly significant increase in the national currency," while the Klondike stampede helped end the depression of the 1890s. But again and again throughout the second half of the nineteenth century, the seekers expended far more money in outfitting themselves, in traveling to and living in the goldfields, and in trying to retrieve the gold than the yield warranted.

What remained, then, was the adventure itself, and it is clear that many whom the gold eluded did not regret their search. California argonaut William M'Collum found his part in the stampede of 1849 "a wild, I may almost acknowledge a hare-brained adventure," but admitted "no regrets, no wishes that it had not been undertaken," concluding, "With California, and all that is in it, I quit even." An even more affirmative view is reflected in the words of a stampeder of half a century later who arrived in Alaska with thirty-five cents and left two years later without even that amount: "I made exactly nothing, but if I could turn time back I would do it over again for less than that."

What lay behind such a statement? First, gold pilgrims enjoyed the opportunity to be adventurer-heroes. They were "embryo Stanleys and Livingstones," Jasons seeking the golden fleece, conquistadors searching out cities of gold. Imagining themselves a great army as their quasi-military units moved across the western landscapes, they shared the camaraderie of camp and claim, devised battle plans, assaulted the earth, tallied losses and triumphs, and dreamed of future glories.

Even without the mythic imagery, the gold rush population showed a pleasing propensity to elevate nearly every accepted fellow adventurer into Somebody in a strange mingling of the hierarchical and the fiercely egalitarian. This trend started in each rush with the naming of companies. One man rolled out his company's name in his journal—"Long Beach, Alaska, Mining and Trading Company"—then noted, "How bulky and pompous that sounds!" But that was the point—each member of a company was a tycoon in the making.

Men further established themselves as Somebody by virtue of any skill or experience at all. Thus, as one observer of gold rush California put it, a fellow who "could blow a fife on training days" at home became transformed into a professor of music, while the builder of a pigsty or kennel would be "a master-builder in California."

Even if a man didn't claim a skill, he was likely to soar in rank, as titles of status and respect were applied with enthusiastic abandon. This practice, too, began in California, where an Englishman upon being called "captain" concluded, "I rank as Captain in California being *nothing;* if I was a real Captain I should of course be a General there."

Another satisfaction was provided by the novel and stimulating surroundings. "This rambling around among the mountain scenery with me is like drinking with the drunkard or gambling with the gambler," admitted one Colorado stampeder. An Alaska adventurer rhapsodized, "Oh, it is all so enjoyable and fascinating to me! It is like reading a book on a new subject, for one interested in nature to visit this country." And the camps and towns were equally exhilarating, rolling along as they did on a tide of intense speculative excitement. Everywhere the stampeders looked, they saw drama. " 'Tis a queer life, this," wrote one Colorado argonaut, likening it to "a Theater on a great, grand scale." This drama was heightened by the fact that camps and towns overflowed with "the wildest schemes of human fancy"— schemes for finding gold, for extracting gold, for parting successful miners from their gold, for developing this new land in any profitable way. An observer's comment on Virginia City, Nevada, in its silver heyday stands as a testament to gold towns as well: "No man who has ever breathed the air of excitement and speculation of Nevada can live and be content in the quiet of his Eastern birthplace." The "bustle and activity" of the mining camp would lure him back even more than the gold itself.

Gold-seekers also reveled in the freedom "to choose . . . between other men's dicta and our own convictions." A Virginia City newspaper editor stated this lure directly: "The one great blessing is perfect freedom. . . . Everyone does what pleases him best." Men moved about at will, lived as they saw fit, and pursued their own schemes. Even the most grueling manual labor became "fun" when seekers could say "we are working for ourselves."

Such a celebration of freedom carried an implicit criticism of the culture from which the stampeders came. "We may miss some of the good things of life by being out here, but we escape some mighty disagreeable experiences," wrote a prospector rejoicing in his freedom from guilt-inducing sermons. Civilization's dissonances and artificial constructs were also noted; to gold camp resident Nellie Cashman, the closeness to nature and simple neighborliness that characterized Klondike existence stood in stark contrast to urban civilization's "banging trolley cars, honking cars, clubs for catty women and false standards of living."

Stampeders also relished their ability to survive in the new lands. A "Pikes Peaker" reported with pride that he camped "right down on the Ground with one Blanket" and enjoyed "a better night's rest than I could on one of the nicest feather Beds in the World" and noted that his arduous effort in killing a mountain goat for food was probably "what made it taste so Good." Such pride in one's abilities is also reflected in a Nome miner's reaction to a letter from a friend about a "pleasure trip" with some "hard luck experiences": "He would last about seventeen minutes up here."

Gold adventurers found particular satisfaction in their ability to do without, commenting repeatedly upon the value of this discipline. "Really, everybody ought to go to the mines, just to see how little it takes to make people comfortable in the world," mused Louise Clappe, while a California prospector preparing bread of snow-and-flour mix on an outdoor spit concluded, "It is astonishing how small a culinary outfit is really needed." An Alaska prospector, feasting on lemonade, potatoes, and onions in the Cape Nome area after a season on the Kobuk, asserted, "It does a fellow good to be without such things a while, if not too long. He knows better how to appreciate them."

Many developed or retained an appreciation of their fellows as well, concluding that people on the gold frontier were more real, more authentic, were "in general, exactly what they appeared to be"—and that was, in general, good. This view has been repeatedly challenged by critics contending that the stampeders brought no values to the frontier or lost what few values they had there, that the feckless and reckless floated westward, rootless, valueless, greedy. The arduousness and broad nature of the stampedes, drawing a wide variety of argonauts, belies such a judgment. The participants themselves felt that the gold rush experience separated weak from strong. They argued in accordance with the social Darwinism of the late nineteenth century that only the stalwart could handle the westering gold quest. California poet Joaquin Miller wrote that "it took men of grit," a "powerful and select lot," to get to the northern California diggings: "The cowards had never started there and the weak died on the way." A Klondike stampeder voiced similar sentiments: "There were no drones in the Klondike. It was no sanitorium for the weary and faint-hearted."

Charges of rootlessness, irresponsibility, and greed are harder to dismiss. Certainly, the rushes offered ample examples of men behaving foolishly and selfishly. They escaped home responsibilities—"Jonahs" in the belly of the elephant they had gone west to see—then shied away from any exercise of citizenship duties on the frontier.

PRECIOUS DUST

At their most benign the stampedes seemed to encourage an extended adolescence, an indefinite rejection of responsibilities and associations valued and expected by eastern culture, from money management to marriage. Argonauts resembled not only Huck Finn but Peter Pan's Lost Boys. They lived in brush arbors, tents, and rude cabins, furnishing those abodes with found items and ingenuity, as a purposeful child would a playhouse. In an odd echo of childhood pastimes, they spent their days on treasure hunts and planning other improbable schemes for wealth. They indefinitely avoided long-term relationships with the opposite sex, set up their own codes of conduct, and exulted in their freedom to do as they pleased. This mentality is illustrated by two gold camp denizens who climbed a mountain and amused themselves by "detaching the largest boulders we could handle," unleashing them on the trees in the valley below, and finding it "a rare satisfaction . . . to know that there was no crusty old fellow living down there, who would come out and shake his fist at us, and threaten to carry a complaint to our mothers."

In a more somber manifestation of adolescent rebellion, many argonauts can be said to resemble the prodigal son of the Bible. Rejecting the path of steady industry at home, they went to a far country, spent their inheritance, often engaged in licentious living, and longed for the comforts and stabilities they had rejected. Alaska stampeder Marshall Bond's comment, while light in tone, echoed the situation of the wayward son in the Bible, reduced to eating the slops he fed the pigs: "I wish . . . that whenever Jags [Bond's dog] waddles up to a plate of food and contemptuously turns his nose up, you'd make him eat it, for it makes me jealous to think he is getting better food than I am." But worse than the living conditions for most was the fear that no fatted calf would be killed in celebration of their return, that they would instead be seen as pathetic ne'er-do-wells.

Yet casting gold-seekers as prodigal sons or Jonahs or Lost Boys obscures the ways in which their experience changed and deepened their understanding of the world. Stampeders felt that the gold frontier changed for the better those strong enough to benefit from it or simply made them wiser in elemental ways. "I never knew nothing until I went to Alaska," wrote one female rush participant. "That's where I learned—it will learn anybody.

There was so much that I couldn't understand. I seen for the first time the nothingness of the small things." Dawsonite Cashman insisted it took "real folks to live by themselves in the lands of the north," where solitude and nature brought out "the soul of folks."

Observers, too, felt that argonauts "got licked into shape, and polished" through their experiences. Future historian H. H. Bancroft, en route to California in 1852, would later remember the California-bound in Panama as having "the confident swagger of greenness yet upon them," demonstrating a perverse selfishness, but the returnees, he wrote, had been "licked into some degree of form and congruity by their rough experiences, rude and ragged . . . but quieter, more subdued, more ready to yield some fancied right for the common good." A reporter judged the western prospectors he met twenty years later to have cultivated a depth and breadth of spirit and intellect that made each "better than his craft."

There were lessons the gold-seekers learned imperfectly or never learned at all. More democratic and egalitarian than their eastern counterparts, the majority nonetheless retained racial and gender prejudices. More aware of the importance of community to individual survival and group order, they often avoided the responsibility that went with this knowledge. Further, miners seemed to remain ignorant of the conflict inherent in the appreciation many showed for nature and the gusto with which they attacked it, for a solitary placer miner was just as much of a ravager of the earth as an industrial mining company—simply on a smaller scale.

They had the evidence right before their eyes. The stripped and scrambled soil and the profusion of prospect holes combined with the careless disposal of the refuse of daily life made mining camps blights upon the landscape, especially as they declined. After the local fluming company had failed and most of the miners had departed one California location for better diggings, the surrounding hills lay stripped of trees, voluminous piles of gravel still dominated the bar, abandoned tents and brush arbors sagged across the forlorn landscape, and the ground was layered with "empty bottles, oyster cans, sardine boxes, and brandied fruit jars." A few years after the Klondike rush, a newcomer found Dawson a jumble of secondhand shops cluttered with the flotsam of the rush, from rubber boots to hand organs. Strewn in the vacant lots and lining the riverfront were "piles of rusting mining machinery—boilers, winches, wheelbarrows and pumps." The land itself lay scarred and reduced, whole hills having crumbled under hydraulic hoses, creek beds having been mangled by the massive dredges first introduced into western mining in 1897.

Unconcerned with the environmental havoc they created, gold-seekers had found their toughest personal challenge to be learning to win or lose with equanimity. It is no accident that the professional gambler often gained more respect, however grudging, than anyone else on the gold frontier. A "sport" could be counted upon to play a game by the rules (even though he may have stacked the cards to begin with), to pay his debts quickly when flush, and to divert a few dollars to a fellow down on his luck. But most important, he won and lost great sums with the same studied indifference.

In doing so, he provided a model to everyone in the gold regions, to all the gamblers fighting heavy odds. The gold frontier was only a slight exaggeration of the nineteenth-century American frontier in general in being a place of high hopes and dashed expectations. It reverberated with false stampedes, with partnerships ended in bitterness or sorrow, with barren winters, with laboriously dug golden dirt and gravel that turned out to be only dirt and gravel, leading to disappointments so great that men could barely bring themselves to speak of them years later.

Such painful failures are behind Deadwood resident Estelline Bennett's perception of a code of conduct in which "being a good loser" was paramount. "I had heard about it all my life," she would write. "I knew that to be a poor loser was almost as bad as being a horsethief." Her father explained why: taking loss in stride required "a combination of courage, philosophy, and self-control."

Frontier westerners cultivated this trait to the extent that Anthony Trollope found them bearing economic loss "as an Indian bears his torture at the stake." Trollope—who judged westerners dirty, uncouth, and prone to heavy drinking—even deduced a "certain manliness" and "dignity" in their stoicism.

Like their counterparts battling drought and grasshoppers on the plains farming frontier, gold-seekers provide one of the best arguments for Frederick Jackson Turner's contention that frontier immigrants had to adapt to the new environment before they could make it adapt to them—if they could at all. The triumphalist Turner saw both adaptations taking place, in the process defining American character and feeding the flow of American progress. Western historians at the end of the twentieth century tend to look with more jaundiced eye at the frontier experience, and, where Turner stressed cooperation and progress, see greed and exploitation and failure.

The stampedes lend themselves to both interpretations, but the truth, I believe, lies somewhere in between. The rushes created a gold frontier that, in its constant movement and renewal and in its rejection of the old and established, maintained the just-born vitality, the democracy, diversity, and individualistic spirit that characterized America's beginnings and its Jacksonian period. Putting large portions of the West and Far North "on the map" with a swiftness unknown to the farming, ranching, and trading frontiers, they also sped westward the assumptions and values of mid- to late-nineteenth-century American culture. Yet at the same time, participants challenged that culture, opting out of a world in which corporate capitalism was replacing small-scale agricultural and entrepreneurial pursuits. Instead, they chose simpler life-styles, more frank and direct social relations over artificial convention.

Further, many expressed ambivalence toward the very cultural goal they seemed to be trying to accomplish: success through the amassing of wealth. Despite his later bitterness, Bob Womack reportedly responded to the news that Win Stratton had become Cripple's first millionaire by exclaiming, "Poor Old Man Stratton! All that money to worry about. . . . I don't envy him one bit!" Klondike gold-seeker Bill Liggett aptly reflected the attitude of many a long-term prospector to a friend: "If I wanted to stay here I could make piles and piles of money. But what would I do with it all? I couldn't spend it. No, I will sell out. . . . My disposition is too wandersome. I shall go to the tropics and try mining there."

Liggett represented a class of stampeders who had found their calling on the gold frontier, men "uneasy" in any settled mode of living. But Liggett's remarks also point up an important element in the gold hunt of many an argonaut from California to the Klondike, however long or short: The gold itself mattered less or came to matter less than the quest for it. "It isn't the gold that I'm wanting/So much as just finding the gold," explained Klondike bard Robert Service.

In this sense, the stampeders themselves wrested a triumphalist interpretation from their journeys and sojourns. They had dreamed of riches, but the experience itself became the real treasure. Those who survived the mining ups and downs found meaning and pride, recognized their part, however small, in the history of westward expansion, celebrated having been alive and free in the grand western vistas, with a fortune always just over the next hill. Theirs was a quixotic quest, but not necessarily an overly self-absorbed and grasping one. Philosopher Josiah Royce, the son of forty-niners, was himself critical of gold rush society but still opined that those who accused Americans of "extraordinary love for gain" failed to understand "how largely we are a nation of idealists."

The rushes amply demonstrated that idealism, that simple optimistic vitality. They were, after all, deliriously hopeful affairs. Despite their hardships and heartbreak, they were fun—at least for a while—and provided the most excitement many stampeders would ever see.* For those able to spin along on the wheel of fortune retaining their faith and optimism, the pleasures outweighed the pain. Stampeders laughed with self-recognition at the Deadwood-era campfire tale of the aged prospector who, upon going to heaven, identified his occupation to Saint Peter. "I'm afraid I can't admit you," Saint Peter said. "We have more of your kind in here now than we can well accommodate; I would much prefer to get rid of some of them if possible, than to allow more to enter." The old prospector promised "to thin them out for you" and spread a rumor of a gold strike in hell. The other prospectors accordingly stampeded

* Hundreds of thousands of Americans shared this excitement vicariously through purchase of (usually worthless) mining stocks, which were cheaper than railroad or industrial stocks.

to the netherworld. But no sooner had they done so than "the saint noticed that the old man was uneasy" and asked what was wrong. "Peter, I guess you'll have to let me out of here," announced the prospector. "You see on thinking it over there may be something in that report of rich diggings in hell."

No wonder it was hard for men to return home and settle down; no wonder even the aging veterans of 1849 began packing for the Klondike when Bonanza and Eldorado creeks blazed to glory. But most stampeders did go home, settled down, became "average citizens" once again. Yet they had lived through too much to ever forget, and the memoirs of such people are shot through with pride and delight in remembrance of the time when a new land with all its wonders and contradictions unfurled before them.

Others chose the rambling life for years or even decades, some returning like Rip Van Winkles only when the family felt sure they were dead or when the family itself had died out.

Others stayed forever—those who died in tents and beside campfires, in saloons and abandoned camps, and those who couldn't move their hearts, souls, and bodies away from the goldfields, instead wandering year after year in golden hopes. Members of this last group filled a hospital veranda in Dawson in the 1920s, "old sourdoughs" of 1897 and '98 eking out their last days contemplating the broad Yukon River. One thought himself on that river on a steamboat and repeatedly looked for confirmation by saying, "I think we're reaching Stewart City now, aren't we, nurse?"

She always assured him they were.

Notes

Note: If more than one quote in a sentence is from the same source, this is indicated by citing only the initial words of the first quote or by including the initial quote and the first part of the one following.

Chapter One: "The Age of Gold"

22 "corroborated by . . .": quoted in J.S. Holliday, *The World Rushed In: The California Gold Rush Experience* (New York: Simon & Schuster, 1981), p. 48.

22 "where you . . .": Elisha Lord Cleaveland, "Hasting to Be Rich. A Sermon . . ." (New Haven, Conn.: J.H. Benham, 1849), p. 15.

23 "see more of . . .": Letter of January 15, 1849, Angell Letters, Beinecke Library, Yale University, New Haven, Conn.

23 "it seemed to . . .": quoted in Richard H. Peterson, *The Bonanza Kings: The Social Origins and Business Behavior of Western Mining Entrepreneurs, 1870–1900* (Lincoln, Nebr.: University of Nebraska Press, 1977), p. 17.

23 "crowded with . . .": Franklin A. Buck, *A Yankee Trader in the Gold Rush: The Letters of Franklin A. Buck* (Boston: Houghton Mifflin, 1930), p. 31.

23 "more or less . . .": Bayard Taylor on California voyagers in *Eldorado: Or Adventures in the Path of Empire* (Lincoln, Nebr.: University of Nebraska Press, 1988), p. 3.

24 "shouldered their . . .": quoted in Holliday, pp. 35–36.

24 "can do . . .": quoted in Eugene Lyon, "Search for Columbus," *National Geographic*, Vol. 181, No. 1 (January 1992), p. 25.

25 "No talk . . .": quoted in T.H. Watkins, *Gold and Silver in the West: The Illustrated History of an American Dream* (New York: Bonanza Books, 1971), p. 21.

25 "as if . . .": Benjamin Parks quoted in Otis E. Young, Jr., "The Southern Gold Rush, 1828–1836," *Journal of Southern History*, XLVII (August 1982), p. 384.

26 "Gold and silver . . .": Watkins, p. 21.
26 "wish[ed] to God . . .": quoted in Frank Bergon and Zeese Papani-kolas, eds., *Looking Far West: The Search for the American West in History, Myth, and Literature* (New York: New American Library, 1978), p. 111.
27 "about half . . .": Watkins, p. 26.
27 "It made . . .": ibid.
27 "Boys, by G[o]d . . .": quoted in Clyde A. Milner II, ed., *Major Problems in the History of the American West* (Lexington, Mass.: D.C. Heath & Co., 1989), p. 313.
27 "miracle . . .": Benjamin Butler Harris, *The Gila Trail: The Texas Argonauts and the California Gold Rush,* ed. Richard H. Dillon (Norman, Okla.: University of Oklahoma Press, 1960), p. 93.
28 "Plenty for . . .": quoted in Holliday, p. 43.
28 "Oh, the gold . . .": quoted in Bergon and Papanikolas, p. 254.
28 "city to come . . .": ibid. p. 255.
28 "perhaps the . . .": Donald Dale Jackson, *Gold Dust* (New York: Knopf, 1980), p. 21.
28 "from the Trinity . . .": Holliday, p. 41.
28 "decided to . . .": George Stewart quoted in Ferol Egan, *The El Dorado Trail: The Story of the Gold Rush Routes Across Mexico* (Lincoln, Nebr.: University of Nebraska Press, 1984), p. 49.
28 "Every day men . . .": quoted in Holliday, p. 49.
29 "Remember Father . . .": Letter of April 4, 1849, Alonzo Hill Papers, Beinecke Library.
29 "that I should . . .": Charles William Churchill, *Fortunes Are for the Few: Letters of a Forty-Niner,* ed. Duane A. Smith and David J. Weber (San Diego: San Diego Historical Society, 1977), p. 26.
29 "at my time . . .": ibid.
29 "Oh, what was . . .": quoted in Donald Jackson, p. 236.
29 "So good bye . . .": quoted in Grant Foreman, *Marcy and the Gold Seekers* (Norman, Okla.: University of Oklahoma Press, 1968), p. 26.
29 "light[ing] out . . .": Mark Twain, *The Adventures of Huckleberry Finn,* Vol. I., *Anthology of American Literature,* ed. George McMichael (New York: Macmillan Publishing Co., 1974), p. 517.
29 "Captain Sutter's . . .": Donald Jackson, p. 66.
30 "no wilderness": quoted in Egan, p. 32.
31 "the more unsettled . . .": John S. Hittell quoted in Watkins, p. 45.
31 "evidently the . . .": Granville Stuart, *The Montana Frontier, 1852–1864* (Lincoln, Nebr.: University of Nebraska Press, 1977), p. 61.
31 "so great . . .": Jane Apostol, "Gold Rush Widow," *Pacific Historian,* Vol. XXVIII, No. 2 (Summer 1984), p. 53.
31 "mountains of . . .": ibid. p. 61.
31 "Somebody had . . .": P. H. Lewis Reminiscences, Bancroft Library, University of California, Berkeley, Calif., p. 10.
32 "IX. Thou shalt . . .": quoted in Marvin Lewis, ed., *The Mining Fron-*

tier: *Contemporary Accounts from the American West in the Nineteenth Century* (Norman, Okla.: University of Oklahoma Press, 1967), p. 41.

32 "without at . . .": *Frazer River Thermometer,* April 1858, Beinecke Library.

32 "lying thick . . .": ibid.

33 "We are again . . .": Lucetta Rogers letter of September 5, 1858, in Ruth B. Moynihan, Susan Armitage, and Christiane Fischer Dichamp, eds., *So Much to Be Done: Women Settlers on the Mining and Ranching Frontier* (Lincoln, Nebr.: University of Nebraska Press, 1990), p. 16.

33 "over $600 . . .": quoted in LeRoy Hafen, ed., *Colorado Gold Rush: Contemporary Letters and Reports 1858–1859* (Philadelphia: Porcupine Press, 1974), p. 32.

34 "every man . . .": ibid. p. 41.

34 "Have it done . . .": ibid. p. 49.

34 "gold, and . . .": fifty-niner quoted in Robert Athearn, "The Fifty-Niners," *The American West,* Vol. XIII, No. 5 (September–October 1976), p. 61.

34 "You might . . .": quoted in Linda Peavy and Ursula Smith, *The Gold Rush Widows of Little Falls: A Story Drawn from the Letters of Pamelia and James Fergus* (St. Paul: Minnesota Historical Society, 1990), p. 36.

34 "wend my . . .": ibid. p. 39.

34 "distances were . . .": Watkins, p. 53.

34 "This brings . . .": *Missouri Democrat,* September 21, 1858, in Hafen, op. cit. p. 56.

35 "Pikes Peak hats . . .": Stanley W. Zamonski and Teddy Keller, *The '59ers: Roaring Denver in the Gold Rush Days* (Frederick, Colo.: Jende-Hagan Bookcorp, 1983), p. 97.

35 "I've got . . .": George Jackson journal entries of January 8 and February 9, 1859, in "Jackson's diary of '59" (Idaho Springs, Ida.: 1929), Barker Texas History Center, Austin, Tex.

35 "saved the rush": William S. Greever, *The Bonanza West: The Story of the Western Mining Rushes 1848–1900* (Moscow, Ida.: University of Idaho Press, 1986), p. 163.

35 "The Rocky Mountains . . .": quoted in Zamonski and Keller, p. 34.

36 "fierce, riotous . . .": William Parsons quoted in Robert Wallace, *The Miners* (New York: Time-Life Books, 1976), p. 22.

36 "Jerusalem, what . . .": quoted in Watkins, p. 55.

36 "By God, I've . . .": ibid. p. 60.

37 "a great city . . .": *Rocky Mountain News,* June 1860, quoted in Glenn Quiett, *Pay Dirt: A Panorama of American Gold Rushes* (New York: D. Appleton-Century Co., 1936), p. 145.

37 "it required . . .": Peterson, p. 88.

37 "the richest . . .": Watkins, p. 65.

37 "pour[ed] their . . .": Charles Crampton, "Gold Rushes Within the

United States, 1800–1900: A General Survey" (M.A. thesis, University of California at Berkeley, 1935), p. 64.

38 "the backwash . . .": Quiett, p. 342.

38 "rarely, if . . .": Loo-Wit Lat-Kla [pseudonym], "Gold Hunting in the Cascade Mountains," facsimile of 1861 edition (New Haven, Conn.: Yale University Library, Western Historical Series No. 3), p. 5.

38 "uninhabitable wasteland": unattributed quote in Roger O. Walker, "One Tall Drink of Water," *True West*, Vol. 27, No. 1 (Fall 1990), p. 29.

39 "It is now . . .": Robert Hauser letter of May 20, 1862, Hauser Papers, Beinecke Library.

39 "nearly every . . .": Julius Merrill, *Bound for Idaho: The 1864 Trail Journal of Julius Merrill*, ed. Irving Merrill (Moscow, Ida.: University of Idaho Press, 1988), p. 26.

40 "too tired . . .": quoted in Quiett, p. 220.

40 "spent the day . . .": Larry Barsness, *Gold Camp: Alder Gulch and Virginia City, Montana* (New York: Hastings House, 1962), p. 11.

40 "I have got . . .": Greever, pp. 219–20.

40 "a regular . . .": Stuart, p. 270.

40 "dug up . . .": ibid. pp. 271–72.

41 "with their blankets . . .": May 1868 report quoted in James Chisolm, *South Pass, 1868; James Chisolm's Journal of the Wyoming Gold Rush*, ed. Lola M. Homsher (Lincoln, Nebr.: University of Nebraska Press, 1975), pp. 44–45.

42 "raised three . . .": quoted in Watson Parker, *Gold in the Black Hills* (Lincoln, Nebr.: University of Nebraska Press, 1982), p. 39.

42 "determine in . . .": quoted in Watkins, pp. 108–109.

42 "reasonably expect . . .": ibid. p. 109.

42 "set comfortable . . .": quoted in Watson Parker, p. 38.

43 "conservative estimate": Greever, p. 291.

43 "to those . . .": ibid. p. 295.

43 "all of . . .": ibid. p. 302.

44 "it was not . . .": Eugene V. Smalley, "1884: The Great Couer d'Alene Stampede," *Idaho Yesterdays*, Vol. 11, No. 3 (Fall 1967), p. 6.

46 "created at . . .": Greever, p. 213.

46 "a springboard . . .": Pierre Berton, *The Klondike Fever* (New York: Carroll and Graf Publishers, 1985), p. 6.

46 "a life independent . . .": Melody Webb, *The Last Frontier* (Albuquerque, N.Mex.: University of New Mexico Press, 1985), p. 96.

47 "the richest placer . . .": Pierre Berton, p. 57.

47 "the all-firedest . . .": [Howard Kelly], *Sourdough Gold: The Log of a Yukon Adventure*, ed. Mary Lee Davis (Boston: W.A. Wilde Co., 1933), p. 168.

47 "all led away . . .": Jeremiah Lynch, *Three Years in the Klondike*, ed. Dale L. Morgan (Chicago: R.R. Donnelley, 1967), p. 104.

48 "piled about . . .": quoted in Terrence Cole, "A History of the Nome

Gold Rush: The Poor Man's Paradise" (Ph.D. dissertation, University of Washington, 1983), p. 5.

48 "Can I . . .": Pierre Berton, p. 121.

48 "stirred from . . .": E. Hazard Wells, *Magnificence and Misery: A First-hand Account of the 1897 Klondike Gold Rush* (Garden City, N.Y.: Doubleday & Co., 1984), pp. 1, 6.

49 "frozen swampland": Pierre Berton, p. 73.

49 "other thousands": Wells, p. 80.

49 "staked from . . .": ibid. p. 113.

49 "flying northward . . .": Joseph Grinnell, "Gold Hunting in Alaska As Told by Joseph Grinnell," ed. Elizabeth Grinnell, *Alaska Journal*, Vol. 13 (Spring 1983), p. 11.

49 "to all points . . .": Wells, 157.

49 "bringing in . . .": Marshall Bond, Jr., *Gold Hunter: The Adventures of Marshall Bond* (Albuquerque, N.Mex.: University of New Mexico Press, 1969), p. 55.

49 "the barrier . . .": Pierre Berton, p. 205.

50 "easier than . . .": Cole, p. 75

50 "half the . . .": quoted in ibid. p. 39.

Chapter Two: Overland to California

53 "land of extremes": Elliott West, *Growing Up with the Country: Childhood on the Far Western Frontier* (Albuquerque, N.Mex.: University of New Mexico Press, 1989), p. 4.

53 "highest and . . .": ibid.

53 "the country's . . .": ibid.

55 "pour[ed] down . . .": Charles Hinman, *"A Pretty Fair View of the Eliphant": or, Ten Letters by Charles G. Hinman . . .*, ed. Colton Storm (Chicago: G. Martin, 1960), letter of May 8, 1849.

55 "What we . . .": ibid.

55 "I would . . .": Lucius Fairchild, *California Letters of Lucius Fairchild*, ed. Joseph Schafer (Madison, Wisc.: State Historical Society of Wisconsin, 1931), p. 38.

55 "You may rest": quoted in Holliday, p. 296.

56 "broad forehead . . .": John Woodhouse Audubon quoted in Egan, p. 66.

56 "lonely woodyards . . .": A.G. Henderson, "My Journey to the Gold Fields: Reminiscences of an Argonaut," *The American West*, Vol. XIII, No. 3 (May–June 1976), p. 5.

56 "a solid . . .": quoted in Foreman, p. 83.

56 "Oh! my God . . .": ibid.

58 "Neither the . . .": quoted in Holliday, p. 59.

58 "they can . . .": ibid. p. 97.

59 "3 good-looking . . .": Brother Gatien Monsimer quoted in Franklin Cullen, C.S.C., "Holy Cross on the Gold Dust Trail," in *Holy Cross on the Gold Dust Trail and Other Western Ventures* (Notre Dame, Ind.: Indiana Province Archives Center, 1989), p. 16.

59 "you had . . .": Holliday, p. 148.

59 "This strangest . . .": Stuart, p. 42.

59 "towering far . . .": Merrill, p. 61.

59 "a broken . . .": William Swain quoted in Holliday, p. 182.

60 "the most miserable . . .": James Abbey quoted in Brigham Madsen, *Gold Rush Sojourners in Great Salt Lake City, 1849 and 1850* (Salt Lake City: University of Utah Press, 1983), p. 26.

60 "The traveling . . .": Arthur St. Clair Denver letter of January 7, 1850, Beinecke Library.

61 "wagon wheels . . .": Holliday, p. 230.

61 "was even . . .": Sarah Royce, *A Frontier Lady: Recollections of the Gold Rush and Early California,* ed. Ralph Henry Gabriel (New Haven, Conn.: Yale University Press, 1932), p. 71.

62 "not much . . .": Egan, p. 109.

63 "meandering line . . .": ibid. p. 157.

63 "covered with . . .": ibid. p. 166.

63 "Faith and . . .": Denver letter of January 7, 1850.

63 "ready to . . .": quoted in David Rich Lewis, "Argonauts and the Overland Trail Experience: Method and Theory," *Western Historical Quarterly,* Vol. XVI, No. 3 (July 1985), p. 295.

63 "fill[ed] the . . .": entry of May 7, 1849, Phineas Blunt journal, Bancroft Library, University of California at Berkeley.

64 "formed in . . .": argonaut quoted in Holliday, p. 105.

64 "commenced . . .": John D. Unruh, Jr., "The Traveling Community of the Overlanders," in Milner, p. 283.

64 "The desert! . . .": quoted in Thomas H. Hunt, "The California Trail: A Personal Quest," *The American West,* Vol. XI, No. 5 (September 1974), p. 19.

64 "Our tongues . . .": Lorenzo Dow Stephens, *Life Sketches of a Jayhawker of '49* (San Jose, Calif.: Nolta Brothers, 1916), p. 21.

65 "a fine . . .": C.C. Cox, ed., "From Texas to California in 1849: Diary of C.C. Cox," ed. Mabelle E. Martin, *Southwestern Historical Quarterly,* Vol. XXIX, No. 1 (July 1925), p. 44.

65 "I never . . .": ibid. pp. 45–46.

65 "I hope . . .": ibid.

65 "a region . . .": Harris, p. 41.

65 "welcome singing": ibid.

65 "hearty and well": entry of July 3, 1849, Blunt journal.

65 "None of . . .": entry of July 7, 1849, ibid.

65 *"Turn back* . . .": Royce, p. 45 (emphasis in original).

66 "an abomination . . .": argonaut quoted in Holliday, p. 253.

66 "rancid bacon . . .": ibid. p. 229.
66 "only as . . .": Harris, p. 89.
66 "precious slice . . . a small . . .": ibid.
66 "terribly inflamed . . .": Egan, p. 105.
66 "most *damniabl* . . .": Fairchild, p. 36.
67 "stony and rough": James Pratt quoted in Holliday, p. 279.
67 "a heightened . . .": Andrew J. Rotter, " 'Matilda for Gods Sake Write': Women and Families on the Argonaut Mind," *California History*, Vol. LVIII (Summer 1979), p. 131.
67 "not even . . .": Royce, p. 87.
67 "the hardest . . .": quoted in Holliday, p. 128.
68 "great cook . . .": quoted in Foreman, p. 74.
68 "Made batter . . .": entry of May 8, 1849, Blunt journal.
68 "mixed beans . . .": quoted in Cullen, p. 16.
68 "doing their . . .": Royce, p. 12.
68 "didn't want . . .": Stephens, p. 20.
68 "as plucky . . .": ibid.
68 "distinguished by . . .": Royce, pp. 8–9.
68 "as much . . .": Cox, p. 39.
69 "that should . . .": Harris, p. 38.
69 "hammered the saddle . . .": ibid.
69 "ludicrous scene": entry of July 11, 1849, Blunt journal.
69 "[Two] men . . .": ibid.
69 "many of our . . .": ibid.
69 "one of the . . .": Stephens, p. 22.
69 "wagons, and . . .": quoted in David Lewis, p. 291.
70 "at the rear . . .": Royce, p. 53.
70 "a monumental . . .": Egan, p. 64.
70 "from being . . .": quoted in Madsen, p. 21.
70 "Dey cosht . . .": quoted by J. Goldsborough Bruff in Bergon and Papanikolas, p. 225.
71 "civilized men": Harris, p. 65.
72 "gourds of water . . .": ibid. p. 80.
72 "peaceful, poor . . .": Holliday, p. 114.
73 "It is true . . .": Billy Holcomb quoted in Dinah Rose, "Billy Holcomb, Hard Luck Miner," *True West*, Vol. 34, No. 8 (August 1987), p. 39.
73 "most miserable . . .": Denver letter of Jan. 7, 1850.
73 "Shoot at . . .": Harris, p. 31.
74 "came powering . . .": Daniel McArthur quoted in Madsen, p. 54.
74 "If there is . . .": unidentified argonaut in Holliday, p. 146.
74 "that, in California . . .": Harris, p. 107.
74 "large tents . . .": Donald Jackson, p. 255.
75 "bowel complaints": Holliday, p. 230.
75 "left . . . to . . .": Brigham Young quoted in Madsen, p. 71.
75 "looked in . . .": quoted in Holliday, p. 230.

75 "They buried . . .": ibid.
75 "without all . . .": Unruh, p. 280.
75 "fat, fresh-killed elk": Harris, p. 102.
76 "two or three . . .": Stephens, p. 22.
76 "Boys, you . . .": ibid.
76 "I knew we . . .": ibid.
76 "Not a buffalo . . .": Holliday, p. 146.
76 "in the midst . . .": ibid.
76 "Burthens of . . . as it is . . .": Hinman letter of May 27, 1849.
77 "as natural . . .": Fairchild, p. 29.
77 "far more . . .": quoted in Holliday, pp. 116–17.
77 "no roaches . . .": ibid.
77 "many a gold . . .": Harris, pp. 30–31.
77 "danced cotillions . . .": ibid. pp. 35–36.
77 "large quantities . . .": entry of May 5, 1849, Blunt journal.
77 "the whole . . .": entry of July 14, 1849, ibid.
77 "had a hard . . .": quoted in Holliday, p. 259.
78 "resembling the . . .": Harris, p. 85.
78 "three or four . . .": ibid. p. 94.
78 "commenced vanishing . . .": ibid.
78 "sweeping the . . .": William Swain quoted in Holliday, p. 123.
78 "a central . . .": John Mack Faragher, "The Midwestern Farming Family, 1850," in Linda K. Kerber and Jane Sherron De Hart, eds., *Women's America: Refocusing the Past* (New York and Oxford: Oxford University Press, 1991), p. 123.
78 "the sweetest . . .": Swain quoted in Holliday, p. 155.
79 "no miners . . .": Harris, p. 108.
79 "It is a . . .": James L. Tyson, *Diary of a Physician in California* (Oakland, Calif.: Biobooks, 1955), p. 75.
79 "lank and brown . . .": Taylor, *Eldorado*, p. 37.
79 "except their . . .": ibid.
79 "a woebegone . . .": quoted in Donald Jackson, p. 297.
79 "looking as . . .": Louise Clappe, *The Shirley Letters* (Santa Barbara, Calif., and Salt Lake City, Utah: Peregrine Smith, 1970), p. 181.
79 "I found . . .": Stephens, p. 25.
79 "so entirely . . .": expedition commander Captain Rucker quoted in Holliday, p. 291.
80 "Any man . . .": quoted in ibid. p. 75.
80 "I will never . . .": Hinman letter of October 18, 1849.

Chapter Three: Sea Travels and California Wanderings

84 "equatorial doldrums": James Morison, *By Sea to San Francisco, 1849–1850: The Journal of Dr. James Morison*, ed. Lonnie J. White and William

R. Gillaspie (Memphis: Memphis State University Press, 1977), p. 8.

84 "lofty hills": Morison, p. 8.

84 "the perpetually . . .": Donald Jackson, p. 103.

85 "one of the . . .": Morison, p. 37.

85 "sometimes halfway . . .": Donald Jackson, p. 108.

85 "tall, gaunt . . .": Taylor, *Eldorado*, p. 8.

86 "sink into . . .": John M. Letts, *California Illustrated: Including a Description of the Panama and Nicaragua Routes. By a Returned Californian* (New York: W. Holdredge, 1852), p. 15.

86 "American hotels . . .": Morison, p. 46.

86 "grass-grown plazas": Taylor, *Eldorado*, p. 22.

86 "a miserable . . .": quoted in George Groh, *Gold Fever* (New York: William Morrow, 1966), p. 33.

86 "steeped in . . .": argonaut quoted in ibid.

87 "All were . . .": Morison, p. 48.

87 "palm, cocoa . . .": Taylor, *Eldorado*, p. 26.

87 "appeared to . . .": ibid. p. 34.

87 "blooming plain": ibid.

88 "not yet having . . .": [Linville Hall], *Around the Horn in '49. Journal of the Hartford Union Mining and Trading Company*. (I.J. Hall, 1898 reprint), p. 17.

88 "chains rattling . . .": ibid. p. 40–41.

88 "one minute . . .": entry of May 1, 1849, John Angell journal, Beinecke Library.

88 "to the damage . . .": Morison, p. 14.

88 "it appeared . . .": ibid. p. 19.

88 "like the trampling . . .": Taylor, *Eldorado*, p. 16.

88 "If I am . . .": Letts, p. 32.

89 "about as bad . . .": Henderson, p. 7.

89 "worn into . . .": Taylor, *Eldorado*, p. 19.

89 "coating the rider . . .": ibid.

89 "We pitied . . .": ibid. p. 21.

89 "a narrow . . .": Emeline Day quoted in JoAnn Levy, *They Saw the Elephant: Women in the California Gold Rush* (Hamden, Conn.: Archon Books, 1990), p. 42.

89 "a lady . . .": Taylor, *Eldorado*, p. 23.

90 "frightful": quoted in John Haskell Kemble, *The Panama Route, 1848–1849* (Columbia, S.C.: University of South Carolina Press, 1990), p. 149.

90 "The monotony . . .": Vincente Perez Rosales quoted in Edwin A. Beilharz and Carlos V. Lopez, trans. and eds., *We Were 49ers! Chilean Accounts of the California Gold Rush* (Pasadena, Calif.: Ward Ritchie Press, 1976), p. 11.

90 "Many of the . . .": Hall, p. 99.

90 "if he lived . . .": ibid.

90 "cast-steel and rose-wood": Hall, pp. 96–97.

91 "musty flour . . . being saved . . .": George Schenck quoted in Kemble, p. 159.

91 "We lived . . .": quoted in Joseph Jackson, *Anybody's Gold: The Story of California's Mining Towns* (New York: D. Appleton-Century Co., 1941), p. 99.

91 "bid against . . .": Donald Jackson, p. 81.

91 "on each . . .": quoted in Kemble, p. 154.

91 "Got up . . .": Levi Stowell quoted in ibid. p. 153.

92 "appear[ed] to . . .": Letts, p. 39.

92 "when they . . .": J.D. Borthwick quoted in Groh, p. 41.

92 "almost a giant . . .": John Steele, *In Camp and Cabin* (New York: Citadel Press, 1962), p. 111.

92 "could hear . . .": Frank Merryat quoted in Groh, p. 50.

92 ". . . pretty soon . . .": Henderson, p. 5.

92 "The sides . . .": quoted in Beilharz and Lopez, p. 5.

93 "We had . . .": Sam Burris quoted in Margaret Henson, "The Cartwrights of San Augustine, Texas: Agrarian Entrepreneurs in the Nineteenth Century," manuscript scheduled for publication, p. 191.

93 "the fruit . . .": Taylor, *Eldorado,* pp. 38–39.

93 "their horses . . .": Morison, p. 55.

93 "just before . . .": ibid.

93 "in a brutal . . .": Angell letter of March 26, 1849.

94 "gentlemanly and . . .": Henderson, p. 11.

94 "never got . . .": ibid.

94 "I never . . .": Taylor, *Eldorado,* p. 27.

94 "no berth . . .": quoted in Kemble, p. 154.

95 "humanity to . . .": ibid.

95 "some making . . .": Rosales quoted in Beilharz and Lopez, p. 7.

95 "hair of ropes . . .": Morison, p. 8.

95 "We are moving . . .": Hall, p. 62.

95 "spreading out . . .": Morison, p. 44.

96 "mad with . . .": Mary Jane Megguier quoted in Levy, p. 38.

96 "They take . . .": Morison, p. 13.

96 "with more . . .": ibid. p. 46.

96 "anarchy and . . .": ibid. p. 22

96 "could be Yankeeized . . .": quoted in Kemble, p. 177.

96 "I heard . . .": Callbreath letter of March 14, 1849, Callbreath letters, Bancroft Library.

97 "retorts, crucibles . . .": Letts, p. 44.

97 "fearful and anxious": Pierre Garnier, *A Medical Journey in California,* ed. Doyce B. Nunis, Jr.; trans. L. Jay Olivia (Los Angeles: Zeitlin and Ver Brugge, 1967), p. 31.

97 "To travel . . .": Robert Louis Stevenson, "El Dorado," quoted in *The Oxford Dictionary of Quotations* (Oxford: Oxford University Press, 1989), p. 522.

97 "laborious life . . . the uncertainty . . .": Taylor, *Eldorado*, p. 40.
97 "gather the . . .": Erving quoted in Hall, p. 13.
98 "a good capital": A. Hill letter of October 7, 1849.
98 "it was the . . .": Harris, p. 129.
98 "had sunk . . .": Henderson, p. 63.
98 "This place . . .": Harris, pp. 129–30.
98 "would settle . . .": Steele, p. 53.
98 "and as the . . .": ibid.
98 "terminated in . . .": ibid.
98 "The Bible says . . .": ibid.
98 "from tree . . .": ibid. p. 65.
99 "to the arms": ibid. p. 68.
99 "like so many . . .": Chisolm, p. 118. The author was commenting on the Wind River Range in Wyoming.
99 "I believe . . .": Steele, p. 87.
99 "We never . . .": ibid.
100 "I feel . . .": Churchill, p. 73.
100 "in comparison . . .": Watkins, p. 48.
100 "a hilly road . . .": Alexander Anderson, *Hand-book and Map to the Gold Region of Frazer's and Thompson's Rivers, with Table of Distances* (San Francisco: J.J. Le Count, 1858).
101 "as it gave . . .": Callbreath letter of January 24, 1859.
102 "almost wishing . . .": quoted in Richard Wright, *Overlanders: 1858 Gold* (Saskatoon, Sask.: Western Producer Prairie Books, 1985), p. 56.
102 "new trails . . .": Warren Sadler memoirs in Sadler papers, Bancroft Library, p. 364.

Chapter Four: One Daunting Trail After Another

105 "Oh how . . .": Wilbur Fiske Parker, " 'The Glorious Orb of Day Has Rose': A Diary of the Smoky Hill Route to Pike's Peak, 1858," ed. Norman Lavers, *Montana, the Magazine of Western History*, Vol. 36, No. 2 (Spring 1986), p. 55.
106 "broke my . . .": quoted in Hafen, *Colorado Gold*, p. 146.
106 "Don't you . . .": ibid.
106 "consisted of . . .": Romanzo Kingman, "Romanzo Kingman's Pike's Peak Journal, 1859," ed. Kenneth F. Millsap, *Iowa Journal of History*, Vol. 48 (1950), p. 69.
106 "two or three": ibid. p. 73.
107 "ten cents . . .": LeRoy Hafen, *Overland Routes to the Gold Fields, 1859, from Contemporary Diaries* (Philadelphia: Porcupine Press, 1974), p. 47.
107 "desolate bluff country": Phyllis F. Dorset, *The New Eldorado: The*

Story of Colorado's Gold and Silver Rushes (New York: Macmillan Co., 1970), p. 35.

107 "through sage . . .": Hafen, *Overland,* p. 47.

107 "We . . . had a . . .": quoted in Hafen, *Colorado Gold,* p. 148.

107 "more a pleasure . . .": ibid. p. 151.

107 "hideously hirsute . . .": Horace Greeley quoted in Quiett, pp. 139–40.

107 "nearly all . . .": Kingman, pp. 77–78.

108 "Oh heavens . . .": Wilbur Parker, p. 58.

108 "the size of . . .": Kingman, p. 76.

108 "Perfect hurricane[s]": William Salisbury, "The Journal of an 1859 Pike's Peak Gold Seeker," ed. David Lindsey, *Kansas Historical Quarterly,* Vol. XXII, No. 4 (Winter 1956), p. 328.

108 "a good looking . . .": Wilbur Parker, p. 58.

109 "well supplied . . .": Hafen, *Overland,* p. 266.

109 "not a word . . .": ibid. p. 268.

109 "broken, barren . . .": Dorset, p. 36.

110 "three days . . .": Hafen, *Overland,* p. 268.

110 "scores": Dorset, p. 36.

110 "breaking them . . .": Blue quoted in Ben Maddow, *A Sunday Between Wars: The Course of American Life from 1865 to 1917* (New York: W.W. Norton & Co., 1979), p. 103.

110 "his mind . . .": Dorset, p. 37.

110 "fifteen inches . . .": quoted in Athearn, p. 23.

111 "dry bread . . .": Robert L. Brown, *The Great Pikes Peak Gold Rush* (Caldwell, Ida.: Caxton Printers, 1985), p. 83.

111 "The first sight . . .": quoted in Zamonski and Keller, p. 95.

111 "to such . . .": Mrs. Nat Collins quoted in Moynihan et al., p. 150.

111 "chilling winds . . .": ibid. p. 154.

112 "carrying the . . .": William Armistead Goulder, *Reminiscences: Incidents in the Life of a Pioneer in Oregon and Idaho* (Moscow, Ida.: University of Idaho Press, 1990), pp. 241–42.

113 "the bad . . .": John Buchanan quoted in John E. Parsons, "Steamboats in the 'Idaho' Gold Rush," *Montana, Magazine of History,* Vol. 10, No. 1 (Winter 1960), p. 57.

113 "organized protests . . .": Parsons, p. 60.

113 "beggar[ed] description . . .": Merrill, p. 27.

113 "little difference . . .": ibid.

113 "several bridges": Peavy and Smith, pp. 117.

114 "none of . . .": Merrill, p. 70.

114 "a faucet . . .": James Knox Polk Miller, *The Road to Virginia City: The Diary of James Knox Polk Miller,* ed. Andrew F. Rolle (Norman, Okla.: University of Oklahoma Press, 1950), p. 7.

114 "no house . . .": Merrill, p. 32.

115 "the largest . . .": Miller, pp. 10–11.

115 "deserted and . . .": ibid. p. 22.

115 "taking a . . .": ibid.

115 "Two of them . . .": Merrill, p. 26.

115 "our danger . . .": Miller, pp. 7–8.

115 "there might . . .": Richard B. Hughes, *Pioneer Years in the Black Hills,* ed. Agnes Wright Spring (Glendale, Calif.: Arthur H. Clark Co., 1957), p. 54.

115 "the road . . .": Merrill, p. 63.

116 "They would . . .": ibid. p. 67.

116 "Deliver me . . .": Pamelia Fergus quoted in Peavy and Smith, p. 185.

116 "work[ing] with . . .": Miller, p. 16.

116 "biting through . . .": ibid. p. 14.

116 "nearly devour[ing]": Merrill, p. 47.

116 "it was rather . . .": ibid. p. 73.

116 "left without . . .": ibid. p. 34.

116 "Into the willows . . .": ibid. p. 80.

116 "This is the . . .": Miller, p. 26.

117 "We are not . . .": Goulder, p. 303.

117 "blow the top": ibid.

117 "Old man . . .": ibid.

117 "None try . . .": Merrill, p. 89.

117 "none of which . . .": ibid. p. 74.

117 "walking to . . .": Barsness, p. 188.

117 "tired of . . .": ibid.

118 "inched forward . . .": Barsness, pp. 189–90.

118 "a poor miner . . .": ibid. p. 190.

118 "not noticing . . .": ibid. p. 184.

118 "Gold is . . .": Greever, p. 384.

119 "like a cow": quoted in Chisolm, p. 96.

119 "sat down . . .": ibid. p. 97.

119 "I commenced . . .": ibid.

119 "for tying . . .": ibid. pp. 60–61.

119 "In fact . . .": ibid.

120 "get valuable . . .": Kansas Pacific Railway Company, *Miners' Guide to the Black Hills and Famous Big Horn Gold Region* . . . (Chicago: Rand, McNally & Co., printers, 1877), n.p.

120 "a well-marked . . .": Watson Parker, p. 47.

120 "a very pretty . . .": Jerry Bryan, *An Illinois Gold Hunter in the Black Hills; the Diary of Jerry Bryan,* ed. Clyde C. Watson (Springfield, Ill.: Illinois State Historical Society, 1960), p. 25.

120 "chuck full . . .": ibid. p. 13.

120 "Indian Outrages": ibid.

120 "that they . . .": ibid.

120 "12 wagons . . .": ibid. p. 16.

120 "Only those . . .": Hughes, p. 59.
121 "as plentiful . . .": "Frontier Sketches," *Denver Field and Farm* (July 1, 1905), p. 8.
121 "tramped all . . .": ibid.
122 "We felt . . .": James Friend quoted in Richard E. Lingenfelter, *The Rush of '89: The Baja California Gold Fever* (Los Angeles: Dawson's Book Shop, 1967), p. 19.
122 "A chance . . .": Marshall Sprague, *Money Mountain: The Story of Cripple Creek Gold* (Lincoln, Nebr.: University of Nebraska Press, 1979), pp. 98–99.
122 "dogs, chickens . . .": ibid. p. 99.
122 "rushing, pelting . . .": Chisolm, p. 71.

Chapter Five: Bound for the Far North

125 "Bring me . . .": Edith Feero Larson quoted in Melanie J. Mayer, *Klondike Women: True Tales of the 1897–1898 Gold Rush* (Athens, Ohio: Ohio University Press, 1989), p. 23.
126 "an older . . .": quoted in Holliday, p. 296.
126 "I am . . .": quoted in Pierre Berton, p. 152.
126 "hanging upon . . .": Wells, pp. 40–41.
126 "I don't think . . .": Alfred McMichael, "Klondike Letters: The Correspondence of a Gold Seeker in 1898," ed. Juliette C. Reinicker, *Alaska Journal,* Vol. 14 (Autumn 1984), p. 9.
127 "set on . . .": Pierre Berton, p. 127.
128 "almost a . . .": ibid. p. 202.
128 "two men . . .": quoted in Mayer, p. 29.
128 "glass jars . . .": Pierre Berton, p. 127.
128 "Great American . . .": Henry Rogers letter of June 27, 1898, Henry Rogers Papers, Beinecke Library.
128 "We stopped . . .": ibid.
129 "red-and-yellow . . .": Martha Louise [Purdy] Black, *My Ninety Years* (Anchorage: Alaska Northwest Publishing Co., 1976), p. 21.
129 "was free . . .": Mayer, p. 52.
130 "vague and hungry interiors": Austin, p. 2.
131 "gold delirium . . .": ibid. p. 22.
131 "people; horses . . .": Mayer, p. 111.
131 "You was . . .": Malinda Jenkins, *Gambler's Wife: The Life of Malinda Jenkins* (Boston: Houghton Mifflin, 1933), p. 193.
132 "boulders, torn-up . . .": Pierre Berton, p. 247.
132 "narrow and . . .": Inga Sjolseth Kolloen quoted in Mayer, p. 78.
132 "about half . . .": McMichael, p. 23.
132 "the last camping . . .": Mayer, p. 78.
132 "and still . . .": Pierre Berton, p. 249.

133 "sliding from . . .": McMichael, p. 27.
133 "by spring . . .": Pierre Berton, p. 255.
133 "long, treeless . . .": Mayer, p. 99.
133 "Look out below!": Jenkins, p. 199.
133 "began to . . .": McMichael, p. 32.
134 "the cooler . . .": Wells, p. 50.
134 "a general . . .": ibid. p. 28.
134 "a steady . . .": Mayer, p. 133.
134 "wild and desolate": Wells, pp. 36–37.
134 "gentle slopes . . .": Mayer, p. 133.
134 "the greatest tent . . .": Pierre Berton, p. 269.
134 "canvas stores . . .": Flora Shaw quoted in Mayer, p. 172.
135 "a fleet . . .": Pierre Berton, p. 269.
135 "There are more . . .": Inga Sjolseth Kolloen quoted in Mayer, p. 172.
135 "witch-watercourse . . .": Laura Berton, *I Married the Klondike* (Boston: Little, Brown & Co., 1954), pp. 162–63.
135 "no one seems . . .": quoted in Mayer, p. 176.
135 "the strange . . .": ibid.
136 "snakelike twists": Laura Berton, p. 180.
136 "quite like . . .": ibid.
136 "a continuous . . .": Mayer, p. 190.
136 "four rock-faced . . .": ibid. p. 196.
136 "The responsibilities . . .": McMichael, p. 42.
137 "frozen swamps . . .": Joanne Hook, "He Never Returned: Robert Hunter Fitzhugh in Alaska," *Alaska Journal,* Vol. 15 (Spring 1985), p. 35.
139 "caught like . . .": Wells, p. 69.
139 "We had been . . .": quoted in Pierre Berton, p. 225.
139 "rose to the . . .": Wells, p. 25.
140 "looked out . . .": Mayer, p. 51.
140 "rolling sea . . .": Arthur Dietz, *Mad Rush for Gold in Frozen North* (Los Angeles: Times-Mirror Printing and Binding House, 1914), p. 106.
140 "The relief . . .": ibid. p. 115.
140 "a hopeless . . .": ibid. p. 125.
141 "so murky . . .": Wells, p. 63.
141 "Well, you . . .": quoted in Mayer, p. 85.
141 "an interminable . . .": Bond, p. 33.
141 "plenty that . . .": Jenkins, p. 194.
141 "Cooking for . . .": Emily Craig quoted in Mayer, p. 35.
141 "dried potatoes . . .": ibid.
142 "How many . . .": quoted in Mayer, p. 97.
142 "About a mile . . .": ibid.
142 "packed like pigs": McMichael, p. 11.
142 "so close . . .": Pierre Berton, p. 138.

142 "failed to . . .": Mayer, p. 135.
143 "rubbin' its . . .": Feero's daughter Edith Feero Larson quoted in ibid. p. 149.
143 "collapsed in . . .": Pierre Berton, p. 166.
143 "slowly roast[ing] . . .": ibid.
143 "The only thing . . .": quoted in Mayer, p. 138.
143 "If you offer . . .": Wells, p. 30.
143 "very scarce . . .": ibid. p. 47.
144 "five times . . .": Pierre Berton, p. 255.
144 "All right . . .": quoted in Mayer, p. 98.
144 "made no . . .": quoted in ibid. p. 97.
144 "blue jean . . .": McMichael, p. 29.
144 "She is . . .": ibid.
144 "small, bright . . .": Wells, p. 58.
144 "Mrs. Fancheon is . . .": ibid.
145 "Great Scott . . .": quoted in Wells, p. 29.
145 "turn back . . .": ibid.
145 "A man can . . .": quoted in Mayer, p. 82.
146 "so named . . .": ibid. p. 47.
146 "highwaymen and . . .": Wells, p. 50.
146 "I thought . . .": quoted in Pierre Berton, p. 262.
147 "stuffed with . . .": ibid. p. 257.
147 "quarreling and . . .": quoted in Mayer, p. 43.
147 "each set[ting] . . .": Pierre Berton, p. 271.
147 "The one . . .": Edith Feero Larson quoted in Mayer, p. 104.
147 "begged and . . .": Wells, p. 63.
147 "not one offer[ing] . . .": ibid. p. 68.
148 "in agony . . .": Pierre Berton, p. 259.
148 "at great expense . . .": McMichael, p. 30.
148 "every man . . .": Black, p. 29.
148 "I have a . . .": Bond, p. 27.
148 "so beautiful . . .": Jenkins, p. 2149
149 "constant passing . . .": J.C. Cantwell quoted in Cole, p. 104.
149 "bloody for . . .": Edward Jesson quoted in ibid. p. 106.
149 "inside an . . .": ibid. p. 105.
149 "White man . . .": quoted in ibid. p. 106.
150 "a 'Nomer' . . .": ibid. p. 115.
150 "it cost . . .": ibid. p. 134.
150 "If ever . . .": Stephen Redgrave quoted in Wright, p. 194.

Chapter Six: Life in the Diggings

151 "an army . . .": Carl Allvar Kullgren quoted in Axel Friman, "Two Swedes in the California Goldfields: Allvar Kullgren and Carl August

Modh, 1850–1856," *Swedish-American Historical Quarterly,* Vol. XXXIV (April 1983), p. 111.

151 "literally turned . . .": quoted in Barsness, p. 55.

152 "immense piles . . .": Clappe, p. 49.

152 "So many men . . .": quoted in Barsness, p. 55.

152 "bees around . . .": ibid.

152 "scenes of . . .": Hughes, p. 107.

152 "crowded thickly . . .": Kelly, p. 186.

152 "it was never . . .": Taylor, p. 192.

152 "moral disgrace": Kelly, p. 106.

152 "clubhouse, news . . .": Barsness, p. 204.

153 "Not a spot . . .": Clappe, p. 49.

153 "had a yellow . . .": Letts, p. 72.

153 "the whole . . .": entry of November 8, 1849, Blunt journal.

153 "for somebody . . .": Goulder, p. 196.

153 "that not . . .": Kelly, p. 185.

153 "sandbars and . . .": Otis E. Young, Jr., *Western Mining: An Informal Account* . . . (Norman, Okla.: University of Oklahoma Press, 1987), p. 110.

154 "surrounded by . . .": quoted in Barsness, p. 11.

154 "Come to camp . . .": Bryan, p. 26.

154 "honest as the day": Young, *Western Mining,* p. 41.

154 "picking into . . .": Holliday, p. 37.

154 "work plenty . . .": Goulder, p. 210.

154 "sifted through . . .": Groh, p. 227.

154 "Digging wells . . .": quoted in Donald Jackson, p. 204.

154 "and wash . . .": quoted in Henson, p. 192.

154 "five thousand . . .": quoted in Goulder, p. 364.

155 "frozen muck . . .": Grinnell, p. 71.

155 "began to . . .": ibid.

155 "I regret . . .": Davidson letter of January 12, 1853, to "Lewis," Davidson Papers, Beinecke Library.

155 "grow discouraged . . .": quoted in Marvin Lewis, p. 40.

155 "the sober . . .": Taylor, *Eldorado,* p. 199.

155 "almost everybody . . .": Dorothy M. Johnson, "The Patience of Frank Kirkaldie," *Montana, the Magazine of Western History,* Vol. XXI, No. 1 (January 1971), p. 16.

155 "is not . . .": Rogers letter of October 1, 1898.

155 "If a person . . .": Clappe, p. 123.

155 "almost all . . .": ibid. (emphasis in original).

156 "drunkards, idlers, and fools": quoted in Donald Jackson, p. 259.

156 "drunken, worthless . . .": Tyson, p. 76.

156 "Wherever Geology . . .": Clappe, p. 117.

156 "from strictly . . .": quoted in Lingenfelter, p. 35.

156 "reflected no . . .": Young, *Western Mining,* p. 3.

156 "d——n fools . . .": quoted in Watson Parker, p. 105.
157 "the wrong . . .": Kelly, p. 89.
157 "in virtually . . .": Young, *Western Mining*, p. 19.
157 "Lots of . . .": Callbreath letter of July 20, 1850.
157 "several pieces . . .": Stuart, p. 202.
157 "I wouldn't . . .": quoted in Pierre Berton, p. 53.
158 "the operation . . .": Charles M. Clark, *A Trip to Pike's Peak and Notes by the Way* (San Jose, Calif.: Talisman Press, 1958), p. 86.
158 "rock and . . .": Charles D. Ferguson, *The Experiences of a Forty-niner During Thirty-four Years' Residence in California and Australia,* ed. Frederick T. Wallace (Cleveland, Ohio: Williams Publishing Co., 1888), p. 120.
158 "When you . . .": Steele, pp. 23–24.
159 "with its . . .": Young, *Western Mining*, p. 109.
159 "an open box . . .": Holliday, p. 38.
159 "sounded like . . .": Prentice Mulford quoted in Marvin Lewis, p. 33.
160 "a connection . . .": Donald Jackson, p. 282.
160 "an almost . . .": ibid. p. 314.
160 "that fine . . .": Barsness, p. 61.
161 "It takes . . .": McMichael, p. 54.
161 "Half of . . .": quoted in Donald Jackson, p. 202.
161 "The boys . . .": Rogers letter of December 4, 1898.
161 "not only your . . .": Henderson, p. 62.
162 "I can get . . .": A. Hill letter of March 14, 1850.
162 "for considerable . . .": Hughes, p. 79.
162 "rinsed the ditch": Friman, p. 118.
163 "flood[ed] the . . .": Harris, p. 124.
163 "It is a . . .": Steele, pp. 114–15.
163 "lay under . . .": Barsness, p. 57.
163 "scattered over . . .": Watson Parker, p. 99.
164 "burning out a shaft": Hook, p. 37.
164 "numerous pillars . . .": Wells, p. 147.
165 "several millions . . .": quoted in Roger D. McGrath, *Gunfighters, Highwaymen and Vigilantes: Violence on the Frontier* (Berkeley, Calif.: University of California Press, 1984), p. 3.
166 "expos[ing] the . . .": Young, *Western Mining*, p. 30.
166 "many a good . . .": ibid.
166 "the billions . . .": quoted in Sally Zanjani, *Goldfield: The Last Gold Rush on the Western Frontier* (Athens, Ohio: Swallow Press/Ohio University Press, 1992), p. 24.
166 "What happened . . .": Donald Jackson, p. 223.
166 "I feel . . .": quoted in Wells, p. 140.
166 "the sudden . . .": Taylor, *Eldorado*, p. 115.
166 "I do not . . .": Hughes, p. 148.

166 "as if propelled . . .": Letts, p. 82.

167 "with renewed . . .": ibid.

167 "almost impossible . . .": Taylor, *Eldorado,* p. 196.

167 "too tired . . .": quoted in Quiett, p. 220.

168 "to limit . . .": Sprague, p. 124.

168 "two or . . .": Taylor, *Eldorado,* p. 70.

169 "two feet . . .": Sprague, p. 20.

169 "Cripple's characteristic . . .": ibid. p. 15.

169 "It was by . . .": Stuart, p. 75.

169 "right over . . .": Henderson, p. 63.

170 "tended to . . .": Barsness, p. 173.

170 "to separate . . .": Hughes, p. 110.

170 "luggage, boots . . .": Thomas Berry, "Gold! But How Much?" *California Historical Quarterly,* Vol. LV, No. 3 (Fall 1976), p. 247.

171 "all of California": Greever, p. 166.

171 "had rather . . .": Clark, p. 83.

171 "It would take . . .": quoted in Grinnell, p. 44.

171 "locate some . . .": McMichael, p. 62.

171 "got uneasy": Sadler, p. 2.

171 "there was no . . .": ibid. p. 215.

171 "thoroughly . . .": Paul Fatout, *Meadow Lake: Gold Town* (Lincoln, Nebr.: University of Nebraska Press, 1969), p. 5.

172 "scales the . . .": ibid.

172 "No beds . . .": quoted in Holliday, p. 359.

172 "a dirty sheet . . .": Lynch, p. 105.

172 "a mere strip . . .": Groh, p. 231.

172 "bowers or . . .": Peavy and Smith, p. 51.

173 "the water . . .": Sadler, p. 188.

173 "after six . . .": McMichael, p. 60.

173 "[a] place to . . .": ibid.

173 "a place for . . .": quoted in Peavy and Smith, p. 131.

173 "pour[ed] in . . .": Grinnell, p. 22.

173 "froze stiff . . .": ibid.

173 "right through . . .": Fred Isbell, *Mining and Hunting in the Far West* (Middleton, Conn.: J.S. Steward, printer, ca. 1871), p. 26.

173 "It seemed . . .": Laura Berton, pp. 193–94.

174 "cut[ting] clear . . .": Webb, pp. 84–85.

174 "gunny-sacking . . .": Grinnell, p. 24.

174 "How I shall . . .": Clappe, p. 55.

174 "people dont . . .": quoted in Peavy and Smith, p. 173.

174 "Damn the . . .": quoted in Kelly, p. 142, referring to the Yukon.

174 "these humbug prices": William Swain in Holliday, p. 331.

174 "I wish . . .": McMichael, p. 36.

175 "It is soon . . .": Hinman, p. 41.

175 "wonders if . . .": quoted in Marvin Lewis, p. 85.

175 "the three B's": Henry Rogers letter of November 15, 1898.
175 "corn-meal . . .": Grinnell, p. 48.
175 "a plentiful supply . . .": McMichael, p. 52.
175 "they could . . .": Lemuel McKeeby quoted in Brown, p. 37.
175 "boiled, roasted and stewed": Grinnell, p. 60.
175 "high in . . .": Elliott West, *The Saloon on the Rocky Mountain Mining Frontier* (Lincoln, Nebr.: University of Nebraska Press, 1979), p. 13.
176 "and a corresponding . . .": Clappe, p. 30.
176 "no easy mode . . .": Callbreath letter of January 24, 1859.
176 "at last . . .": Rogers letter of November 19, 1898.
176 "to eat . . .": Grinnell, p. 47.
176 "This little pig . . .": Kelly, p. 142.
176 "the purchaser . . .": Groh, p. 229.
177 "baskets, tin . . .": quoted in Holliday, p. 39.
177 "so badly . . .": Pierre Berton, p. 75.
177 "there was plenty . . .": Holliday, p. 41.
177 "so heavy . . .": Donald Jackson, p. 234.
177 "died from . . .": Kelly, p. 143.
177 "[a] Chicago . . .": ibid. p. 119.
177 "I would have . . .": Churchill, p. 58.
177 "a great army . . .": Goulder, pp. 215–16.
178 "bills are . . .": ibid.
178 "to be ready . . .": Peavy and Smith, p. 148.
178 "I never more . . .": quoted in Fatout, p. 15.
178 "steep[ing] their . . .": quoted in Quiett, p. 88.
178 "empty bottles . . . ": Holliday, p. 364.
178 "perfect Saturnalia": Clappe, pp. 93–94.
178 "remorseless, persevering . . .": ibid.
178 "temporary relationships . . .": Barsness, p. 204.
178 "warranted to . . .": quoted in Fatout, p. 70.
178 "demoralized and . . .": Bond, p. 49.
179 "Wherever there . . .": Taylor, *Eldorado*, p. 67.
179 "work, meals, sleep": Donald Jackson, p. 303.
179 "two old-timers . . .": Pierre Berton, p. 386.
179 "Oh! joy . . .": Stuart, p. 196.
179 "this is the . . .": Charles V. Hume, "The Gold Rush Actor: His Fortunes and Misfortunes in the Mining Camps," *The American West*, Vol. IX, No. 3 (May 1972), p. 17.
180 "change[d] at . . .": Goulder, p. 373.
180 "orchestra . . .": Grinnell, p. 34.
180 "by far the . . .": Clappe, p. 70.
181 "a whole library . . .": Grinnell, p. 21.
181 "with three . . .": Stuart, p. 160.
181 "Did you know . . .": Estelline Bennett, *Old Deadwood Days* (Lincoln, Nebr.: University of Nebraska Press, 1982), p. 268.

181 "things he . . .": ibid.
181 "strain[ing] to . . .": Callbreath letter of January 24, 1859.
181 "Hell of . . .": Quiett, p. 342.
181 "revelers lost . . .": Black, p. 47.
181 "We had no . . .": Kelly, pp. 193–94.
182 "I must cheer . . .": entry of November 13, 1849, Blunt journal.
182 "I came here . . .": quoted in Peavy and Smith, p. 55.
182 "You can bet . . .": quoted in Donald Jackson, p. 284.
182 "I can see . . .": A. Hill letter of December 14, 1851.
182 "If I were . . .": quoted in Hook, p. 38.
182 "it would look . . .": quoted in Johnson, p. 23.
182 "I have come . . .": ibid. p. 22.
182 "have a hell . . .": quoted in Stuart, p. 62.
183 "rather a hard . . .": Hauser, p. 36.
183 "Sometimes I get . . .": Callbreath letter of January 24, 1859.
183 "In 999 cases . . .": Chisolm, p. 75.
183 "practising a . . .": ibid.

Chapter Seven: "Cities of the Magic Lantern"

185 "a mad . . .": Kelly, p. 100.
185 "about the oddest . . .": quoted in Cole, p. 138.
185 "more like . . .": Walter Haas, "Extracts from Alaskan Letters," Haas Scrapbook, Beinecke Library, p. 7.
185 "odd . . .": quoted in Roger W. Lotchkin, *San Francisco 1846–1856: From Hamlet to City* (Lincoln, Nebr.: University of Nebraska Press, 1974), p. 291.
186 "throng of . . .": quoted in Lotchkin, p. 278.
186 "too crowded . . .": Kelly, p. 99.
186 "a perfect . . .": quoted in Cole, p. 140.
186 "frozen swampland": Pierre Berton, p. 73.
187 "It may be . . .": William Bronson, "Nome," *The American West*, Vol. VI, No. 4 (July 1969), p. 21.
187 "five thousand . . .": Laura Berton, p. 119.
187 "I have known . . .": Tyson, p. 60.
188 "at random . . .": Zamonski and Keller, p. 68.
188 " 'Sutter' has . . .": A. Hill letter of June 12, 1850.
188 "popular vote": Cole, p. 89.
189 "Of all the . . .": Blanchard, p. 69.
189 "seemed to . . .": Taylor, *Eldorado*, p. 226.
189 "Can the world . . .": ibid. p. 164.
189 "it would . . .": Holliday, p. 32.
189 "This place . . .": quoted in Zamonski and Keller, pp. 24–25.
189 "six good . . .": ibid.

189 "at the rate . . .": Sprague, p. 94.
189 "twenty-six saloons . . .": ibid.
190 "old settlers . . .": quoted in Mayer, p. 121.
190 "fully thirty . . .": Wells, p. 139.
190 "a city . . .": Cole, p. 3.
190 "seemed actually . . .": ibid. p. 84.
190 "a house . . .": ibid.
190 "city blocks . . .": Paula Petrik, *No Step Backward: Women and Family on the Rocky Mountain Mining Frontier, Helena, Montana, 1865–19191* (Helena, Mont.: Montana Historical Society Press, 1987), p. 4.
191 "especially stupid . . .": Bennett, p. 94.
191 "to make traffic . . .": ibid. p. 19.
191 "a perfect . . .": quoted in Mayer, p. 120.
191 "Gone to . . .": quoted in Zamonski and Keller, p. 97.
191 "were living . . .": Pierre Berton on Yukon rush, p. 75.
192 "in tents . . .": A. Hill letter of September 24, 1849.
192 "from blowing . . .": Lotchkin, p. 295.
192 "on a few . . .": entry of November 30, 1849, Blunt journal.
192 "of entirely . . .": Hough quoted in Barsness, p. 76.
192 "mostly small . . .": quoted in Mayer, p. 120.
192 "rough board . . .": Black, p. 38.
192 "Oh! you can . . .": Sarah Walsworth quoted in Levy, p. 54.
192 "had neither . . .": Zamonski and Keller, p. 33.
192 "about 20 . . .": McMichael, p. 20.
193 "cut-glass chandeliers . . .": Mulrooney quoted in Mayer, p. 156.
193 "a saloon . . .": Clark, p. 95.
194 "the rails . . .": Cole, p. 124.
194 "knock-down theatres . . .": ibid.
194 "even with favorable . . .": Bennett, pp. 92–93.
194 "That's a train . . .": quoted in West, *Saloon,* p. 106.
194 "oysters from . . .": McGrath, p. 111.
194 "fresh beef . . .": Taylor, *Eldorado,* pp. 86–87.
195 "the first potatoes . . .": Jenkins, p. 214.
195 "Expenses 'go . . .": Miller, p. 77.
195 "with the nuggets . . .": Webb, p. 130.
196 "old and . . .": Wells, p. 80.
196 "individual stores . . .": ibid. p. 86.
196 "Henceforth I . . .": Bond, p. 48.
197 "smells like . . .": Wells, p. 101.
197 "order six . . .": ibid. pp. 154–55.
197 "Thank God . . .": quoted in Bond, p. 45.
197 "as large . . .": A. Hill letter of November 15, 1850.
197 *"good butter . . .":* ibid. (emphasis in original).
198 "you can't . . .": Arthur St. Clair Denver letter of October 16, 1850.
198 "You . . . think . . .": A. Hill letter of April 27, 1851.

198 "They will . . .": A. Hill letter of January 7, 1853.

198 "within a week . . .": Pierre Berton, p. 291.

198 "an ancient . . .": ibid.

198 "six hundred pairs . . .": Stuart, p. 261.

199 "with a sack . . .": Fatout, pp. 41–42.

199 "at once went . . .": ibid.

199 "the parts . . .": Barsness on Virginia City, p. 83.

199 "most men . . .": Pierre Berton, p. 76.

199 "with the possible . . .": C.S.A. Frost quoted in Cole, p. 141.

199 "The expenses . . .": ibid.

199 "had some . . .": Miller, p. 73.

200 "odd, grotesque . . .": Bayard Taylor, *Colorado: A Summer Trip* (New York: E.P. Putnam and Son, 1867), p. 57.

200 "a very dull . . .": Miller, p. 74.

200 "motley collection . . .": quoted in Webb, p. 91.

200 "On landing . . .": Black, p. 38.

200 "treat[ing] all . . .": Tyson, p. 71.

200 "poured out . . .": Sacramento observer quoted in Holliday, p. 323.

200 "pan[ning] their . . .": Pierre Berton, p. 367.

201 "so that . . .": ibid.

201 "a trustworthy . . .": Barsness, p. 65.

201 "shining and arrogant": Bennett, p. 84.

201 "statistics of . . .": quoted in Fatout, p. 43.

201 "The Skedaddle . . .": ibid. p. 44.

201 "The richest . . .": Bennett, p. 5.

201 "something exhilarating . . .": quoted in Lotchkin, p. 296.

202 "Though I . . .": ibid.

202 "anything edible . . .": Zamonski and Keller, p. 99.

202 "egg shells . . .": quoted in Levy, p. 179.

202 "empty bottles . . .": Fatout, p. 14.

202 "This is not . . .": quoted in Levy, p. 179.

202 "of a rich . . .": quoted in Watson Parker, p. 147.

202 "piled high . . .": Pierre Berton, p. 290.

203 "colloid quagmires . . .": Kelly, p. 99.

203 "rivers of . . .": quoted in Cole, p. 88.

203 "from one . . .": Taylor, *Eldorado*, pp. 204–205.

203 "still worse . . . stumps . . .": ibid.

203 "one great . . .": quoted in Holliday, p. 321.

203 "lined with . . .": quoted in Lotchkin, p. 188.

203 "almost three . . .": ibid. p. 111.

203 "pools of . . .": ibid. p. 151.

204 "carloads of . . .": Taylor, *Eldorado*, p. 229.

204 "accumulated refuse . . .": Barsness, p. 73.

204 "ocean[s] . . .": Pierre Berton, p. 292.

204 "a nasty, dirty . . .": quoted in Groh, p. 205.
205 "perhaps one-fourth . . .": Lotchkin, p. 175.
205 "piles of combustibles": ibid. p. 176.
205 "for quite . . .": Kelly, p. 187.
205 "every man . . .": Samuel H. Willey quoted in Lotchkin, p. 176.
206 "Even when . . .": Albert Bernard de Russailh quoted in Groh, p. 205.
206 "could have gotten . . .": Rogers letter of February 14, 1899.
206 "young fellows . . .": Hughes, p. 109.
206 "nearly walked . . .": quoted in Lotchkin, p. 84.
207 "so much raw . . .": Bennett, p. 132.
207 "there was plenty . . .": Kelly, p. 148.
207 "crowed about . . .": Barsness, p. 75.
208 "represented by . . .": Pierre Berton, p. 377.
208 "plunged into . . .": Hume, p. 17.
208 "for the purpose . . .": Taylor, *Eldorado*, p. 207.
208 "to whom . . .": ibid.
208 "American melodrama . . .": Barsness, p. 225.
208 "of the blood-curdling . . .": Kelly, p. 148.
208 "some of the . . .": Black, p. 49.
209 "mock trials . . .": Fatout, p. 60.
209 "the grand excitement . . .": Bennett, p. 25.
209 "civilized hearth fires": Barsness, p. 265.
210 "smashing and . . .": Laura Berton, p. 199.
210 "put[ting] summer . . .": ibid. p. 200.
210 "They have built . . .": quoted in Fatout, p. 91.
210 "narrow-brimmed . . .": Taylor, *Eldorado*, p. 231.
211 "symboliz[ing] a . . .": West, *Saloon*, p. 77.
211 *"arriviste* . . . different . . .": Petrik, p. 57.
211 "Leadville! No . . .": quoted in West, *Saloon*, p. 130.
211 "much of the . . .": quoted in Petrik, p. 17.
211 "was young . . .": Bennett, p. 4.
211 "Well, we'll . . .": ibid. p. 300.
211 "we were surrounded . . .": ibid. p. 284.

Chapter Eight: Self, Society, and the "Battle of Life"

223 "Ye have . . .": A. Hill letter of September 24, 1849.
224 "None caring . . .": Entry of October 23, 1849, Blunt journal.
224 "sharing nothing . . .": Garnier, p. 65.
224 "Their hearts . . .": quoted in Holliday, p. 370.
224 "[a] rough . . .": quoted in West, *Saloon*, p. 8.
224 "a free . . .": quoted in Holliday, p. 329.
224 "we don't like . . .": Ferguson, p. 72.

224 "almost as . . .": ibid.

224 "There is no . . .": quoted in Fatout, pp. 16–17.

225 "California will . . .": John Woodhouse Audubon quoted in Donald Jackson, p. 260.

225 "All combine . . .": William H. Brewer quoted in McGrath, p. 12.

225 "vice seems . . .": California observer quoted in Holliday, p. 336.

225 "who never . . .": Lynch, p. 31.

225 "seemed to . . .": ibid.

225 "No organized . . .": Kelly, p. 197.

225 "a stupendous . . .": A. Hill letter of December 29, 1849.

225 "It is looked . . .": quoted in Holliday, p. 355.

225 "a most worshipful . . .": quoted in West, *Saloon,* p. 9.

225 "the only benches . . .": George Pringle quoted in Quiett, p. 347.

226 "the young men . . .": quoted in Barsness, p. 256.

226 "We think . . .": 1854 *Pioneer* quoted in Levy, pp. 187–88.

226 "he don't . . .": Stuart, p. 69.

226 "show-shoe itinerant": West, *Saloon,* p. 78.

226 "here we . . .": quoted in Lotchkin, p. 292.

226 "straightened up . . .": Bennett, p. 174.

226 "a harder task . . .": quoted in Lotchkin, p. 326.

227 "There will . . .": quoted in Fatout, p. 3.

227 "daring, unprincipled": Clappe, p. 143.

227 "any drunken . . .": quoted in Fatout, pp. 16–17.

227 "aristocracy of vice": Hughes, p. 199.

227 "use[d] past . . .": West on Colorado argonauts, *Saloon,* p. 7.

227 "Nothing less . . .": Ferguson, p. 134.

227 "the general . . .": Taylor, *Eldorado,* p. 46.

227 "mining-camp man . . . liberal-minded . . .": quoted in McGrath, p. 110.

227 "no hypocrisy . . .": Kelly, p. 193.

228 "men were . . .": ibid.

228 "disregard for . . .": Taylor, *Eldorado,* p. 46.

228 "business looked . . .": Sadler, p. 356.

228 "some rascals . . .": quoted in Mayer, p. 224.

228 "We come . . .": quoted in Levy, p. 108.

228 "Folks up . . .": Rogers letter of April 10, 1899.

228 "a man . . .": Kelly, p. 193.

228 "clubbed away": Reinhart, p. 58.

228 "I have seen . . .": quoted in Hook, p. 34.

229 "charged with . . .": Hughes, p. 229.

229 "It is not . . .": Lynch, p. 119.

229 "If you . . .": quoted in Henderson, p. 62.

229 "nobody was . . .": Laura Berton, p. 218.

229 "perfect stranger . . .": Robert Hunter Fitzhugh quoted in Hook, p. 34.

229 "A man never . . .": Henry Rogers letter of August 27, 1899.

229 "those who . . .": Stuart, p. 263.

229 "any real . . .": Kelly, p. 198.

229 "like so much . . .": ibid. p. 168.

230 "Take 'em in . . .": quoted in Goulder, p. 307.

230 "There is more . . .": quoted in Donald Jackson, p. 299.

230 "infusion of . . .": Taylor, *Eldorado*, p. 193.

230 "rotten to . . .": Hughes, p. 200.

230 "the roaring hells . . .": Chisolm, p. 100.

230 "some grave . . .": ibid. pp. 11–12.

231 "begging for . . .": quoted in Holliday, p. 322.

231 "business and . . .": quoted in Lotchkin, p. 300.

231 "a square . . .": Herman Francis Reinhart, *The Golden Frontier: The Recollections of Herman Francis Reinhart 1851–1869*, ed. Doyce B. Nunis, Jr. (Austin, Tex.: University of Texas Press, 1962), p. 58.

231 "hundreds who . . .": Pierre Berton, p. 311.

231 "no money . . .": entry of November 13, 1849, Blunt journal.

231 "gentlemanly bank-clerk . . .": Kelly, p. 195.

231 "standing before . . .": ibid. p. 194.

232 "viewed with . . .": Rogers letter of July 3, 1899.

232 "rough clothes . . .": ibid.

232 "Doesn't he . . .": Rogers letter of August 27, 1899.

232 "work for . . .": Kelly, p. 122.

232 "a very large . . .": ibid. p. 101.

233 "The expenses . . .": quoted in Peavy and Smith, p. 149.

233 "ended their . . .": Royce, p. 87.

233 "the hardships . . .": Dobson Prest quoted in Wright, p. 247.

233 "center[s] of sickness": Lotchkin, p. 183.

233 "With all . . .": Steele, p. 57.

233 "I have . . .": Clappe, p. 127.

233 "the very dust . . .": quoted in Rose, p. 42.

233 "sickness in . . .": Tyson, p. 77.

233 "more broken-down . . .": ibid. p. 80.

233 "I would pity . . .": Charles Ferguson quoted in Groh, p. 238.

234 "the winter scourge": Goulder, p. 229.

234 "almost none . . .": Groh, pp. 170–71.

234 "in a mild . . .": Dr. J.J. Chambers quoted in Wells, p. 131.

234 "blood in . . .": Cole, p. 89.

234 "extremely filthy . . . mountain . . .": Clark, p. 96.

235 "I think . . .": Moynihan et al., p. 12.

235 "Excitement makes . . .": A. Hill letter of September 24, 1849.

235 "half raw . . .": Kelly, p. 134.

235 "sickening depression . . .": Royce, p. 87.

235 "[drinking] their meals": West, *Saloon*, p. 23.

235 "promising young . . .": Harris, p. 116.

235 "miners whose . . .": argonaut quoted in Holliday, p. 397.

235 "sudden increases . . .": Taylor, *Eldorado,* p. 230.

236 "smart enough . . .": Barsness, p. 61.

236 "would go . . .": Sprague, p. 45.

236 "no man . . .": Clark, p. 92.

237 "by the use . . .": Hughes, p. 231.

237 "in a blanket . . .": Stuart, p. 76.

237 "doctored himself . . .": Groh, p. 235.

237 "scurvy trails": Grinnell, p. 51.

237 "up to their . . .": Groh, p. 239.

238 "got mad . . .": Adolphus Windeler quoted in ibid. p. 234.

238 "To be sick . . .": quoted in Friman, p. 120.

238 "Neither shalt . . .": quoted in Marvin Lewis, p. 45.

238 "as a rule . . .": Kelly, p. 131.

238 "the men on . . .": Ferguson, p. 123.

238 "some of them . . .": Clappe, pp. 31–32.

238 "sheltered, doctored . . .": Harris, p. 123.

238 "tends to . . .": Grinnell, p. 54.

239 "utter loneliness": Steele, pp. 119–20.

239 "hard-worked, hurried . . .": Wilson quoted in Levy, p. 76.

239 "found him . . .": ibid.

239 "Seems good . . .": Rogers letter of April 10, 1899.

239 "You may . . .": A. Hill letter of December 24, 1849.

240 "I am not . . .": A. Hill letter of October 16, 1850.

240 "If you get . . .": quoted in Holliday, p. 384.

240 "the moment I . . .": entry of November 26, 1849, Blunt journal.

240 "I am not . . .": entry of November 27, 1849, ibid.

240 "getting fast hold": entry of September 16, 1850, ibid.

240 "lime-juice . . .": ibid.

240 "Doctors are . . .": quoted in Peavy and Smith, p. 79.

241 "They will hardly . . .": quoted in Holliday, p. 314.

241 "in their . . .": quoted in Groh, p. 180.

241 "predilection for . . .": Garnier, p. 64.

241 "Scurvy prevented . . .": quoted in Mayer, p. 33.

242 "in profound . . .": Ferguson, pp. 121–22.

242 "perhaps he . . .": ibid.

242 "That settled . . .": ibid.

242 "the ear and . . .": Harris, p. 142.

242 "a mixture . . .": ibid. p. 143.

242 "There has been . . .": quoted in Holliday, p. 372.

242 "to fry . . .": Edwin Morse quoted in Groh, p. 235.

242 "bringing them . . .": ibid.

243 "pest houses": Lotchkin, p. 185.

243 "spent weeks . . .": Pierre Berton, p. 296.

244 "those of vigorous . . .": Groh, p. 174.

244 "crawled to . . .": Hall, p. 182.
244 "John, shall . . .": ibid.
244 "The men came . . .": ibid.
244 "John, it's . . .": ibid. p. 187.
245 "die like . . .": Bennett, p. 185.
245 "400 + is . . .": A. Hill letter of May 20, 1857.
245 "dig[ging] for . . .": Cole, p. 72.
245 "were returned . . .": Glenda J. Choate, *Skagway, Alaska Gold Rush Cemetery History and Guide Book* (Skagway, Alaska: Lynn Canal Publishing, 1989), p. 21.
245 "Burning and . . .": Jenkins, p. 210.
245 "made a pathway . . .": Joaquin Miller, "The Men of Forty-nine" in Bergon and Papanikolas, p. 237.

Chapter Nine: "No Law but Miners' Law"

247 "Gentlemen, any . . .": quoted in Kelly, p. 164.
247 "What the hell": ibid.
247 "Who the hell . . .": ibid.
247 "what in hell": ibid.
247 "to convince . . .": ibid.
248 "the only . . .": Holliday, p. 4249
249 "neither the . . .": Hughes, p. 167.
249 "were declared . . .": Barsness, p. 125.
249 "one policeman . . .": Keith A. Murray, *Reindeer and Gold* (Bellingham, Wash.: Center for Pacific Northwest Studies, 1988), pp. 3–4.
250 "very infrequent": quoted in Holliday, p. 40.
250 "buckskin purses . . .": Royce, p. 80.
250 "life & property . . .": A. Hill letter of December 24, 1849.
250 "the bully . . .": California argonaut quoted in Marvin Lewis, p. 4.
250 "The extent . . .": Holliday, p. 40.
250 "good order . . . every . . .": Henderson, p. 12.
250 "a single . . .": ibid.
251 "more than . . .": quoted in McGrath, p. 190.
251 "especially at . . .": Garnier, p. 60.
251 "no petty . . .": Bennett, p. 36.
251 "knew that . . .": Royce, p. 81.
252 "well protected": William Swain quoted in Holliday, p. 317.
252 "accepted and . . .": argonaut quoted in Marvin Lewis, p. 4.
252 "knew this . . .": Stephens, p. 28.
252 "I do not . . .": Steele, p. 118.
252 "all differ[ed] . . .": Clark, p. 111.
253 "three hundred . . .": Hughes, p. 106.
253 "not much . . .": Murray, p. 185.

253 "no one could . . .": ibid. p. 189.

253 "many innocent . . .": ibid. p. 196.

253 "the character . . .": Hughes, p. 106.

254 "it took . . .": Murray, p. 154.

254 "covered with . . .": Grinnell, p. 74.

254 "You can . . .": quoted in Marvin Lewis, p. 5.

254 "the fust man . . .": ibid.

254 "it would be . . .": Hughes, p. 126.

255 "friendly, neighborly . . .": Chisolm, p. 103.

255 "the sole . . .": Wright, p. 6.

256 "one-half . . .": Fatout, p. 64.

256 "no duty . . .": quoted in Zanjani, p. 200.

256 "all those . . .": quoted in Jan S. Stevens, "Stephen J. Field: A Gold Rush Lawyer Shapes the Nation," *Journal of the West*, Vol. XXIX, No. 3 (July 1990), p. 43.

256 "men in . . .": ibid.

256 "competing confectioners . . . lawyers . . .": Lotchkin, p. 189.

256 "was not . . .": McGrath, p. 186.

257 "many and bitter . . .": Hughes, p. 119.

257 "inordinate blackguard . . . erects . . .": quoted in Barsness, p. 130.

257 "stabbed very . . .": Clappe, p. 114.

257 "You see . . .": quoted in McGrath, p. 214.

257 "both his . . .": quoted in Holliday, p. 317.

257 "After a man . . .": Steele, p. 191.

257 "a kind, steady . . .": ibid. p. 190.

258 "so bold . . .": Jerusha Merrill quoted in Moynihan et al., p. 14.

258 "anything and . . .": Zamonski and Keller, p. 73.

258 "robbery of . . .": Walden quoted in Bronson, pp. 24–25.

258 "if you should . . .": Kelly, p. 137.

258 "It was pretty . . .": Jenkins, p. 213.

259 "placing a . . .": Hughes, pp. 153–54.

259 "big bully": ibid.

259 "an attempt to . . .": ibid.

259 "became entangled . . .": ibid.

259 "disinterested parties . . . that . . .": ibid. p. 155.

259 "unanimous voice": ibid. p. 156.

259 "there was no desire . . .": ibid.

259 "to escort . . .": ibid. p. 158.

259 "eight hundred . . .": H.J. Hawley quoted in Lynn I. Perrigo, "Law and Order in Early Colorado Mining Camps," *Mississippi Valley Historical Review*, Vol. XXVIII (1941–42), p. 44.

260 "his previous . . .": Clappe, p. 68.

260 "age and . . .": Stuart, p. 216.

260 "harmless . . .": Clappe, p. 88.

260 "in the most . . .": ibid. p. 86.

260 "tried to . . .": Webb, p. 80.
260 "all in camp . . .": Austin, p. 55.
261 "I never saw . . .": J.D. Borthwick quoted in Groh, p. 245.
261 "The people make . . .": Callbreath letter of June 30, 1849.
261 "appeal to . . .": Chisolm, p. 113.
261 "Thus . . . an accomplished . . .": ibid.
261 "This is rough . . .": quoted in Perrigo, p. 44.
261 "almost everybody . . .": Clappe, p. 86.
261 "blubbering and . . .": James Fergus quoted in Peavy and Smith, p. 153.
261 "The gallows . . .": ibid.
262 "despite good . . .": Webb, p. 95.
262 "so great . . .": Clappe, p. 151.
262 "advocated . . .": Harris, p. 138.
263 "the court . . .": Hughes, p. 158.
263 "The people here wish . . .": Clappe, p. 62.
263 "the people here . . . Humbug! . . .": Holliday, p. 4264
264 "any political . . .": Lotchkin, p. 142.
264 "Gold was . . .": Tyson, p. 95.
264 "noncooperative communities": Lotchkin, p. 142.
264 "If good . . .": quoted in McGrath, p. 128.
264 "inconsequent shootings": Bennett, p. 112.
264 "had spent . . .": McGrath, p. 176.
264 "miners, laborers . . .": Marvin Lewis, p. 58.
265 "for storage": Kelly, p. 178.
265 "posed as . . .": Hughes, p. 196.
265 "If word . . .": quoted in Pierre Berton, p. 355.
266 "there is no . . .": quoted in McGrath, p. 209.
266 "Blood & crime . . .": A. Hill letter of April 7 or 17, 1850.
267 "the only real . . .": Zamonski and Keller, p. 178.
267 "legal fees . . .": Lotchkin, p. 160.
267 "the law . . .": ibid.
267 "If three men . . .": quoted in McGrath, p. 194.
268 "routinely discharged . . .": McGrath, p. 216.
268 "a burdensome . . .": Zamonski and Keller, p. 199.
268 "I didn't . . .": quoted in Marvin Lewis, p. 21.
268 "The court don't . . .": quoted in Quiett, p. 153.
268 "making a farce . . .": Zamonski and Keller on early Denver, p. 216.
268 "tolerat[ing] considerable . . .": Lotchkin on San Francisco, p. 142.
268 "that by good . . .": entry of November 27, 1850, Blunt journal.
268 "not justifiable . . .": ibid.
268 "operated on speculation": Zamonski and Keller, p. 199.
269 "that 'side-door tips' . . .": Kelly, p. 178.
269 "set about . . .": Bronson, pp. 25–26.

269 "even commanded . . .": ibid.

269 "made his . . .": Arthur Treadwell Walden quoted in Bronson, pp. 24–25.

269 "There was often . . .": McGrath, p. 208.

270 "the police . . .": ibid.

270 "all the sheriffs . . .": Walden quoted in Bronson, pp. 24–25.

270 "able leadership . . . the respectable . . .": Stuart, p. 237.

270 "kill the . . .": quoted in Marvin Lewis, p. 25.

271 "Hunt your . . .": Steele, p. 171.

271 "I'm powerful . . .": ibid. p. 173.

271 "most shameful . . .": Goulder, p. 232.

271 "so numerous . . .": Steele, p. 1271

271 "the banished . . .": ibid. p. 101.

271 "laugh in . . .": Hughes, p. 196.

271 "swaggered": ibid.

272 "exactly what . . .": Kelly, p. 193.

272 "no pretense . . .": Hughes, p. 196.

272 "could drive . . .": Zamonski and Keller, pp. 127–28.

272 "flowing black moustaches": Black, p. 23.

273 "degenerated . . .": Lotchkin, p. 191.

273 "swooped down . . .": ibid.

273 "in a state . . .": Groh, p. 217.

274 "accused of . . .": Lotchkin, p. 201.

274 "they kept . . .": Barsness, p. 52.

274 "howling, shouting . . .": Clappe, pp. 155–56.

274 "poor, exhausted miners": ibid.

274 "You will . . .": ibid.

274 "There is not . . .": Barsness, p. 53.

275 "mob violence . . .": Mrs. Lee Whipple-Haslam quoted in Moynihan et al., p. 30.

275 "This was . . .": Arthur St. Clair Denver letter of October 16, 1850.

275 "noninterference . . .": McGrath, p. 225.

275 "If our Acting . . .": quoted in Barsness, p. 146.

275 "dreadful government": Jonathan Blanchard quoted in Robert H. Keller, Jr., "A Puritan at Alder Gulch and the Great Salt Lake: Rev. Jonathan Blanchard's Letters from the West, 1864," *Montana, the Magazine of Western History*, Vol. 36, No. 3 (Summer 1986), p. 71.

275 "most perfect . . .": ibid. p. 66.

275 "The air . . .": quoted in Barsness, p. 249.

276 "I did not . . .": ibid.

276 "A man in . . .": ibid. p. 250.

276 "I have been . . .": Goulder, p. 279.

276 "were very . . .": Beilharz and Lopez, p. xix.

276 "we have just . . .": quoted in Perrigo, p. 46.

276 "found ample . . .": Charles Howard Shinn, *Mining Camps, A Study* *in American Frontier Government* (New York: Alfred A. Knopf, 1948), p. 6.

Chapter Ten: Natives and Strangers

279 "[seek] the . . .": West, *Saloon*, p. 90.
280 "a boiling . . .": Carl Kullgren quoted in Friman, pp. 108–109.
280 "some 260 . . .": Friman, p. 102.
280 "disastrous depression": Paul Friggens, "The Curious 'Cousin Jacks': Cornish Miners in the American West," *The American West*, Vol. XV, No. 6 (November–December 1978), p. 6.
280 "Poles, Italians . . .": Wright, p. 130.
281 "several thousand": Laura Berton, p. 209.
281 "taboo": Sprague, p. 97.
282 "Wherever there . . .": quoted in Friggens, p. 5.
282 "quiet homes . . .": Steele, p. 146.
283 "a great . . .": quoted in James J. Rawls, "Gold Diggers: Indian Miners in the California Gold Rush," *California Historical Quarterly*, Vol. LV (Spring 1976), p. 31.
283 "well treated . . .": Stuart, p. 55.
283 "the Indians . . .": quoted in Rawls, p. 31.
283 "the invisible . . .": Samuel Ward quoted in ibid. p. 33.
284 "there were . . .": Holliday, p. 39.
284 "whites seemed . . .": Rawls, p. 34.
284 "no more . . .": ibid. p. 36.
284 "war of . . .": Tyson, p. 78.
284 "had rather . . .": quoted in Holliday, p. 328.
284 "class of . . .": Goulder, p. 263.
285 "full restitution . . .": Zamonski and Keller, p. 85.
285 "on the ground . . .": Fergus quoted in Peavy and Smith, p. 139.
285 "He who . . .": quoted in Lotchkin, p. 110.
285 "were barred . . .": Murray, p. 181.
285 " 'catching on' . . .": Grinnell, p. 18.
286 "chronic jealousy": quoted in McGrath, p. 146.
286 "depended for . . .": McGrath, p. 146.
286 "remained friendly . . .": Goulder, p. 192.
286 "always ready . . .": Grinnell, p. 19.
286 "the best . . .": quoted in Chisolm, p. 124.
286 "these same . . .": ibid.
287 "dying like . . .": Fred Kimball quoted in William H. Wilson, "To Make a Stake: Fred G. Kimball in Alaska, 1899–1909," *Alaska Journal*, Vol. 13 (Winter 1983), p. 110.
287 "it seemed . . .": Steele, p. 204.
287 "confiding, generous . . .": Grinnell, p. 19.

287 "Scarcely six . . .": Wells, p. 161.

288 "2,288 black . . .": Ray R. Albin, "The Perkins Case: The Ordeal of Three Slaves in Gold Rush California," *California History,* Vol. 67 (December 1988), p. 215.

288 "free white . . .": Walter Colton quoted in Levy, p. 210.

289 "with their . . .": Rudolph Lapp, *Blacks in Gold Rush California* (New Haven, Conn.: Yale University Press, 1977), p. 64.

289 "retain slave . . .": Levy, p. 214.

289 "to tell . . .": Lapp, p. 67.

290 "with drink . . .": ibid. p. 141.

290 "I saw . . .": quoted in ibid. p. 21.

290 "fragile and . . .": ibid. p. 77.

290 "more tolerant . . .": J.D. Borthwick quoted in ibid. p. 78.

290 "pluck, tenacity . . .": black abolitionist argonaut Mifflin Wistar Gibbs quoted in ibid. p. 16.

290 "There are . . .": quoted in ibid. p. 13.

291 "Do you think . . . ": quoted in ibid. p. 23.

291 "This is . . .": quoted in ibid. p. 23.

291 " 'decent' white . . .": ibid. p. 59.

291 "tried to . . .": ibid.

292 "not just . . .": ibid. p. 13.

292 "organized black . . .": Albin, p. 220.

292 "in tremendous . . .": Lapp, p. 22.

292 "an excellent . . .": quoted in Levy, p. 93.

293 "owned two . . .": Lapp, p. 102.

293 "suggestive of . . .": ibid. p. 239.

293 "a trio . . .": Clark, p. 103.

293 "had had . . .": ibid.

293 "hoping that . . .": ibid.

294 "and turned . . .": Bond, p. 32.

294 "a successful . . .": William Sherman Savage, *Blacks in the West* (Westport, Conn.: Greenwood Press, 1976), p. 84.

294 "well adapted . . .": Beilharz and Lopez, p. xv.

294 "grind[ing] resistant materials": ibid.

294 "murder[ed] . . . that it . . .": Clappe, p. 110.

294 "They ain't . . .": ibid.

295 "volunteers to . . .": Harris, p. 130.

295 "took up . . .": Steele, p. 166.

295 "lost their . . .": ibid. p. 167.

296 "decidedly . . .": ibid. p. 168.

296 "I would . . .": Levy, p. 85.

296 "a very . . .": Clappe, pp. 150–51.

296 "murder every . . .": ibid.

297 "gained control": Randall E. Rohe, "Chinese Miners in the Far West," in Milner, p. 340.

297 "dominate": Rohe, p. 340.

297 "Mexicans . . .": McGrath on Bodie, p. 140.

298 "most exotic . . .": Lotchkin, pp. 123–24.

298 "which paid . . .": Rohe, p. 330.

298 "The Chinamen . . .": quoted in ibid. p. 335.

298 "held the . . .": ibid. p. 337

298 "Your true . . .": Kelly, p. 103.

298 "overwhelmingly white": McGrath, p. 5.

298 "the largest . . .": Bennett, p. 28.

299 "mysterious Chinese inscriptions": Barsness, p. 237.

299 "shabby little shops": Bennett, p. 28.

299 "accounted for . . .": Rohe, p. 341.

299 "were often . . .": Kelly, p. 103.

299 "[make] money . . .": quoted in Barsness, pp. 241–42.

299 "long after . . .": Rohe, p. 333.

299 "ungodly": John R. Wunder, "Law and Chinese in Frontier Montana," *Montana, the Magazine of Western History*, Vol. XXX (July 1980), p. 21.

299 "resented even . . .": Barsness, pp. 241–42.

300 "The contemplated . . .": Goulder, pp. 354–55.

300 "We don't mind . . .": quoted in Barsness, p. 239.

300 "the Chinese . . .": Barsness, p. 239.

301 "lugged it . . .": Fatout, p. 73.

301 "a drain . . .": quoted in McGrath, p. 137.

301 "dozens . . .": McGrath, p. 139.

301 "with complacency . . .": quoted in Barsness, pp. 241–42.

302 "The white . . .": Goulder, p. 257.

302 "could hardly . . .": ibid.

302 "unofficial counsel . . .": Bennett, p. 27.

302 "had the face . . .": ibid. p. 29.

302 "became anti-Chinese, too": Wunder, p. 18.

303 "one of the . . .": quoted in Marvin Lewis, pp. 133–35.

303 "clear a . . .": ibid. p. 135.

303 "attend[ing] strictly . . .": Kelly, p. 103.

303 "nearly every . . .": McGrath, p. 120.

304 "chauvinism . . . at . . .": Lotchkin, p. 115.

304 "a strong . . .": Brian McGinty, "The Green and the Gold," *American West*, Vol. XV, No. 2 (April 1978), p. 66.

304 "access to . . .": Petrik, p. 9.

305 "assessed the . . .": Shirley Ewart, "Cornish Miners in Grass Valley: The Letters of John Coad, 1858–1860," *Pacific Historian*, Vol. 25, No. 4 (Winter 1981), p. 40.

305 "sober . . .": ibid. p. 39.

305 "till the . . .": quoted in William Weber Johnson, *The Forty-Niners* (New York: Time-Life Books, 1974), p. 68.

305 "played-out silver mine": Wallace, p. 162.
305 "the first . . .": Laura Berton, p. 148.
305 "Not knowing . . .": quoted in Webb, p. 90.
306 "were simply . . .": Murray, p. 194.
306 "with their . . .": quoted in Rohe, p. 333.
306 "Though the . . .": Clappe, p. 36 (emphasis in original).
307 "Oh! I must . . .": quoted in Friman, p. 119.

Chapter Eleven: Home Ties

309 "I would . . .": Hinman, p. 40.
309 "write me . . .": John Callbreath letter of July 20, 1850.
309 "hoorahing . . .": quoted in Holliday, p. 310.
309 "We all . . .": Rogers letter of September 19, 1898.
309 "enraptured," "transfigured": Lynch, p. 23.
310 "treasures . . .": Goulder, p. 216.
310 "the most . . .": McMichael, p. 19.
310 "I live . . .": quoted in Holliday, p. 310.
310 "a letter . . .": A. Hill letter of March 15, 1850.
310 "and read . . .": Lynch, p. 42.
311 "more than . . .": Holliday, p. 311.
311 "mountains of . . .": Kelly, p. 173.
311 "nervous individuals . . .": Hughes, p. 111.
311 "tons of . . .": ibid.
312 "six abreast . . .": Lotchkin, p. 102.
312 "a delivery . . .": Steele, p. 107.
312 "long strings . . . at all . . .": Clark, p. 115.
312 "within 20 . . .": Wells, pp. 94–95.
312 "around the . . .": Cole, p. 143.
312 "to the amazement . . .": ibid. p. 144.
313 "inquire for . . .": Hughes, pp. 111–12.
313 "unsophisticated . . .": ibid.
313 "by way . . .": Stuart, p. 239.
313 "between sixty . . .": Goulder, p. 253.
313 "break the . . .": ibid. p. 254.
314 "the hardest . . .": Grinnell, p. 43.
314 "addressees would . . .": Zamonski and Keller, p. 29.
314 "a notice . . .": ibid.
314 "This does . . .": Wells, p. 96.
314 "sent two . . .": Hinman, p. 39.
314 "for I can . . .": ibid. p. 40.
315 "spook meetings . . . drifting in . . .": Jenkins, p. 226.
315 "passengers . . .": Bennett, p. 25.
316 "I find . . .": Holliday, p. 359.

316 "Each letter . . .": Rogers letter of February 4, 1899.
316 "We are . . .": quoted in Peavy and Smith, p. 151.
316 "The reason . . .": ibid.
316 "I am just . . .": quoted in Donald Jackson, p. 283.
316 "not be equaled . . .": Jasper S. Hill, *The Letters of a Young Miner Covering the Adventures of Jasper S. Hill . . .* ed. Doyce B. Nunis, Jr. (San Francisco: John Howell Books, 1964), p. 39.
316 "While I . . .": Bond, p. 29.
316 "It is . . .": McMichael, p. 45.
316 "Do not feel . . .": Rogers letter of September 2, 1898.
317 "Well, I . . .": J. Hill, *Letters,* p. 30.
317 "Glad to . . .": ibid. p. 36.
317 "more [money] . . .": A. Hill letter of March 14, 1850.
317 "I have done . . .": A. Hill letter of April 14, 1850.
317 "struggling . . .": A. Hill letter of September 27, 1850.
317 "I do not . . .": A. Hill letter of October 16, 1850.
317 "Do not be . . .": Rogers letter of October 20, 1898.
317 "When a man . . .": quoted in Holliday, pp. 370–71.
318 "If I could . . .": Rogers letter of March 16, 1899.
318 "that had . . .": Clark on Colorado argonauts, p. 119.
318 "I hope that if . . .": Churchill, p. 89.
318 "I hope that you . . .": Holliday, p. 320.
318 "I intend . . .": A. Hill letter of September 27, 1850.
319 "that it . . .": Churchill, p. 101.
319 "the longer . . .": Peavy and Smith, p. 101.
319 "in food . . .": quoted in Marvin Lewis, p. 45.
319 "I have wrote . . .": A. Hill letter of May 15, 1853.
319 "Now she . . .": Peavy and Smith, p. 111.
319 "I understand . . .": quoted in Holliday, p. 353.
320 "You say . . .": A. Hill letter of August 1, 1850.
320 "I suppose you . . .": A. Hill letter of June 12, 1850.
320 "Well, at . . .": quoted in Holliday, p. 396.
320 "was thought . . .": J. Hill, p. 34.
320 "all the time . . .": Rogers letter of September 2, 1898.
320 "he is . . .": Fairchild, p. 120.
320 "wished me . . .": J. Hill, p. 64.
320 "Mr. [Daniel] Bosworth . . .": Peavy and Smith, pp. 49–50.
320 "Tell Mrs. Bosworth . . .": ibid. p. 56.
321 "I really . . .": quoted in Holliday, pp. 352–53.
321 "If some . . .": Peavy and Smith, p. 97.
321 "I have so . . .": Bond, p. 45.
321 "friends and . . .": J. Hill, p. 79.
321 "If you see . . .": McMichael, p. 45.
321 "stay at . . .": quoted in Holliday, p. 350.
321 "I wouldn't . . .": Grinnell, p. 73.

321 "under the . . .": quoted in Holliday, p. 361.

321 "It is the . . .": Fairchild, p. 38.

321 "learnt me . . .": quoted in Holliday, p. 361.

321 "[in] this . . .": McMichael, p. 52.

322 "I shall read . . .": William Swain quoted in Holliday, p. 337.

322 "I am sorry . . .": Buck, p. 68.

322 "In this country . . .": quoted in West, *Saloon*, p. 4.

322 "leave this . . .": California argonaut quoted in Holliday, p. 335.

322 "I wish . . .": Rogers letter of May 27, 1899.

322 "You must . . .": Callbreath letter of August 18, 1849.

323 "got a . . .": Cornelius Griswold letter of July 2, 1860, Griswold Letters, Beinecke Library.

323 "enjoying yourself . . .": J. Hill, pp. 57–58.

323 "on account . . .": A. Hill letter of August 1, 1850.

323 "If you was . . .": quoted in Peavy and Smith, p. 84.

324 "any of the . . .": J. Hill, p. 60.

324 "would like . . .": ibid.

324 "Now Some . . .": A. Hill letter of February 22, 1850.

324 "It is cruel . . .": A. Hill letter of April 26, 1850.

324 "You say it . . .": A. Hill letter of December 19, 1852.

324 "I see that . . .": J. Hill, p. 73.

324 "Society will . . .": Callbreath letter of October 1, 1850.

324 "Little Eliza . . .": Holliday, p. 339.

325 "some thirty . . .": J. Hill, p. 71.

325 "be kind . . .": Kirkaldie quoted in Dorothy Johnson, "The Patience of Frank Kirkaldie," p. 14.

325 "not to go . . .": Peavy and Smith, p. 76.

325 "You mentioned . . .": quoted in Dorothy Johnson, p. 18.

325 "I fear . . .": Hinman, p. 40.

325 "pressing matters . . . it is . . .": Peavy and Smith, p. 49.

325 "I ought . . .": ibid. p. 138.

325 "their own . . .": ibid.

325 "You don't . . .": quoted in Lotchkin, p. 101.

325 "it is a . . .": ibid. p. 287.

326 "took full charge": Black, p. 45.

326 "fresh-baked . . .": ibid.

326 "miners, prospectors . . .": ibid.

326 "had made . . .": quoted in William Johnson, p. 139.

326 "sometimes stood . . .": Pierre Berton, p. 379.

326 "God bless . . .": quoted in ibid. p. 379.

326 "They have . . .": quoted in Chisolm, p. 158.

327 "Jane I left . . .": quoted in Groh, pp. 232–33.

327 "My thoughts . . .": J. Hill, p. 31.

327 "You say . . .": A. Hill letter of December 1, 1851.

327 "that I was . . .": quoted in Donald Jackson, p. 305.

327 "cold welcome . . .": Grinnell, p. 47.

327 "I thank . . .": J. Hill, p. 62.

327 "Just think . . .": Fairchild, p. 121.

328 "But I am . . .": quoted in Holliday, p. 353.

328 "that I mite . . .": quoted in Groh, pp. 232–33.

328 "I do not . . .": quoted in Holliday, p. 396.

328 "a bull-necked . . .": Robert Hunter Fitzhugh quoted in Hook, p. 38.

328 "if I ever . . .": A. Hill letter of August 29, 1852.

328 "less than . . .": Bond, p. 56.

328 "a bitter . . .": ibid.

328 "Poetry doesn't . . .": Grinnell, p. 52.

329 "May you . . .": quoted in Holliday, p. 344.

329 "If you . . .": ibid.

329 "take a good . . .": quoted in Holliday, p. 387.

329 "come directly . . .": quoted in Groh, p. 226.

329 "Don't give . . .": Rogers letter of April 10, 1899.

329 "as fast as . . .": Henderson, p. 12.

330 "You seem . . .": Callbreath letter of December 18, 1854.

330 "could only . . .": ibid.

330 "a spreeing . . .": Peavy and Smith, p. 80.

330 "distance and . . .": ibid.

330 "that while . . .": quoted in Holliday, p. 338.

330 "O! that . . .": ibid. p. 386.

330 "I am very . . .": quoted in Dorothy Johnson, p. 18.

331 "As to a . . .": quoted in Holliday, p. 339.

331 "Mother and . . .": ibid.

331 "in the very . . .": Peavy and Smith, p. xv.

331 "education, discipline . . .": ibid. p. 40.

331 "To be so . . .": quoted in Dorothy Johnson, p. 21.

332 "need not . . .": quoted in Peavy and Smith, p. 111.

332 "very destitute": ibid. p. 63.

332 "Their neighbours . . .": ibid. p. 63.

332 "Why you *might* . . .": ibid. p. 62.

332 "You done . . .": ibid. p. 78.

332 "and make a . . .": ibid. p. 85.

332 "shift[ed] some . . .": ibid. p. 104.

332 "the water . . .": quoted in Dorothy Johnson, p. 17.

332 "Perhaps I . . .": ibid.

333 "I should suppose . . .": Churchill, p. 98.

333 "If you fail . . .": quoted in Holliday, p. 384.

333 "bound to . . .": ibid. p. 347.

334 "about $125 . . .": Rogers letter of July 26, 1899.

334 "Mrs. [Margaretha] Ault . . .": Peavy and Smith, p. 159.

334 "Poor John . . .": ibid. pp. 212–13.

335 "worse than . . .": Grinnell, p. 52.
335 "Has anyone . . .": Choate on the Yukon, p. 16.
335 "We have gotten . . .": ibid. p. 17.
335 "my mother's . . .": Ferguson, pp. 201–202.
335 "would tell . . .": ibid.
335 "ceased years . . .": quoted in Marvin Lewis, p. 33.
336 "in a far-away . . .": ibid.

Chapter Twelve: Gender and Gold

337 "You have no . . .": quoted in Lotchkin, p. 293.
337 "got nearer . . .": quoted in Holliday, p. 355.
337 "scarcer than gold": Lynch, p. 162.
337 " 'Society' girls": Robert Hunter Fitzhugh quoted in Hook, p. 37.
338 "doorways filled . . .": quoted in Levy, p. 178.
338 "like 'boys . . .' ": West, *Saloon,* p. 48.
338 "a great attraction . . .": Ferguson, pp. 148–49.
338 "a godforsaken . . .": Lee, p. 7.
339 "I said . . .": quoted in Levy, p. xxi.
339 "a double . . .": ibid.
339 "I have become . . .": Clappe, p. 74.
340 "exactly like . . .": Elizabeth Gunn quoted in Levy, p. 112.
340 "rock the cradle . . .": quoted in ibid. p. 176.
340 "worked side . . .": Barsness, p. 157.
340 "found several . . .": Levy, p. xxi.
340 "two intelligent . . .": quoted in ibid. p. 112.
341 "shabby little cabin": Bennett, p. 13.
341 "tight knee . . .": McMichael, p. 50.
341 "daily, after . . .": Levy, p. 113.
341 "hik[ing] inland . . .": Murray, p. 190.
341 "a Severe . . .": quoted in Peavy and Smith, p. 2341
341 "Four Hundred . . .": ibid.
341 "first lady prospector": Sprague, p. 73.
342 "she figured . . .": ibid.
342 "a capable . . .": Wells, p. 136.
342 "did not yield . . .": Black, p. 57.
342 "the society . . .": quoted in Levy, p. 174.
343 "there is a wide . . .": quoted in Petrik, p. 74.
343 "library, church . . .": Mayer, p. 240.
343 "impoverished fortunes": Lynch, p. 16.
343 "take in . . .": ibid.
343 "I don't . . .": ibid.
343 "comfortless as . . .": quoted in Levy, p. 61.

343 "If we had . . .": quoted in Peavy and Smith, p. 180.
344 "they would . . .": Chisolm, p. 81.
344 "You can go . . .": quoted in Levy, p. 29.
344 "as though . . .": quoted in Levy, p. 195.
344 "She would . . .": Bennett, p. 16.
344 "no place . . . hell . . .": Jenkins, pp. 221, 228.
344 "the men . . .": Laura Berton, p. 116.
345 "I had been brought . . .": ibid.
345 "This strange . . .": Clappe, p. 54.
345 "vast new . . .": Black, p. 59.
345 "wherever there . . .": Bennett, p. 80.
345 "modest cabin . . . the presence . . .": Chisolm, p. 80.
346 "a two-storey . . .": Lynch, pp. 110–11.
346 "laid the table . . .": Black, p. 40.
346 "all trying . . .": Stuart, p. 213.
346 "decreed . . .": ibid. p. 256.
346 "explaining . . .": ibid.
347 "self-imposed": Moynihan et al., p. 4.
347 "alike impossible . . .": Kelly, p. 148
347 "anxious to . . .": Ferguson, pp. 148–49.
347 "angels of . . .": ibid.
348 "every virtuous . . .": quoted in Levy, p. 222.
348 "Do you speak . . .": Bennett, p. 59.
349 "going independent . . .": quoted in Levy, p. 35.
349 "a plain person . . .": Royce, p. 83.
349 "was not . . .": ibid.
349 "dance, sing . . .": quoted in Mayer, p. 81.
350 "met a man . . .": Levy, p. 91.
350 "shocked and . . .": Bennett, p. 96.
350 "Why didn't . . .": Jenkins, p. 208.
350 "You could a-put . . .": ibid.
350 "absolutely indelicate": Clappe, p. 3.
350 "The men . . .": McMichael, p. 49.
351 "We had a . . .": quoted in Mayer, p. 115.
351 "She'll be . . .": ibid. pp. 77–78.
351 "Honey, Honey . . .": ibid.
351 "awkward to . . .": Lynch, p. 163.
351 "They were very . . .": quoted in ibid. p. 164.
352 "said justice . . .": quoted in Levy, p. 83.
352 "All good women . . .": Bennett, p. 22.
352 " 'square' or . . .": Kelly, p. 185.
352 "a drunk . . .": McGrath, p. 184.
352 "an awful . . .": Clappe, p. 41.
353 " 'Course, where . . .": quoted in Mayer, p. 126.
353 "She was one . . .": Bennett, p. 16.

353 "it is the . . .": quoted in Levy, p. 91.
353 "plentiful . . . who can . . .": Barsness, p. 156.
353 "Womens labor . . .": Peavy and Smith, p. 193.
353 "a good, faithful . . .": Grinnell, p. 12.
353 "She has . . .": quoted in Levy, p. 106.
353 "just dipped . . .": quoted in Moynihan et al., p. 9.
353 "female labour . . .": ibid.
354 "warning the . . .": Petrik, p. 11.
354 "that the good . . .": quoted in Wunder, p. 20.
354 "doubled his offer": Levy, p. 91.
354 "bread made . . .": ibid. p. 92.
354 "doing full . . .": quoted in Holliday, p. 355.
354 "tramping along . . .": Margaret Frink quoted in Levy, p. 93.
354 "there is nothing . . .": Grinnell, p. 30.
354 "to determine . . .": Ferguson, pp. 148–49.
355 "pulled her . . .": Webb, p. 96.
355 "a motherly . . .": Wells, p. 119.
355 "starved for . . .": Dorley quoted in Mayer, p. 220.
355 "as much as . . .": Barsness, p. 156.
355 "natural progression": Levy, p. 95.
355 "nine out . . .": ibid.
355 "from a precious . . .": Wilson quoted in ibid. p. 102.
356 "every boarding . . .": quoted in Margaret S. Woyski, "Women and Mining in the Old West," *Journal of the West,* Vol. XX (April 1981), p. 44.
356 "If I had . . .": quoted in Levy, p. 98.
356 "When feminines . . .": quoted in Barsness, p. 157.
356 "an Illinois . . .": Wells, p. 127.
356 "made enough . . .": Mayer, p. 156.
357 "she had . . .": Levy, p. 122.
357 "awful toney bed": Bennett, p. 97.
357 "set her husband . . .": quoted in Lotchkin, p. 307.
357 "They are making . . .": ibid.
357 "Pay me . . .": Jenkins, p. 220.
357 "He sure . . .": ibid.
357 "were often . . .": Moynihan et al., p. xvii.
357 "The better . . .": quoted in ibid. p. 15.
357 "almost day . . .": Mrs. Nat Collins quoted in ibid. p. 156.
357 "were unsuccessful . . .": ibid.
358 "was useless . . .": Thomas Berger, *Little Big Man* (Greenwich, Conn.: Fawcett Publications, 1964), p. 133.
358 "Magnificent woman . . .": Clappe, p. 39.
358 "nine hundred . . .": ibid.
358 "a man . . .": ibid.
358 "as if . . .": ibid.

358 "ENLARGEMENT . . .": quoted in Levy, p. 150.
358 "We must . . .": ibid. p. 151.
358 "offer[ing] the . . .": Bond, p. 50.
359 "the most successful . . .": Pierre Berton, p. 384.
359 "I leave . . .": ibid. p. 384.
359 "some had . . .": Lynch, pp. 58–59.
359 "entertaining companions . . .": Black on Dawson dance hall girls, p. 48.
359 "stolid and . . .": Reporter on boomtown Summit City quoted in Fatout, p. 39.
359 "the stigma . . .": Black, p. 48.
359 "the freedom . . .": Pierre Berton, p. 385.
360 "more in . . .": Barsness, p. 208.
360 "As soon as . . .": Louisa Cook quoted in Moynihan et al., pp. 167–68.
360 "appear[ed] to . . .": Petrik, p. 26.
360 "old hands . . .": Sprague, p. 95.
360 "summer women": Petrik, p. 33.
360 "the elegance . . .": Kelly on Dawson prostitutes, p. 190.
361 "white slave girls": Black, p. 48.
361 "Chinese men . . .": McGrath, p. 131.
361 "as valuable . . .": Barsness, p. 234.
361 "common prostitutes": Pierre Berton, p. 385.
361 "a small-town . . .": stampeder quoted in ibid. p. 385.
361 "frame shacks . . . to a . . .": ibid. p. 385.
362 "questionable boardinghouse": Black, p. 69.
362 "Me plead . . .": quoted in ibid.
362 "the lovely . . .": Bennett, p. 6.
362 "too close . . .": Kelly, p. 190.
363 "said she . . .": Holliday, p. 355.
363 "ease, taste . . .": quoted in Levy, p. 153.
363 "advanced fifteen . . .": ibid. p. 171.
363 "legitimate economy . . . 265 property . . .": ibid. pp. 26–27.
363 "in some . . .": ibid. p. 56.
363 "occasional prostitute": ibid. p. 35.
363 "economic elite": ibid. p. 36.
364 "The very air . . .": quoted in Levy, p. 109.
364 "loafed around . . .": Barsness, p. 157.
364 "Was there . . .": Elizabeth Fisk quoted in Petrik, p. 76.
364 "two women . . .": quoted in Woyski, p. 42.
364 "spatted . . . concluded . . .": Grannis quoted in Maddow, p. 90.
365 "had to put . . .": Barsness, p. 158.
365 "There were a . . .": quoted in ibid. p. 158.
365 "It is a . . .": Elizabeth Fisk quoted in Petrik, p. 96.
365 "geared to . . .": Petrik, p. 136.

365 "admit[ted] in . . .": Lotchkin on San Francisco, p. 308.
365 "doing the . . .": Bennett, p. 49.
366 "shelter and . . .": Black quoted in Mayer, p. 227.
366 "I liked . . .": Black, p. 68.

Chapter Thirteen: Perspectives

368 "Expense! . . .": quoted in Pierre Berton, p. 419.
368 "If you have . . .": ibid.
368 "The poor ginks . . .": quoted in ibid. p. 382.
369 "for its size . . .": ibid. p. 79.
369 "peddling water . . .": ibid. p. 418.
369 "to feed . . .": Sprague, p. 164.
369 "There ain't . . .": quoted in "Frontier Sketches," *Denver Field and Farm,* January 7, 1905, p. 8.
370 "untimely . . .": McGrath, p. 2.
370 "grabbers": quoted in Donald Jackson, p. 333.
370 "they have lawed . . .": quoted in "Frontier Sketches," *Denver Field and Farm,* January 7, 1905, p. 8.
370 "that was . . .": Ferguson, p. 125.
370 "poverty and . . .": French laundress Fanny Loviot on her fellow French lottery immigrants in Donna Evleth, "The Lottery of the Golden Ingots," *American History Illustrated,* Vol. XXIII, No. 3 (May 1988), p. 37.
370 "scarcely broke even": Young, *Western Mining,* p. 288.
370 "a highly . . .": ibid. p. 124.
370 "a wild . . .": quoted in Donald Jackson, n.p.
371 "I made . . .": quoted in Pierre Berton, p. 429.
371 "embryo . . .": Bond on his Dawson-bound company, p. 22.
371 "Long Beach . . .": Grinnell, p. 42.
371 "could blow . . .": quoted in Levy, p. 109.
371 "I rank . . .": ibid.
371 "This rambling . . .": quoted in Peavy and Smith, p. 61.
371 "Oh, it is . . .": Grinnell, p. 39.
372 " 'Tis a queer . . .": Wilbur Parker, p. 61.
372 "the wildest . . .": Holliday, p. 338.
372 "No man . . .": quoted in Moynihan et al., p. 117.
372 "to choose . . .": 1854 *Pioneer* quoted in Levy, pp. 187–88.
372 "The one great . . .": quoted in Barsness, p. 80.
372 "we are working . . .": Grinnell, p. 23.
372 "We may miss . . .": Stuart, pp. 251–52.
372 "banging trolley . . .": quoted in Mayer, p. 224.
372 "right down . . .": Griswold letter of October 2, 1860.
372 "what made . . .": ibid.

373 "pleasure trip . . . hard luck . . .": Haas, p. 152.
373 "Really, everybody . . .": Clappe, p. 196.
373 "It is astonishing . . .": Steele, p. 58.
373 "It does . . .": Grinnell, p. 72.
373 "in general . . .": Kelly, p. 193.
373 "it took . . .": quoted in Marvin Lewis, p. 52.
373 "There were no . . .": Lynch, pp. 53–54.
374 "detaching the . . .": Chisholm, p. 86.
374 "I wish . . .": Bond, p. 46.
374 "I never knew . . .": Jenkins, p. 200.
375 "real folks . . .": quoted in Mayer, p. 224.
375 "got licked . . .": John D. Borthwick on California stampeders, quoted in Marvin Lewis, p. xiii.
375 "the confident . . .": quoted in Donald Jackson, p. 325.
375 "better than . . .": Chisolm, pp. 111–12.
375 "empty bottles . . .": Clappe, pp. 194–95.
376 "piles of . . .": Laura Berton, p. 37.
376 "being a good loser": Bennett, p. 146.
376 "I had heard . . .": ibid.
376 "a combination . . .": ibid.
376 "as an Indian . . .": quoted in Bergon and Papanikolas, p. 299.
377 "certain manliness . . . dignity": ibid.
377 "Poor Old Man . . .": quoted in Sprague, p. 107.
377 "If I wanted . . .": quoted in Wells, p. 144.
378 "It isn't . . .": Robert Service, "The Spell of the Yukon," in *Collected Poems of Robert Service* (New York: Dodd, Mead & Co., 1940), p. 5.
378 "extraordinary love . . .": quoted in Sarah Royce, p. 78.
378 "I'm afraid . . .": Hughes, p. 136.
378 "to thin . . .": ibid.
379 "the saint . . .": ibid.
379 "Peter, I . . .": ibid.
379 "I think we're . . .": quoted in Laura Berton, pp. 144–45.

Selected Bibliography

Anyone interested in pursuing the topic of American gold rushes will find a vast and varied literature. The following is not meant to be a comprehensive bibliography, but a guide to some of the published sources I found helpful and a stimulus to the reader in discovering others.

BOOKS

Anderson, Alexander. *Hand-book and Map to the Gold Region of Frazer's and Thompson's Rivers, with Table of Distances.* San Francisco: J.J. Le Count, 1858.

Austin, Basil. *The Diary of a Ninety-Eighter.* Mount Pleasant, Mich.: J. Cumming, printer, 1968.

Bancroft, Hubert H. *California Inter Pocula.* San Francisco: History Company, 1888.

Barsness, Larry. *Gold Camp: Alder Gulch and Virginia City, Montana.* New York: Hastings House, 1962.

Beilharz, Edwin A., and Carlos V. Lopez, trans. and eds. *We Were 49ers! Chilean Accounts of the California Gold Rush.* Pasadena, Calif.: Ward Ritchie Press, 1976.

Bennett, Estelline. *Old Deadwood Days.* Lincoln, Nebr.: University of Nebraska Press, 1982, reprint of 1928 edition.

Bergon, Frank, and Zeese Papanikolas, eds. *Looking Far West: The Search for the American West in History, Myth, and Literature.* New York: New American Library, 1978.

Berton, Laura. *I Married the Klondike.* Boston: Little, Brown & Co., 1954.

Berton, Pierre. *The Klondike Fever.* New York: Carroll and Graf Publishers, 1985, reprint of 1958 edition.

Selected Bibliography

Black, Martha Louise. *My Ninety Years.* Anchorage: Alaska Northwest Publishing Co., 1976.

Bond, Marshall, Jr. *Gold Hunter: The Adventures of Marshall Bond.* Albuquerque, N.Mex.: University of New Mexico Press, 1969.

Brown, Robert L. *The Great Pikes Peak Gold Rush.* Caldwell, Ida.: Caxton Printers, 1985.

Bruff, J. Goldsborough. *Gold Rush: The Journals, Drawings and Other Papers of J. Goldsborough Bruff,* ed. Georgia W. Read and Ruth Gains. New York: Columbia University Press, 1944.

Bryan, Jerry. *An Illinois Gold Hunter in the Black Hills; the Diary of Jerry Bryan.* Springfield, Ill.: Illinois State Historical Society, 1960.

Buck, Franklin A. *A Yankee Trader in the Gold Rush.* Boston: Houghton Mifflin, 1930.

Caughey, John W. *Gold Is the Cornerstone.* Berkeley, Calif.: University of California Press, 1948.

Chisolm, James. *South Pass, 1868; James Chisolm's Journal of the Wyoming Gold Rush,* ed. Lola M. Homsher. Lincoln, Nebr.: University of Nebraska Press, 1975; reprint of 1960 edition.

Choate, Glenda J. *Skagway, Alaska Gold Rush Cemetery History and Guide Book.* Skagway, Alaska: Lynn Canal Publishing, 1989.

Churchill, Charles William. *Fortunes Are for the Few: Letters of a Forty-Niner,* eds. Duane A. Smith and David J. Weber. San Diego Historical Society, 1977.

Clappe, Louise A.K.S. *The Shirley Letters.* Santa Barbara, Calif., and Salt Lake City, Utah: Peregrine Smith, 1970.

Clark, Charles M. *A Trip to Pike's Peak and Notes by the Way.* San Jose, Calif.: Talisman Press, 1958.

Coates, Ken S. and William R. Morrison. *Land of the Midnight Sun: A History of the Yukon.* Edmonton, Alberta: Hurtig Publishers, 1988.

Dietz, Arthur. *Mad Rush for Gold in Frozen North.* Los Angeles: Times-Mirror Printing and Binding House, 1914.

Dorset, Phyllis F. *The New Eldorado: The Story of Colorado's Gold and Silver Rushes.* New York: Macmillan Co., 1970.

Egan, Ferol. *The El Dorado Trail: The Story of the Gold Rush Routes Across Mexico.* Lincoln, Nebr.: University of Nebraska Press, 1984; reprint of 1970 edition.

Fairchild, Lucius. *California Letters of Lucius Fairchild,* ed. Joseph Schafer. Madison, Wisc.: State Historical Society of Wisconsin, 1931.

Fatout, Paul. *Meadow Lake: Gold Town.* Lincoln, Nebr.: University of Nebraska Press, 1969.

Selected Bibliography

Ferguson, Charles D. *The Experiences of a Forty-niner During Thirty-four Years' Residence in California and Australia*, ed. Frederick T. Wallace. Cleveland, Ohio: Williams Publishing Co., 1888.

Fisher, Vardis, and Opal L. Holmes. *Gold Rushes and Mining Camps of the Early American West*. Caldwell, Ida.: Caxton Printers, 1968.

Foreman, Grant. *Marcy and the Gold Seekers*. Norman, Okla.: University of Oklahoma Press, 1968; reprint of 1939 edition.

Garnier, Pierre. *A Medical Journey in California*, ed. Doyce B. Nunis, Jr., and trans. L. Jay Olivia. Los Angeles: Zeitlin and Ver Brugge, 1967.

Goulder, William Armistead. *Reminiscences: Incidents in the Life of a Pioneer in Oregon and Idaho*. Moscow, Ida.: University of Idaho Press, 1990; reprint of 1909 edition.

Greever, William S. *The Bonanza West: The Story of the Western Mining Rushes 1848–1900*. Moscow, Ida.: University of Idaho Press, 1986; reprint of 1963 edition.

Groh, George. *Gold Fever*. New York: William Morrow, 1966.

Hafen, LeRoy, ed. *Colorado Gold Rush: Contemporary Letters and Reports 1858–1859*. Philadelphia: Porcupine Press, 1974; reprint of 1941 edition.

————. *Overland Routes to the Gold Fields, 1859, from Contemporary Diaries*. Philadelphia: Porcupine Press, 1974; reprint of 1942 edition.

————. *Pike's Peak Gold Rush Guidebooks of 1859*. Philadelphia: Porcupine Press, 1974; reprint of 1941 edition.

Hall, Linville [supposed author]. *Around the Horn in '49. Journal of the Hartford Union Mining & Trading Company*. Wethersfield, Conn.: Printed by L.J. Hall, on board the Henry Lee, 1849; reprinted by Rev. I.J. Hall, 1898.

Harris, Benjamin Butler. *The Gila Trail: The Texas Argonauts and the California Gold Rush*, ed. Richard H. Dillon. Norman, Okla.: University of Oklahoma Press, 1960.

Hill, Jasper S. *The Letters of a Young Miner Covering the Adventures of Jasper S. Hill During the California Gold Rush, 1849–1852*, ed. Doyce B. Nunis, Jr. San Francisco: John Howell Books, 1964.

Hine, Robert V. *Community on the American Frontier: Separate but Not Alone*. Norman, Okla.: University of Oklahoma Press, 1960.

Hinman, Charles G. *'A Pretty Fair View of the Eliphant': or, Ten Letters by Charles G. Hinman Written During His Trip Overland from Groveland, Illinois, to California in 1849 and His Adventures in the Gold Fields in 1849 and 1850*, ed. Colton Storm. Chicago: Printed for Everett D. Graff by G. Martin, 1960.

Selected Bibliography

Holliday, J.S. *The World Rushed In: The California Gold Rush Experience.* New York: Simon & Schuster, 1981.

Hughes, Richard B. *Pioneer Years in the Black Hills,* ed. Agnes Wright Spring. Glendale, Calif.: Arthur H. Clarke Co., 1957.

Isbell, Fred. *Mining and Hunting in the Far West.* Middletown, Conn.: J.S. Steward, printer, ca. 1871.

Jackson, Donald Dale. *Gold Dust.* New York: Knopf, 1980.

Jackson, Joseph. *Anybody's Gold: The Story of California's Mining Towns.* New York: D. Appleton-Century Co., 1941.

Jenkins, Malinda. *Gambler's Wife: The Life of Malinda Jenkins.* Boston: Houghton Mifflin, 1933.

Johnson, William Weber. *The Forty-Niners.* New York: Time-Life Books, 1974.

Kansas Pacific Railway Company. *Miners' Guide to the Black Hills and Famous Big Horn Gold Region . . .* Chicago: Rand, McNally Co., printers, 1877.

Kelly, Howard [supposed author]. *Sourdough Gold: The Log of a Yukon Adventure,* ed. Mary Lee Davis. Boston: W.A. Wilde Co., 1933.

Kemble, John Haskell. *The Panama Route, 1848–1849.* Columbia, S.C.: University of South Carolina Press, 1990; reprint of 1943 edition.

Lapp, Rudolph. *Blacks in Gold Rush California.* New Haven, Conn.: Yale University Press, 1977.

Lee, Mabel Barbee. *Cripple Creek Days.* Lincoln, Nebr.: University of Nebraska Press, 1984; reprint of 1958 edition.

Letts, John M. *California Illustrated: Including a Description of the Panama and Nicaragua Routes. By a Returned Californian.* New York: W. Holdredge, 1852.

Levy, JoAnn. *They Saw the Elephant: Women in the California Gold Rush.* Hamden, Conn.: Archon Books, 1990.

Lewis, Marvin. *The Mining Frontier: Contemporary Accounts from the American West in the Nineteenth Century.* Norman, Okla.: University of Oklahoma Press, 1967.

Limerick, Patricia Nelson. *The Legacy of Conquest: The Unbroken Past of the American West.* New York: W.W. Norton & Co., 1987.

Lingenfelter, Richard E. *The Rush of '89: The Baja California Gold Fever.* Los Angeles: Dawson's Book Shop, 1967.

Lotchkin, Roger W. *San Francisco, 1846–1856: From Hamlet to City.* Lincoln, Nebr.: University of Nebraska Press, 1974.

Lynch, Jeremiah. *Three Years in the Klondike,* ed. Dale L. Morgan. Chicago: R.R. Donnelley, 1967.

Selected Bibliography

Maddow, Ben. *A Sunday Between Wars: The Course of American Life from 1865 to 1917*. New York: W.W. Norton & Co., 1979.

Madsen, Brigham. *Gold Rush Sojourners in Great Salt Lake City, 1849 and 1850*. Salt Lake City: University of Utah Press, 1983.

Malone, Michael P., and Richard B. Roeder. *Montana, A History of Two Centuries*. Seattle: University of Washington Press, 1976.

Mayer, Melanie J. *Klondike Women: True Tales of the 1897–1898 Gold Rush*. Athens, Ohio: Ohio University Press, 1989.

McGrath, Roger D. *Gunfighters, Highwaymen and Vigilantes: Violence on the Frontier*. Berkeley, Calif.: University of California Press, 1984.

Merrill, Julius. *Bound for Idaho: The 1864 Trail Journal of Julius Merrill*, ed. Irving Merrill. Moscow, Ida.: University of Idaho Press, 1988.

Miller, James Knox Polk. *The Road to Virginia City: The Diary of James Knox Polk Miller*, ed. Andrew F. Rolle. Norman, Okla.: University of Oklahoma Press, 1950.

Milner, Clyde A. II, ed. *Major Problems in the History of the American West*. Lexington, Mass.: D.C. Heath, 1989.

Monaghan, Jay. *Chile, Peru, and the California Gold Rush of 1849*. Berkeley, Calif.: University of California Press, 1973.

Morison, James. *By Sea to San Francisco, 1849–1850: The Journal of Dr. James Morison*, ed. Lonnie J. White and William R. Gillaspie. Memphis: Memphis State University Press, 1977.

Morrell, W.P. *The Gold Rushes*. London: Adam and Charles Black, 1940.

Moynihan, Ruth B., Susan Armitage, and Christiane Fischer Dichamp, eds. *So Much to Be Done; Women Settlers on the Mining and Ranching Frontier*. Lincoln, Nebr.: University of Nebraska Press, 1990.

Murray, Keith A. *Reindeer and Gold*. Bellingham, Wash.: Center for Pacific Northwest Studies, 1988.

Parker, Watson. *Gold in the Black Hills*. Lincoln, Nebr.: University of Nebraska Press, 1982; reprint of 1966 edition.

Paul, Rodman W. *California Gold: The Beginning of Mining in the Far West*. Lincoln, Nebr.: University of Nebraska Press, 1965; reprint of 1947 edition.

———. *The California Gold Discovery: Sources, Documents, Accounts and Memoirs Relating to the Discovery of Gold at Sutter's Mill*. Georgetown, Calif.: Talisman Press, 1967.

———. *The Far West and the Great Plains in Transition 1859–1900*. New York: Harper and Row, 1988.

———. *Mining Frontiers of the Far West, 1848–1880*. New York: Holt, Rinehart, and Winston, 1963.

Selected Bibliography

Peavy, Linda, and Ursula Smith. *The Gold Rush Widows of Little Falls: A Story Drawn from the Letters of Pamelia and James Fergus*. St. Paul: Minnesota Historical Society, 1990.

Peterson, Richard H. *The Bonanza Kings: The Social Origins and Business Behavior of Western Mining Entrepreneurs, 1870–1900*. Lincoln, Nebr.: University of Nebraska Press, 1977.

———. *Bonanza Rich: Lifestyles of the Western Mining Entrepreneurs*. Moscow, Ida.: University of Idaho Press, 1991.

Petrik, Paula. *No Step Backward: Women and Family on the Rocky Mountain Mining Frontier, Helena, Montana, 1865–1900*. Helena, Mont.: Montana Historical Society Press, 1987.

Pomfret, John Edwin. *California Gold Rush Voyages*. San Marino, Calif.: Huntington Library, 1954.

Quiett, Glenn C. *Pay Dirt: A Panorama of American Gold Rushes*. New York: D. Appleton-Century Co., 1936.

Reinhart, Herman Francis. *The Golden Frontier: The Recollections of Herman Francis Reinhart 1851–1869*, ed. Doyce B. Nunis, Jr. Austin, Tex.: University of Texas Press, 1962.

Royce, Sarah. *A Frontier Lady: Recollections of the Gold Rush and Early California*, ed. Ralph Henry Gabriel. New Haven, Conn.: Yale University Press, 1932.

Savage, William Sherman. *Blacks in the West*. Westport, Conn.: Greenwood Press, 1976.

Service, Robert. *Collected Poems of Robert Service*. New York: Dodd, Mead & Co., 1940.

Shinn, Charles Howard. *Mining Camps: A Study in American Frontier Government*. New York: Alfred A. Knopf, 1948; reprint of 1885 edition.

Smith, Duane. *Rocky Mountain Mining Camps: The Urban Frontier*. Niwot, Colo.: University Press of Colorado, 1992.

Sprague, Marshall. *Money Mountain: The Story of Cripple Creek Gold*. Lincoln, Nebr.: University of Nebraska Press, 1979; reprint of 1953 edition.

Steele, John. *In Camp and Cabin*. New York: Citadel Press, 1962.

Stephens, Lorenzo Dow. *Life Sketches of a Jay-hawker of '49*. San Jose, Calif.: Nolta Brothers, 1916.

Stuart, Granville. *The Montana Frontier 1852–1864*. Lincoln, Nebr.: University of Nebraska Press, 1977; reprint of 1925 edition.

Taylor, Bayard. *Colorado: A Summer Trip*. New York: E.P. Putnam and Son, 1867.

———. *Eldorado: Or Adventures in the Path of Empire*. Lincoln, Nebr.: University of Nebraska Press, 1988; reprint of 1949 edition.

Tyson, James L. *Diary of a Physician in California*. Oakland, Calif.: Biobooks, 1955.

Wallace, Robert, and editors of Time-Life Books. *The Miners*. New York: Time-Life Books, 1976.

Watkins, T.H. *Gold and Silver in the West: The Illustrated History of an American Dream*. New York: Bonanza Books, 1971.

Webb, Melody. *The Last Frontier*. Albuquerque, N.Mex.: University of New Mexico Press, 1985.

Weiss, Richard. *The American Myth of Success: From Horatio Alger to Norman Vincent Peale*. Urbana, Ill.: University of Illinois Press, 1988.

Wells, E. Hazard. *Magnificence and Misery: A Firsthand Account of the 1897 Klondike Gold Rush*. Garden City, N.Y.: Doubleday & Co., 1984.

Wells, Merle W. *Gold Camps and Silver Cities*, 2nd ed. Moscow, Ida.: Idaho Department of Lands, 1983.

West, Elliott. *Growing Up with the Country: Childhood on the Far Western Frontier*. Albuquerque, N.M.: University of New Mexico Press, 1989.

————. *The Saloon on the Rocky Mountain Mining Frontier*. Lincoln, Nebr.: University of Nebraska Press, 1979.

Wright, Richard. *Overlanders: 1858 Gold*. Saskatoon, Sask.: Western Producer Prairie Books, 1985.

Young, Otis E., Jr. *Western Mining: An Informal Account of Precious Metals Prospecting, Placering, Lode-Mining, and Milling on the American Frontier*. Norman, Okla.: University of Oklahoma Press, 1987; reprint of 1970 edition.

Zamonski, Stanley W., and Teddy Keller. *The '59ers: Roaring Denver in the Gold Rush Days*. Frederick, Colo.: Jende-Hagen Bookcorp, 1983; reprint of 1961 edition.

ARTICLES

Albin, Ray R. "The Perkins Case: The Ordeal of Three Slaves in Gold Rush California." *California History*, Vol. 67 (December 1988), pp. 215–27.

Apostol, Jane. "Gold Rush Widow." *Pacific Historian*, Vol. XXVIII, No. 2 (Summer 1984), pp. 49–55.

Athearn, Robert. "The Fifty-Niners." *The American West*, Vol. XIII, No. 5 (September–October 1976), pp. 22–25ff.

Berry, Thomas. "Gold! But How Much?" *California Historical Quarterly*, Vol. LV, No. 3 (Fall 1976), pp. 247–55.

Bigando, C. Robert. "Tom Miner's Folly: The Twisted Tale of the

Mogollon Mining and Prospecting Company, 1871." *Journal of Arizona History,* Vol. 31 (Spring 1990), pp. 61–78.

Bronson, William. "Nome." *The American West,* Vol. VI, No. 4 (July 1969), pp. 20–31.

Chang, Clenn K.M. "Leading the Way: The Prospecter in the Trans-Mississippi West." *Journal of the West,* Vol. XX (April 1981), pp. 31–37.

Cox, C.C. "From Texas to California in 1849: Diary of C.C. Cox," ed. Mabelle E. Martin. *Southwestern Historical Quarterly,* Vol. XXIX, No. 1 (July 1925), pp. 36–50.

Cullen, Franklin, C.S.C. "Holy Cross on the Gold Dust Trail: The 1850 Expedition to California." *Holy Cross on the Gold Dust Trail and Other Western Ventures.* Notre Dame, Ind.: Indiana Province Archives Center, 1989. (No. 5, Preliminary Studies in the History of the Congregation of Holy Cross in America)

Doyle, Susan Badger. "Journeys to the Land of Gold: Emigrants on the Bozeman Trail, 1863–1866." *Montana, the Magazine of Western History,* Vol. 41, No. 4 (Autumn 1991), pp. 54–67.

Dutka, Barry L. "New York Discovers GOLD! in California." *California History,* Vol. LXIII (Fall 1984), pp. 313–19.

Egan, Ferol. "Twilight of the Californios." *The American West,* Vol. VI, No. 2 (March 1969), pp. 34–42.

Evleth, Donna. "The Lottery of the Golden Ingots." *American History Illustrated,* Vol. XXIII, No. 3 (May 1988), pp. 34–37ff.

Ewart, Shirley. "Cornish Miners in Grass Valley: The Letters of John Coad, 1858–1860." *Pacific Historian,* Vol. 25, No. 4 (Winter 1981), pp. 38–45.

Farr, William E. "Germans in Montana Gold Camps: Two Views." *Montana, the Magazine of Western History,* Vol. 32, No. 4 (Autumn 1982), pp. 58–73.

Friggens, Paul. "The Curious 'Cousin Jacks': Cornish Miners in the American West." *The American West,* Vol. XV, No. 6 (November–December 1978), pp. 4–7, 62ff.

Friman, Axel. "Two Swedes in the California Goldfields: Allvar Kullgren and Carl August Modh, 1850–1856." *Swedish-American Historical Quarterly,* Vol. XXXIV (April 1983), pp. 102–29.

Genini, Ronald. "The Fraser-Cariboo Gold Rushes: Comparisons and Contrasts with the California Gold Rush." *Journal of the West,* Vol. XI, No. 3 (July 1972), pp. 470–87.

Grinnell, Joseph. "Gold Hunting in Alaska As Told by Joseph Grinnell," ed. Elizabeth Grinnell. *Alaska Journal,* Vol. 13 (Spring 1983), book insert.

Henderson, A.G. "My Journey to the Gold Fields: Reminiscences of an Argonaut." *The American West,* Vol. XIII, No. 3 (May–June 1976), pp. 4–12.

Hook, Joanne. "He Never Returned: Robert Hunter Fitzhugh in Alaska." *Alaska Journal,* Vol. 15 (Spring 1985), pp. 33–38.

Hume, Charles V. "The Gold Rush Actor: His Fortunes and Misfortunes in the Mining Camps." *The American West,* Vol. IX, No. 3 (May 1972), pp. 14–19.

Johnson, Dorothy M. "The Patience of Frank Kirkaldie." *Montana, the Magazine of Western History,* Vol. XXI, No. 1 (January 1971), pp. 12–27.

Keller, Robert H., Jr. "A Puritan at Alder Gulch and the Great Salt Lake: Rev. Jonathan Blanchard's Letters from the West, 1864." *Montana, the Magazine of Western History,* Vol. 36, No. 3 (Summer 1986), pp. 62–75.

Kingman, Romanzo. "Romanzo Kingman's Pike's Peak Journal, 1859," ed. Kenneth F. Millsap. *Iowa Journal of History,* Vol. 48 (1950), pp. 55–85.

Kittredge, William, and Steven M. Krauzer. " 'Mr. Montana' Revised: Another Look at Granville Stuart." *Montana, the Magazine of Western History,* Vol. 36, No. 4 (Autumn 1986); pp. 14–23.

Levy, JoAnn. "Eliza Farnham and the Golden State." *True West,* Vol. 27, No. 1 (Fall 1990), pp. 37–41.

Lewis, David Rich. "Argonauts and the Overland Trail Experience: Method and Theory." *Western Historical Quarterly,* Vol. XVI, No. 3 (July 1985), pp. 285–305.

McGinty, Brian. "The Green and the Gold." *The American West,* Vol. XV, No. 2 (April 1978), pp. 18–21, 65ff.

McMichael, Alfred. "Klondike Letters: The Correspondence of a Gold Seeker in 1898," ed. Juliette C. Reinicker. *Alaska Journal,* Vol. 14 (Autumn 1984), book insert.

Mueller, George D. "Real and Fancied Claims: Joseph Richard 'Skookum Joe' Anderson, Miner in Central Montana, 1880–1897." *Montana, the Magazine of Western History,* Vol. 35, No. 2 (Spring 1985), pp. 50–59.

Nelson, Arnold and Helen. "Bringing Home the Gold." *Alaska Journal,* Vol. 9 (Summer 1979), pp. 52–59.

Olmsted, Roger. "San Francisco and the Vigilante Style: I." *The American West,* Vol. 7, No. 1 (January 1970), pp. 6–11ff.

———. "San Francisco and the Vigilante Style: II." *The American West,* Vol. 7, No. 2 (March 1970), pp. 20–26ff.

Parker, Wilbur Fiske. " 'The Glorious Orb of Day Has Rose': A Diary of the Smoky Hill Route to Pike's Peak, 1858," ed. Norman Lavers.

Selected Bibliography

Montana, the Magazine of Western History, Vol. 36, No. 2 (Spring 1986), pp. 50–61.

Parsons, John E. "Steamboats in the 'Idaho' Gold Rush. *Montana, the Magazine of Western History,* Vol. 10, No. 1 (Winter 1960), pp. 51–61.

Perez-Venero, Alejandro. "The 'Forty-Niners Through Panama." *Journal of the West,* Vol. XI, No. 3 (July 1972), pp. 460–69.

Perrigo, Lynn I. "Law and Order in Early Colorado Mining Camps." *Mississippi Valley Historical Review,* Vol. XXVIII (1941–42), pp. 41–62.

Rawls, James J. "Gold Diggers: Indian Miners in the California Gold Rush." *California Historical Quarterly,* Vol. LV (Spring 1976), pp. 28–45.

Reid, Bernard J. "Life in the California Goldfields in 1850: The Letters of Bernard J. Reid," ed. Mary McDougall Gordon. *Southern California Quarterly,* Vol. LXVII, No. 1 (Spring 1985), pp. 51–69.

Richmond, Robert W., and Robert W. Mardock, eds. "The Mining Frontier." *A Nation Moving West: Readings in the History of the American Frontier.* Lincoln, Nebr.: University of Nebraska Press, 1966, pp. 190–211.

Riley, Glenda. "Women on the Panama Trail to California, 1849–1869." *Pacific Historical Review,* Vol. 55, No. 4 (November 1986), pp. 531–48.

Rohe, Randall E. "Chinese Miners in the Far West." *Major Problems in the History of the American West,* ed. Clyde A. Milner II. Lexington, Mass.: D.C. Heath & Co., 1989.

Rose, Dinah. "Billy Holcomb, Hard Luck Miner." *True West,* Vol. 34, No. 8 (August 1987), pp. 39–43.

Roske, Ralph J. "The World Impact of the California Gold Rush, 1849–1857." *Arizona and the West,* Vol. 5 (1963), pp. 187–232.

Rotter, Andrew J. " 'Matilda for Gods Sake Write': Women and Families on the Argonaut Mind." *California History,* Vol. LVIII (Summer 1979), pp. 128–41.

Salisbury, William. "The Journal of an 1859 Pike's Peak Gold Seeker," ed. David Lindsey. *Kansas Historical Quarterly,* Vol. XXII, No. 4 (Winter 1956), pp. 321–41.

Smalley, Eugene V. "1884: The Great Coeur d'Alene Stampede." *Idaho Yesterdays,* Vol. 11, No. 3 (Fall 1967), pp. 2–9.

Steffen, Jerome O. "The Mining Frontiers of California and Australia: A Study in Comparative Political Change and Continuity." *Pacific Historical Review,* Vol. LII, No. 4 (November 1983), pp. 429–40.

Stevens, Jan S. "Stephen J. Field: A Gold Rush Lawyer Shapes the Nation." *Journal of the West,* Vol. XXIX, No. 3 (July 1990), pp. 40–53.

Thompson, Gerald. " 'Is There a Gold Field East of the Colorado?' The La Paz Rush of 1862." *Southern California Quarterly*, Vol. 67 (Winter 1985), pp. 345–60.

Torrez, Robert J. "The San Juan Gold Rush of 1860 and Its Effect on the Development of Northern New Mexico." *New Mexico Historical Review*, Vol. 63 (July 1988), pp. 257–72.

Walker, Roger O. "One Tall Drink of Water." *True West*, Vol. 27, No. 1 (Fall 1990), pp. 28–32.

Wilson, William H. "To Make a Stake: Fred G. Kimball in Alaska, 1899–1909." *Alaska Journal*, Vol. 13 (Winter 1983), pp. 108–14.

Woyski, Margaret S. "Women and Mining in the Old West." *Journal of the West*, Vol. XX (April 1981), pp. 39–45.

Wunder, John R. "Law and Chinese in Frontier Montana." *Montana, the Magazine of Western History*, Vol. XXX (July 1980), pp. 18–31.

Young, Otis E., Jr. "The Southern Gold Rush, 1828–1836." *Journal of Southern History*, Vol. XLVII (August 1982), pp. 373–92.

UNPUBLISHED DOCUMENTS AND EPHEMERA

I made use of the large western history holdings at the Beinecke Library, Yale; at the Bancroft Library, University of California at Berkeley; and at the Denver Public Library (Western Americana Collection). Among the Beinecke materials, I found the letters of forty-niner Alonzo Hill and those of Alaska gold-hunter Henry Rogers to be particularly rich. At the Bancroft, the journal of forty-niner Phineas Blunt and the memoirs of another California argonaut, Warren Sadler, were especially useful, although the latter appears to mix dime-novel fiction with fact. At the Denver Public Library, a particularly informative source was Terrence Cole's "A History of the Nome Gold Rush: The Poor Man's Paradise" (University of Washington, Ph.D. dissertation, 1983).

Index

Index

Index

Index

4/19

5·01 MK 15 8/00
6/03 MR 15 5/01
8/05 ⑰ 4/05
07/07 ⑱ 05/06 ⓢ